David A. Phoenix,
Sarah R. Dennison,
and Frederick Harris

Antimicrobial Peptides

Related Titles

Gualerzi, C. O., Brandi, L., Fabbretti, A., Pon, C. L. (eds.)

Antibiotics

Targets, Mechanisms and Resistance

2013
ISBN: 978-3-527-33305-9

Castanho, M., Santos, N. (eds.)

Peptide Drug Discovery and Development

Translational Research in Academia and Industry

2012
ISBN: 978-3-527-32891-8

Thormar, H. (ed.)

Lipids and Essential Oils as Antimicrobial Agents

2011
ISBN: 978-0-470-74178-8

Jorgenson, L., Nielson, H. M. (eds.)

Delivery Technologies for Biopharmaceuticals

Peptides, Proteins, Nucleic Acids and Vaccines

2009
ISBN: 978-0-470-72338-8

Jensen, K. J. (ed.)

Peptide and Protein Design for Biopharmaceutical Applications

2009
ISBN: 978-0-470-31961-1

Sewald, N., Jakubke, H.-D.

Peptides: Chemistry and Biology

Second, Revised and Updated Edition
2009
ISBN: 978-3-527-31867-4

Jakubke, H.-D., Sewald, N.

Peptides from A to Z

A Concise Encyclopedia

2008
ISBN: 978-3-527-31722-6

David A. Phoenix, Sarah R. Dennison, and Frederick Harris

Antimicrobial Peptides

WILEY-VCH

WILEY-VCH Verlag GmbH & Co. KGaA

The Authors

Prof. David A. Phoenix
University of Central Lancashire
Pharmacy & Biomedical Sciences
Preston PR1 2HE
United Kingdom

Dr. Sarah R. Dennison
University of Central Lancashire
Pharmacy & Biomedical Sciences
Preston PR1 2HE
United Kingdom

Dr. Frederick Harris
University of Central Lancashire
School of Forensic and
Investigative Sciences
Preston PR1 2HE
United Kingdom

Cover design:
The cover shows aurein 2.3, a typical amphibian antimicrobial peptide in the presence of a lipid bilayer. The molecular dynamics simulation was undertaken by Dr. Manuela Mura, University of Central Lancashire. This MD simulation shows the potential of aurein 2.3 to use a pore-type mechanism of membrane interaction.

■ All books published by **Wiley-VCH** are carefully produced. Nevertheless, authors, editors, and publisher do not warrant the information contained in these books, including this book, to be free of errors. Readers are advised to keep in mind that statements, data, illustrations, procedural details or other items may inadvertently be inaccurate.

Library of Congress Card No.: applied for

British Library Cataloguing-in-Publication Data
A catalogue record for this book is available from the British Library.

Bibliographic information published by the Deutsche Nationalbibliothek
The Deutsche Nationalbibliothek lists this publication in the Deutsche Nationalbibliografie; detailed bibliographic data are available on the Internet at <http://dnb.d-nb.de>.

© 2013 Wiley-VCH Verlag GmbH & Co. KGaA, Boschstr. 12, 69469 Weinheim, Germany

All rights reserved (including those of translation into other languages). No part of this book may be reproduced in any form – by photoprinting, microfilm, or any other means – nor transmitted or translated into a machine language without written permission from the publishers. Registered names, trademarks, etc. used in this book, even when not specifically marked as such, are not to be considered unprotected by law.

Composition   Toppan Best-set Premedia Limited, Hong Kong

Printing and Binding   Markono Print Media Pte Ltd, Singapore

Cover Design   Simone Benjamin, McLeese Lake

Print ISBN: 978-3-527-33263-2
ePDF ISBN: 978-3-527-65288-4
ePub ISBN: 978-3-527-65287-7
mobi ISBN: 978-3-527-65286-0
oBook ISBN: 978-3-527-65285-3

Printed in Singapore
Printed on acid-free paper

Contents

Preface *IX*
List of Abbreviations *XIII*

1 Antimicrobial Peptides: Their History, Evolution, and Functional Promiscuity *1*
Summary *1*
1.1 Introduction: The History of Antimicrobial Peptides *1*
1.2 AMPs: Evolutionarily Ancient Molecules *4*
1.3 AMPs: Multifunctional Molecules *11*
1.3.1 Defensins as Effectors of Immunity *12*
1.3.2 Defensins and Wound Healing *16*
1.3.3 Defensins and Canine Coat Color *17*
1.4 Discussion *17*
References *19*

2 Cationic Antimicrobial Peptides *39*
Summary *39*
2.1 Introduction *39*
2.2 CAMPs and Their Antimicrobial Action *42*
2.3 CAMPs That Adopt an α-Helical Structure *43*
2.4 CAMPs That Adopt a β-Sheet Structure *51*
2.5 CAMPs That Adopt Extended Structures Rich in Specific Residues *55*
2.6 Discussion *60*
References *62*

3 Anionic Antimicrobial Peptides *83*
Summary *83*
3.1 Introduction *83*
3.2 AAMPs in the Respiratory Tract *85*
3.3 AAMPs in the Brain *87*
3.4 AAMPs in the Epidermis *91*
3.5 AAMPs in the Epididymis *94*

3.6	AAMPs in Blood Components	96
3.7	AAMPs in the Gastrointestinal Tract and Food Proteins	97
3.8	AAMPs and Their Structure–Function Relationships	99
3.9	Discussion	101
	References	102

4 Graphical Techniques to Visualize the Amphiphilic Structures of Antimicrobial Peptides 115

Summary 115

4.1	Introduction	115
4.2	Amphiphilic Structures Adopted by AMPs	116
4.3	Qualitative Methods for Identifying Amphiphilic Structure	118
4.4	Quantitative Techniques for Analyzing Amphiphilic Structure	121
4.4.1	Techniques Based on Hydropathy Plot Analysis	121
4.4.2	Techniques Based on Fourier Transforms	123
4.4.3	Amphipathic Index	124
4.4.4	Hydrophobic Moment Analysis	124
4.4.4.1	Choice of Hydrophobicity Scales	125
4.4.4.2	Effect of Sequence Length on Amphiphilicity	126
4.4.5	Classification of Amphiphilic α-Helices Using the Approach of Segrest	127
4.4.6	Amphiphilicity Profiling Analysis of Tilted α-Helices	129
4.4.7	Extended Hydrophobic Moment Plot Analysis of Tilted α-Helices	130
4.4.8	Amphiphilicity Quantified Using the Approach of Keller	130
4.4.9	Amphiphilicity Quantified Using the Approach of Brasseur	132
4.5	Discussion	133
	References	135

5 Models for the Membrane Interactions of Antimicrobial Peptides 145

Summary 145

5.1	Introduction	145
5.2	CM-Associated Factors That Affect the Antimicrobial Action of α-CAMPs	149
5.3	Mechanisms Used by CAMPs for Microbial Membrane Interaction	154
5.4	Established Models for the Membrane Interactions of α-AMPs	155
5.4.1	Barrel-Stave and Toroidal Pore Models	155
5.4.2	Carpet Mechanism and the Shai–Huang–Matsazuki Model	158
5.5	Recent Novel Models for the Membrane Interactions of α-AMPs	159
5.6	Tilted Peptide Mechanism	161
5.7	Amyloidogenic Mechanisms	163
5.8	Discussion	166
	References	168

6	**Selectivity and Toxicity of Oncolytic Antimicrobial Peptides** *181*	
	Summary *181*	
6.1	Introduction *181*	
6.2	Peptide-Based Factors That Contribute to the Anticancer Action of Anticancer Peptides *183*	
6.2.1	Sequence Length *187*	
6.2.2	Net Positive Charge *187*	
6.2.3	Hydrophobicity *191*	
6.2.4	Amphiphilicity *195*	
6.3	Membrane-Based Factors That Contribute to the Anticancer Action of ACPs *196*	
6.3.1	Membrane Receptors *199*	
6.3.2	Cholesterol *199*	
6.3.3	Anionic Membrane Components *200*	
6.3.4	Glycoproteins and Glycolipids *201*	
6.3.5	Proteoglycans *202*	
6.3.6	Phospholipids *203*	
6.4	Discussion *205*	
	References *207*	

Index *223*

Preface

The indiscriminate and widespread use of antibiotics for both medical and non-medical purposes [1] has led to the emergence of pathogenic bacteria with multi-drug resistance, with reports of these pathogens increasing at an alarming rate [2, 3]. This is especially true in the hospital environment where microbes with resistance towards conventional antibiotics are becoming increasingly common [4, 5], with recent high-profile examples including vancomycin-resistant enterococci [6] and methicillin-resistant *Staphylococcus aureus* (MRSA) [7]. There is clearly an urgent need for new antibiotics with novel mechanisms of antimicrobial action and the focus of this book is antimicrobial peptides (AMPs), which show high potential to serve in this capacity. In Chapter 1 we chart the discovery of AMPs, which is generally taken as the late 1980s when major research on these peptides began. This research revealed that AMPs are evolutionarily ancient molecules that are endogenous antibiotics produced by nearly all living organisms. However, it is now also becoming clear that in addition to their antimicrobial function, AMPs serve a variety of other roles, including modulation of the innate and adaptive immune systems via their ability to function as chemotactic agents. Over 2000 AMPs are now known and characterization has shown that the vast majority of these peptides are cationic, which are discussed in Chapter 2, while the remaining AMPs are generally anionic and are considered in Chapter 3. These chapters show that the capacity of AMPs to kill microbes depends upon a number of their structural and physiochemical characteristics such as charge and amphiphilicity, which facilitates their ability to partition into the membranes of the target organism. In most cases, this action leads to membrane permeabilization and death of the host microbe, although in some cases AMPs are translocated across the membrane to attack intracellular targets such as DNA.

The most researched AMPs are those that form α-helices (α-AMPs), which may be their inherent secondary structure, although in most cases these peptides are unfolded in solution and require the membrane interface to adopt α-helical conformations. These α-helices are generally amphiphilic, which allows the apolar face of the peptide to interact with the membrane hydrophobic core while concomitantly permitting its polar face to engage in electrostatic interactions with the membrane lipid head-group region [8]. Based on the spatial regularity of the residues within these amphiphilic structures, a number of techniques have been developed that can predict the potential of peptides to form membrane interactive α-AMPs [9, 10]. A number of prediction techniques for other types of AMPs have also been presented [11, 12] and an overview of this area of research is discussed in Chapter 4.

In addition to peptide-based properties, a number of membrane-based factors also contribute to the ability of AMPs to interact with membranes. Major examples of these factors include the transmembrane potential, lipid-packing characteristics, and the net negative charge generally carried by microbial membranes, which are targeted by cationic AMPs and thereby play fundamental roles in both the activity and selectivity of these peptides. Based on this research, a number of models to describe the antimicrobial activity of these peptides have been proposed, which are discussed in Chapter 5. In addition, these models appear to describe the ability of AMPs to target and kill cancer cells, the membranes of which also carry a net negative charge, which is discussed in Chapter 6. Currently, there is currently little evidence of microbial resistance to AMPs [13] and, taken with the research described in the foregone chapters, this has led to the view that AMPs are attractive propositions as lead compounds to serve in a number of scenarios [14]. Major examples of this use include the treatment of cancer [15], along with infection control in the food industry, agriculture [16, 17], and healthcare [5, 18–20]. For example, the fungal defensin, plectasin, and its derivative, NZ2114, are currently under development by Novozymes as lead compounds for use against MRSA and *S. aureus* with resistance to vancomycin [21].

Although much has been learnt about AMPs and their various biological roles since they were first discovered, the factors underpinning their microbial selectivity and toxicity are still poorly understood. In a sense, the antimicrobial mechanisms of these peptides draw parallels to the "lock and key" model postulated for enzyme activity [22], where the "key" refers to characteristics of the peptide and the "lock" refers to those of the target membrane. For the action of AMPs to occur efficiently, the "lock" and "key" need to be fully engaged, and it is hoped that the discussion of these peptides and our current understanding of their function will further stimulate research into fully elucidating these structure–function relationships as well as draw attention to the importance of AMPs in the pharmaceutical industry.

University of Central Lancashire, Preston, UK David A. Phoenix
November 2012 Sarah R. Dennison
  Frederick Harris

References

1 Costelloe, C., Metcalfe, C., Lovering, A., Mant, D., and Hay, A.D. (2010) Effect of antibiotic prescribing in primary care on antimicrobial resistance in individual patients: systematic review and meta-analysis. *British Medical Journal*, **340**, c2096.

2 Arias, C.A. and Murray, B.E. (2009) Antibiotic-resistant bugs in the 21st century – a clinical super-challenge. *The*

New England Journal of Medicine, **360**, 439–443.
3 Livermore, D.M. (2009) Has the era of untreatable infections arrived? *The Journal of Antimicrobial Chemotherapy*, **64** (Suppl. 1), i29–i36.
4 Spellberg, B., Guidos, R., Gilbert, D., Bradley, J., Boucher, H.W., Scheld, W.M., Bartlett, J.G., Edwards, J., Jr, and Infectious Diseases Society of America (2008) The epidemic of antibiotic-resistant infections: a call to action for the medical community from the Infectious Diseases Society of America. *Clinical Infectious Diseases*, **46**, 155–164.
5 Zucca, M., Scutera, S., and Savoia, D. (2011) Antimicrobial peptides: new frontiers in the therapy of infections, in *Drug Development – A Case Study Based Insight into Modern Strategies* (ed. C. Rundfeldt), InTech, New York, pp. 123–162.
6 Arias, C.A. and Murray, B.E. (2012) The rise of the *Enterococcus*: beyond vancomycin resistance. *Nature Reviews Microbiology*, **10**, 266–278.
7 Gould, I.M., David, M.Z., Esposito, S., Garau, J., Lina, G., Mazzei, T., and Peters, G. (2012) New insights into methicillin-resistant *Staphylococcus aureus* (MRSA) pathogenesis, treatment and resistance. *International Journal of Antimicrobial Agents*, **39**, 96–104.
8 Pasupuleti, M., Schmidtchen, A., and Malmsten, M. (2012) Antimicrobial peptides: key components of the innate immune system. *Critical Reviews in Biotechnology*, **32**, 143–171.
9 Phoenix, D.A. and Harris, F. (2002) The hydrophobic moment and its use in the classification of amphiphilic structures [review]. *Molecular Membrane Biology*, **19**, 1–10.
10 Phoenix, D.A., Harris, F., Daman, O.A., and Wallace, J. (2002) The prediction of amphiphilic alpha-helices. *Current Protein & Peptide Science*, **3**, 201–221.
11 Porto, W.F., Silva, O.N., and Franco, O. (2012) Prediction and rational design of antimicrobial peptides, in *Protein Structure* (ed. E. Faraggi), InTech, New York, pp. 377–396.
12 Juretic, D., Vukicevic, D., Petrov, D., Novkovic, M., Bojovic, V., Lucic, B., Ilic, N., and Tossi, A. (2011) Knowledge-based computational methods for identifying or designing novel, non-homologous antimicrobial peptides. *European Biophysics Journal with Biophysics Letters*, **40**, 371–385.
13 Peschel, A. and Sahl, H.-G. (2006) The co-evolution of host cationic antimicrobial peptides and microbial resistance. *Nature Reviews Microbiology*, **4**, 529–536.
14 Hadley, E.B. and Hancock, R.E.W. (2010) Strategies for the discovery and advancement of novel cationic antimicrobial peptides. *Current Topics in Medicinal Chemistry*, **10**, 1872–1881.
15 Ashley, L.H., Melanie, R.P.C., and David, W.H. (2012) Obstacles and solutions to the use of cationic antimicrobial peptides in the treatment of cancer, in *Small Wonders: Peptides for Disease Control* (ed. K. Rajasekaran, J.W. Cary, J. Jaynes, and E. Montesinos), American Chemical Society, Washington, DC, pp. 61–78.
16 Tiwari, B.K., Valdramidis, V.P., O'Donnell, C.P., Muthukumarappan, K., Bourke, P., and Cullen, P.J. (2009) Application of natural antimicrobials for food preservation. *Journal of Agricultural and Food Chemistry*, **57**, 5987–6000.
17 Meng, S., Xu, H., and Wang, F. (2010) Research advances of antimicrobial peptides and applications in food industry and agriculture. *Current Protein & Peptide Science*, **11**, 264–273.
18 Afacan, N.J., Yeung, A.T.Y., Pena, O.M., and Hancock, R.E.W. (2012) Therapeutic potential of host defense peptides in antibiotic-resistant infections. *Current Pharmaceutical Design*, **18**, 807–819.
19 Mayer, M.L., Easton, D.M., and Hancock, R.E.W. (2010) Fine tuning host responses in the face of infection: emerging roles and clinical applications of host defence peptides. *Advances in Molecular and Cellular Microbiology*, **18**, 195–220.
20 Park, S.-C., Park, Y., and Hahm, K.-S. (2011) The role of antimicrobial peptides in preventing multidrug-resistant bacterial infections and biofilm formation. *International Journal of Molecular Sciences*, **12**, 5971–5992.

21 Brinch, K.S., Tulkens, P.M., Van Bambeke, F., Frimodt-Møller, N., Høiby, N., and Kristensen, H.-H. (2010) Intracellular activity of the peptide antibiotic NZ2114: studies with *Staphylococcus aureus* and human THP-1 monocytes, and comparison with daptomycin and vancomycin. *Journal of Antimicrobial Chemotherapy*, **65**, 1720–1724.

22 Fischer, E. (1894) Einfluss der Configuration auf die Wirkung der Enzyme. *Berichte der deutschen chemischen Gesellschaft*, **27**, 2985–2993.

List of Abbreviations

α-ACPI	ACP with inactivity against cancer cells
[G–]	Active against Gram negative bacteria
[G+, G–]	Active against Gram positive and Gram negative bacteria
[G+, G–, F]	Active against Gram positive and Gram negative bacteria and fungi
[G+]	Active against Gram positive bacteria
ASP	agouti signal peptide
α-MSH	alpha-melanocyte stimulating hormone
AD	Alzheimer's Disease
ACE	angiotensin converting enzyme
AAMPs	Anionic antimicrobial proteins and peptides
ACPs	Anticancer peptides
APCs	antigen-presenting cells
AMPs	Antimicrobial peptides
ANN	Artificial neural networks
ATD	Atopic dermatitis
α-ACPs	α-helical anticancer peptides
α-AMPs	α-helical antimicrobial peptides
α-CAMPs	α-helical cationic antimicrobial peptides
α-LA	α-lactalbumin
ALA	5-aminolevulinic acid
BDs	Big defensins
β-LG	β-lactoglobulin
β-ACPs	β-sheet anticancer peptides
β-CAMPs	β-sheet cationic antimicrobial peptides
CL	Cardiolipin
CAMPs	Cationic antimicrobial peptides
CVC	Central venous catheters
CT	Chemotherapy
CS	Chondroitin sulfate
CF	Cystic fibrosis
CM	Cytoplasmic membranes
DCs	Dendritic cells

List of Abbreviations

DCD	Dermcidin
EGF	epidermal growth factor
E-ACPs	extended structures
EPS	Extracellular polymeric substance
FPA	fibrinopeptide A
FPB	fibrinopeptide B
FDA	Food and Drug Administration
$\langle\mu G\rangle$	Glycine moment
GAG	Glycosaminoglycan
[G+, G−, F, P]	Gram positive and Gram negative bacteria, and fungi and parasites
GPCR	G-protein-coupled receptor
HS	Heparan sulfate
HMMs	Hidden Markov models
HBDs	Human β-defensins
HIV	Human immunodeficiency virus
$\langle\mu_H\rangle$	Hydrophobic moment
$\langle H \rangle$	Hydrophobicity
HAs	hylids of Australia
IFS	Incremental feature selection
ACPAO	Ineffective against non-cancerous cells and erythrocytes
LFM	Lactoferricin
LD50	lethal dose 50%
LPS	Lipopolysaccharide
LPC	Lysophosphatidylcholine
LysylPG	Lysylated PG
MIP-3α	macrophage inflammatory protein-3α
Mc1r	melanocortin 1 receptor
M-enk	methionineenkephalin
M-enk-RF	methionine-enkephalin-arginine phenylalanine
MRSA	Methicillin-resistant Staphylococcus aureus
MICs	Minimum inhibition concentration
mRM	Minimum redundancy method
MD	Molecular dynamics
MHP	Molecular hydrophobic potential
MLP	Molecular lipophilic potential
MDR	Multi-drug resistant
MDRPA	Multidrug-resistant *Pseudomonas aeruginosa*
PR-CAMPs	multiple arginine residues-CAMPs
MyD88	myeloid differentiation primary response gene 88
NEP	neutral endopeptidase
OM	Outer membrane
OA	Ovalbumin
PLUNC	Palate, lung, nasal epithelium clone
PC	Phosphatidylcholine

PE	Phosphatidylethanolamine
PG	Phosphatidylglycerol
PI	Phosphatidylinositol
PS	Phosphatidylserine
PACT	Photodynamic antimicrobial chemotherapy
PLS	Principle latent structures
PCA	Principal component analysis
PEA	Proenkephalin A
PGs	Proteoglycans
QSAR	Quantitative structure activity relationships
RT	Radiation therapy
RANAEs	ranids from Asia, North America, and Europe
RBD3	rat β-defensin 3
ROS	reactive oxygen species
SHM	Shai, Huang and Matsazuki
SP	Sphingomyelin
SAR	Structure-activity relationships
SAAPs	Surfactant-associated anionic peptides
TLRs	Toll-like receptors
ACPT	Toxic to both cancerous and non-cancerous cells
TM	Transmembrane
$\Delta\psi$	Transmembrane potential
TA-CAMPs	tryptophan and arginine residues-CAMPs
WHO	World Health Organization

1
Antimicrobial Peptides: Their History, Evolution, and Functional Promiscuity

Summary

Eukaryotic antimicrobial peptides (AMPs) first became a research focus in the middle decades of the twentieth century with the description of cecropins from moths and magainins from frogs. Since then, the number of reported AMPs has burgeoned to over 2000 with representatives in virtually all eukaryotic organisms. The availability of databases of AMPs has facilitated phylogenetic analyses, which have shown that the origins of β-defensins are traceable to an ancestral gene over half a billion years old, indicating the evolutionarily ancient nature of AMPs. An emerging theme from research on AMPs is their multifunctional nature, which we review here for β-defensins, and show that these peptides play roles in wound healing and modulation of both the innate and adaptive immune systems via the dual ability to promote and suppress the proinflammatory response to microbial infection.

1.1
Introduction: The History of Antimicrobial Peptides

Antimicrobial peptides (AMPs) have been recognized in prokaryotic cells since 1939 when antimicrobial substances, named gramicidins, were isolated from *Bacillus brevis*, and were found to exhibit activity both *in vitro* and *in vivo* against a wide range of Gram-positive bacteria [1, 2]. Gramicidins were later shown to successfully treat infected wounds on guinea-pig skin, indicating their therapeutic potential for clinical use [3], and were the first AMPs to be commercially manufactured as antibiotics [4]. In the case of humans and other living creatures, which are constantly exposed to the threat of microbial infection, it had long been known that protection against these infections was provided by the adaptive immune system. However, this left the question as to why plants and insects, which lack an adaptive immune system, also remain free from infections for most of the time. The answer to this question is now known to be that similarly to prokaryotes, eukaryotes also produce AMPs and, historically, some sources attribute the discovery of eukaryotic AMPs to early work on plants [5] when in 1896 it was shown

Antimicrobial Peptides, First Edition. David A. Phoenix, Sarah R. Dennison, and Frederick Harris.
© 2013 Wiley-VCH Verlag GmbH & Co. KGaA. Published 2013 by Wiley-VCH Verlag GmbH & Co. KGaA.

that a substance lethal to bread yeast was present in wheat flour [6]. At the end of the 1920s, lysozyme was identified by Alexander Fleming and is considered by some authors to be the first reported instance of a peptide with antimicrobial activity [7]. However, the mechanism of action used by lysozyme is now known to be enzymatic destruction of the bacterial cell wall, placing it at its time of discovery in a different category to AMPs, which utilize non-enzymatic mechanisms of antimicrobial activity [8, 9]. In 1928, Fleming discovered penicillin [10] and in the 1940s, along with Howard Florey and Ernst Chain, he brought the therapeutic use of penicillin to fruition, which led these three men to share the 1945 Nobel Prize for Medicine [11]. With the advent of penicillin and streptomycin in 1943, began the "Golden Age of antibiotics," which led to a rapid loss of interest in the therapeutic potential of natural host antibiotics such as lysozyme and the importance of this immune defense strategy [12, 13]. However, in 1942, the antimicrobial substance that had previously been detected in wheat flour [6] was isolated from wheat endosperm (*Triticum aestivum*) and found to be a peptide that inhibited the growth of a variety of phytopathogens, such as *Pseudomonas solanacearum* and *Xanthomonas campestris* [14]. Later named purothionin in the mid-1970s [15, 16], this peptide is now known to be a member of the family of thionins, which are AMPs distributed across the plant kingdom [5]. At the time this work was undertaken there was also the realization that the "Golden Age of antibiotics" had ended and with the rise of multidrug-resistant microbial pathogens in the early 1960s, an awakened interest in host defense molecules was prompted [17, 18]. It is this point in time that some sources consider to be the true origin of research into AMPs [19], beginning with studies that were conducted in the 1950s and 1960s, when it was shown that cationic proteins were responsible for the ability of human neutrophils to kill bacteria via oxygen-independent mechanisms – clearly not activity associated with the adaptive immune system [20, 21]. In 1962, in what some consider to be the first description of an animal AMP [22], bombinin was reported in the orange speckled frog *Bombina variegate* [23]. Also in the 1960s, the antimicrobial protein, lactoferrin, was isolated from milk [24] and small antimicrobial molecules were observed to be induced in the hemolymph of wax moth larvae after challenging with *Pseudomonas aeruginosa* [25]. In the late 1970s and 1980s several groups reported a number of AMPs and antimicrobial proteins from leukocytes [26], including what are now known to be α-defensins from rabbits [27–29] and humans [30]. Along with purothionin described above, these defensins were among the first cysteine-stabilized AMPs to be reported (Chapter 2). In 1981, in what are now generally considered as landmark studies, Boman *et al.* injected bacteria into the pupae of the silk moth, *Hyalophora cecropia*, and isolated the inducible cationic antimicrobial proteins, P9A and P9B, from the hemolymph of these pupae (Chapter 2) [31]. Soon after, these peptides were sequenced, characterized, and renamed as the more familiar "cecropins," thereby constituting the first major α-helical AMPs to be reported [32]. In 1987, another landmark study occurred when Zasloff *et al.* (1987) isolated and characterized cationic AMPs from the African clawed frog, *Xenopus laevis*, and, reflecting their defense role, named these peptides magainins after the Hebrew word for "shield" [33]. A few years later, β-defensins and θ-defensins, which differ from α-defensins with respect to their cysteine pairings, were characterized after isolation, respectively, from bovine granulocytes [34] and leukocytes of the rhesus monkey [35]. In the mid-1990s, Brogden et al. identified the first anionic AMPs in *X. Laevis* [36] and characterized several other such peptides in ruminants, including sheep and cattle [37]. Ironically, also in the early 1990s, evidence began to accumulate that led to the current view that lysozyme possesses antimicrobial activity involving non-enzymatic mechanisms that are similar to AMPs, thereby substantiating the view that it was one of the first of these peptides to be discovered [8]. Based on these results, a number of investigators considered the possibility that AMPs may play a role in the defense systems of organisms lacking an adaptive immune system [38]. In the mid-1990s, this was confirmed for the fruit fly, *Drosophila melanogaster*, when it was shown that the deletion of a gene encoding an AMP rendered the insect susceptible to a massive fungal infection [39]. Since these earlier studies, AMPs have been extensively studied not only in plants [40, 41] and insects [42–44], but also other invertebrate organisms that lack an adaptive immune system [45–48] although most of the current understanding of AMPs has been obtained from studies on those isolated from amphibian skin secretions, which is a rich source of these peptides [49–52]. In combination, these studies have established that AMPs exist in virtually all multicellular organisms [53] and it is increasingly being recognized that that these peptides play an important role in the immune system of mammals, including humans [54–57]. These peptides have been identified at most sites of the human body normally exposed to microbes such as the skin and mucosae [54, 55], and are produced by a number of blood cell types, including neutrophils, eosinophils, and platelets [58–60]. However, as research into the expression of AMPs progressed, it became clear that the production of these peptides may be either constitutive or induced by inflammation or injury [38]. Typically, for example, α-defensins and dermcidin (the precursor of AMPs involved in skin defense) tend to be produced constitutively, whereas the majority of β-defensins are inducible [61–63]. Moreover, although particular AMPs may predominate at specific body sites only a small minority are exclusively produced by a certain cell type or tissue and each tissue has its own spectrum of AMPs that may vary in composition depending upon the prevailing physiological conditions [55]. For example, peptides derived from dermcidin are the major AMPs in human sweat but show differing profiles between the body sites of a given individual in response to exercise [64]. One question that has puzzled investigators since the discovery of AMPs is the fact that the minimal inhibitory concentrations (MICs) required for their *in vitro* antimicrobial activity are generally much higher than the physiological concentrations of these peptides found *in vivo* [65]. Two major explanations proposed for this observation are that at sites of inflammation, AMPs can accumulate at high local concentrations sufficiently above their MIC to exert their antimicrobial effect or that these peptides may act synergistically with other AMPs [65]. Over the last decade, these synergistic effects have been demonstrated for a variety of AMPs [66], including those that are structurally similar and from the same host organism, such as magainin and PGLa, which is another α-helical peptide from X. Laevis [67, 68], and those that are structurally dissimilar and from differing host organisms, such as LL-37, an α-helical human peptide, and indolicidin, an extended bovine peptide (Chapter 2) [69]. Studies over the last decade have also established that some organisms produce AMPs as suites of closely related peptides that synergize to produce a broad spectrum of antimicrobial activity, such as maximins, which are α-helical AMPs produced in the brains of amphibians [70], and cyclotides, which are cyclic cystine knot AMPs produced in the leaves, flowers, stems, and roots of various plants [71, 72]. As more has been learnt about AMPs, it has become somewhat arbitrary as to their precise definition. For example, perforin and the complement component C9 are large proteins of approximately 60 kDa that under physiological conditions insert into membranes and form pores as result of highly regulated immune processes, and are thus not classified as AMPs [73, 74]. In contrast, lactoferrin, which is around 80 kDa, is generally included as an AMP, based on the fact that it is ubiquitous in various body fluids and utilizes a non-specific mode of antimicrobial action similar to other AMPs (Chapter 5) [75, 76]. Moreover, some "AMPs" do not appear to exert direct antimicrobial activity such as the PLUNC (palate, lung, nasal epithelium clone) proteins that appear to primarily play a role in neutralizing endotoxins, promoting the agglutination of bacteria, and modulating cytokine production [77]. Nonetheless, it is now well established that the production of AMPs is a defense strategy used across eukaryotes, evidenced by the list of databases dedicated to these peptides that have appeared almost every year over the last decade (Table 1.1). Examination of these databases shows that in excess of 2000 AMPs have now been listed and the number of these peptides being reported is increasing rapidly [87, 89]. The availability of these databases has allowed comparisons to be made between AMPs based on a variety of criteria, most often structure–function relationships and mechanisms of antimicrobial action, which are discussed in later chapters of this book. However, two less well discussed aspects of AMPs are reviewed in the remainder of this chapter, namely their ancient origins and evolution along with their functional promiscuity, encompassing a number of biological roles in addition to antimicrobial activity.

1.2
AMPs: Evolutionarily Ancient Molecules

Since the discovery of magainins in X. laevis [33], it has become clear that the skin secretions of many anurans include a spectrum of AMPs that ranges between 10 and 20 members [92–94] along with a variety of other bioactive peptides, including neuropeptides, pheromones, and neuronal nitric oxide synthase inhibitors [95–97], which has led to the availability of sequence data for these molecules at the levels of protein [91] and DNA [98–100]. Taken with the ubiquity of AMPs across the anuran suborders [94], this has facilitated investigation into the evolutionary history of these peptides, particularly those from frogs of the Ranidae and Hylidae families [96, 101–107]. In a seminal work, one of the earliest investigations

Table 1.1 Representative databases dedicated to AMPs.

Year	Database	Website	Content	Key reference
2002	AMSDb	http://www.bbcm.univ.trieste.it/~tossi/amsdb.html	Animal/plant AMPs	–
2002	SAPD	http://oma.terkko.helsinki.fi:8080/~SAPD	Synthetic AMPs	[78]
2003	NAD	http://www.nih.go.jp/~jun/NADB/search.html	General AMPs	–
2004	A/OL	http://www.atoapps.nl/AOLKnowledge/	Antimicrobial compounds	–
2004	Peptaibol	http://www.cryst.bbk.ac.uk/peptaibol/home.shtml	Fungal AMPs	[79]
2006	CyBase	http://research.imb.uq.edu.au/cybase	Plant AMPs	[80]
2006	PenBase	penbase.immunaqua.com	Shrimp AMPs	[81]
2007	BACTIBASE	bactibase.pfba-lab.org	Bacterial AMPs	[82]
2007	Defensins	defensins.bii.a-star.edu.sg	Defensins across eukarya	[83]
2007	AMPer	http://marray.cmdr.ubc.ca/cgi-bin/amp.pl	Animal/plant AMPs	[84]
2008	PhytAMP	phytamp.pfba-lab-tun.org	Plant AMPs	[85]
2008	RAPD	http://faculty.ist.unomaha.edu/chen/rapd/index.php	Recombinant AMPS	[86]
2009	APD2	http://aps.unmc.edu/AP	Natural AMPs	[87]
2010	CAMP	http://www.bicnirrh.res.in/antimicrobial	General AMPs	[88]
2012	YADAMP	www.yadamp.unisa.it	General AMPs	[89]
2012	DAMPD	http://apps.sanbi.ac.za/dampd	General AMPs	[90]
2012	DADP	http://split4.pmfst.hr/dadp/	Amphibian AMPS	[91]

into the phylogeny of these peptides considered the evolutionary relationships between AMPs from hylids of South America (HSAs) and hylids of Australia (HAs) along with ranids from Asia, North America, and Europe (RANAEs) [102]. Essentially, this latter study derived the amino acid sequences for precursor proteins of caerins, which are AMPs from *Litoria caerulea*, a member of the HAs [96, 102], and then aligned these sequences with those of precursor proteins from various HSAs and RANAEs [102], which have previously been shown to belong to a single family, the pre-prodermaseptins [52]. This alignment revealed that the precursor proteins from these three groups of frogs possessed highly conserved

N-terminal pre-prosequences of approximately 50 residues, including a 22-residue signal peptide and an acidic spacer region, that was linked to a hyper variable C-terminal domain. This domain corresponded to progenitor AMPs with great diversity in length, sequence, net positive charge, and antimicrobial spectra [102]. The strong conservation of these N-terminal pre-prosequences allowed molecular phylograms to be constructed, which showed that nucleotide sequences of pre-prodermaseptins from HSAs, HAs, and RANAEs formed three separate clusters. Further analysis of these phylograms indicated a key result: the genes encoding the pre-prodermaseptins in these clusters arose from a common ancestral locus, which subsequently diversified by several rounds of duplication and divergence of loci. Most of these duplication events appeared to have occurred in a species ancestral to the ranids and hylids, although gene duplication in *L. caerulea* appeared to have taken place after the divergence of HSAs and HAs [102]. To provide a temporal framework for the origins and evolution of the pre-prodermaseptin genes, data from phylogenetic analyses [101, 102] were used in conjunction with the historical biogeography of hylids and ranids, which is strongly linked with tectonic events that occurred during fragmentation of the supercontinent Gondwana to eventually form the modern day continents [108, 109]. This historical reconstruction showed that these genes arose before the isolation of India and South America from Africa in a pan-Gondwanan land mass, and therefore originated from an ancestral gene in excess of 150 million years old [101, 102]. Moreover, given that hylids and ranids belong to the Neobatrachia, which diverged from Archeobatrachia in early Jurassic times [110, 111], and that pre-prodermaseptins have not been detected in this latter suborder [112, 113], these observations suggested that the ancestral gene of HSAs, HAs, and RANAEs may be up to around 200 million years old [101]. Further analysis of cDNA from pre-prodermaseptins suggested that duplications of this ancestral gene accompanied by accelerated mutations in the AMPs progenitor region and the action of positive selection all appeared to be mechanisms that contributed to the hypervariability of their C-terminal domain, and hence the great diversity of modern day AMPs from HSAs, HAs, and RANAEs [102]. Interestingly, there was evidence to suggest that the diversity of these AMPs may have in part resulted from random substitutions involving the operation of a mutagenic error-prone DNA polymerase [102] similar to that reported for some bacteria [114–116]. Since the initial study [102], these results have been supported and extended by later studies, which have been facilitated by the growing repository of cDNA for AMPs of anuran species [101, 106, 107, 113]. For example, recent phylogenetic analyses have shown that the signal sequences of AMPs are highly conserved not only within lineages of the Neobatrachia, but also within those of Bombinatoridae and Pipidae from the Archeobatrachia. Although high divergence between the signal sequences of these three lineages was observed, there was evidence to suggest that the genes encoding AMPs in anurans had evolved convergently on at least three occasions in evolutionary time [113]. In another study, which compared the signal sequences of a range of bioactive peptides from anurans, phylogenetic analyses showed that caerulein neuropeptides produced by *Litoria* spp. have a different evolutionary origin to the pre-prodermaseptins found in these frogs [105]. This latter study also showed that the profile of bioactive peptides produced by individual frogs from a range of species was sufficiently characteristic to form the basis of a diagnostic technique that was able to differentiate between subspecies and different population clusters of the same species and thereby provide insight into anuran evolutionary relationships. Use of this technique showed that the bioactive peptide profile of *L. caerulea* from mainland Australia differed strongly to that of the same species present on offshore islands that had been isolated for around 10 000 years, indicating that evolutionary change can be effected in a relatively short time in evolutionary terms [105]. Taken overall, these studies have led to the consensus view that the panopoly of AMPs produced by hylids, ranids, and other frogs represents the successful evolution of a defense system that maximizes host protection against rapidly changing microbial biota while minimizing the potential development of microbial resistance to these peptides [52, 101, 102].

In contrast to the peptides discussed above, which are produced by organisms of a single taxonomic order, defensins are AMPs that are produced by creatures across the eukaryotic kingdoms [48, 117–120]. In vertebrates these peptides have been identified in fish [121, 122], amphibians [92, 93, 96], reptiles [123, 124], rodents [125, 126], monotremes [127, 128], birds [129–132], marsupials [133], and mammals [134–136], including humans and other primates [137–139]. The defensins of vertebrates fall within the α-, β-, and θ-defensin groups described above, and are generally cationic, amphiphilic peptides that contain around 15–50 amino acid residues, including a conserved motif based on six cysteine residues that form three intramolecular disulfide bonds (Chapter 2). Both α- and β-defensins adopt triple-stranded antiparallel β-sheet configurations but whereas the former peptides form disulfide bonds via the linking of Cys1–Cys6, Cys2–Cys4, and Cys3–Cys5, the latter peptides form these bonds through links between Cys1–Cys5, Cys2–Cys4, and Cys3–Cys6 (Figure 1.1) [137, 138]. In contrast, θ-defensins are cyclized molecules of 18 residues that appear to be the product of a head-to-tail ligation of two truncated α-defensins and are the only known circular peptides of mammalian origin [140]. Defensins are also found in plants [141–143], fungi [144, 145], and invertebrates, such as insects [120, 146, 147], arachnids, including spiders [148, 149], ticks [150, 151], and scorpions [152, 153], crustaceans, including crabs [154–156] and lobsters [157], and bivalvia, including clams [158, 159] and mollusks [160–163]. These invertebrate peptides commonly adopt the cysteine-stabilized α-helical and β-sheet (CS$\alpha\beta$) fold, which consists of a single α-helix that is connected to a β-sheet formed from multiple antiparallel strands depending upon the number of disulfide bridges in the molecule [141, 164–167].

It has been proposed that all defensins evolved from a single precursor based on similarities in sequences, structures, modes of action, and the inter-functionality of these peptides derived from different kingdoms [120, 146, 168]. For example, plant defensins were found to be structurally similar to their insect counterparts [169] while some fungal defensins displayed high levels of sequence homology to those found in invertebrates [164, 166]. The identification of defensins in lower eukaryotes by these latter studies led to the suggestion that the ancestral gene of

(a) 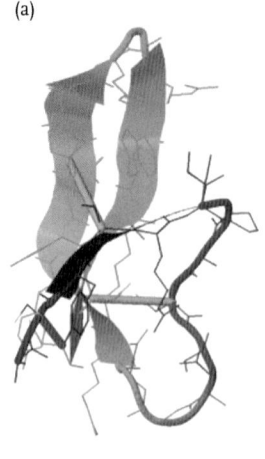   (b)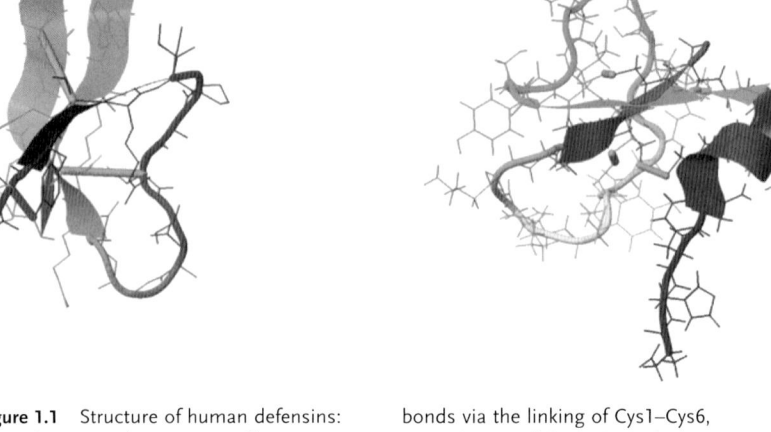

Figure 1.1 Structure of human defensins: three-dimensional structures of human α-defensin 1 (a) and β-defensin 1 (b). Both α- and β-defensins adopt triple-stranded antiparallel β-sheet configurations; however, whereas the former peptides form disulfide bonds via the linking of Cys1–Cys6, Cys2–Cys4, and Cys3–Cys5, the latter peptides form these bonds through links between Cys1–Cys5, Cys2–Cys4, and Cys3–Cys6.

these peptides existed before the fungal and insect lineages of the eukaryotic domain diverged, and may be therefore at least 1 billion years old [164, 166]. Consistent with these observations a conserved structural motif, the γ-core motif, was identified in cysteine-stabilized AMPs, including defensins, from organisms across the phylogenetic spectrum [170–172]. This motif was composed of two antiparallel β-sheets with an interposed short turn region and, based on its evolutionary lineage, it was proposed that defensins from mammals, plants, fungi, and invertebrates emerged from a common ancestor that can be traced back to prokaryotic origins and was therefore in the region of at least 2.6 billion years old [172, 173]. Strongly supporting this proposal, AMPs were identified in the myxobacteria *Anaeromyxobacter dehalogenans* and *Stigmatella aurantiaca*, which showed strong similarities in sequence and structure to fungal defensins, and it was suggested that these bacterial peptides may represent the ancestors of eukaryotic defensins [174, 175]. It has previously been hypothesized that myxobacteria played a central role in the endosymbiotic origins of the early eukaryotic nuclear genome [176, 177], and it was proposed that the transfer of myxobacterial genes encoding these ancestral defensins may have mediated the immune defense of early eukaryotes and thereby the lineage of modern day defensins [174, 175].

The evolutionary relationship between defensins from vertebrates and invertebrates is far from clear, but based on common sequence and structural features,

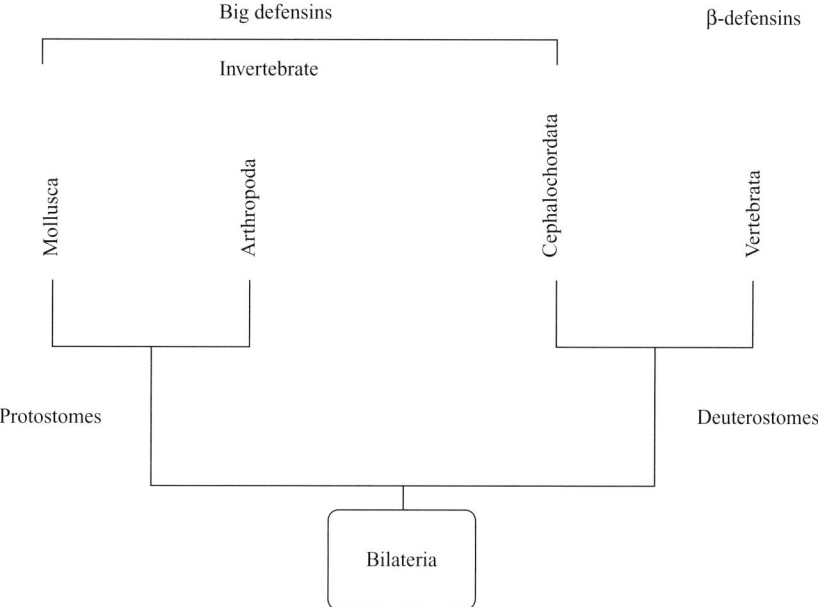

Figure 1.2 Evolution of β-defensins: phylogenetic tree indicating evolutionary relationships of bilateral animals that led to the emergence of β-defensins in the vertebrate lineage. This tree shows that the evolutionary history of these peptides to be traced back to Bilateria, indicating that they are in the region of at least 500 million years old [178–180].

there is evidence to suggest that some of the former peptides may have their origins in the "big defensins" (BDs), which have been identified in invertebrates, but not in vertebrates (Figure 1.2) [138, 181]. BDs are a family of evolutionarily conserved AMPs that were initially isolated from crustaceans [154–156], later identified in mussels [159, 160, 182–184], and most recently described in amphioxus [185], which are the closest invertebrate ancestors to vertebrates [178, 186, 187]. The presence of BDs in mollusks, which are Protostomes, and amphioxus, which are Deuterostomes (Cephalochordata), enables the evolutionary history of these peptides to be traced back to their common ancestor, Bilateria (Figure 1.2), indicating that they are in the region of at least 500 million years old [178–180]. BDs are between 79 and 94 residues in length, and comprise an N-terminal region that mediates activity against Gram-positive bacteria along with a C-terminal region that is active against Gram-negative bacteria [154, 181]. This C-terminal region has been shown to form a three-stranded β-sheet stabilized by three disulfide bridges that exhibits close structural homology to a number of a mammalian β-defensins, which also show selective activity against Gram-negative bacteria [138, 156, 181]. Taken with the fact that BDs and mammalian β-defensins show similar gene structures, the similarities in structure and bioactivity possessed by these two groups of AMPs suggested an evolutionary relationship with β-defensins emerging from BDs through exon shuffling or intronization of exonic sequences [181, 188]. In relation to other mammalian defensins, α-defensins have only been detected in certain animals [189], such as humans [138] and horses [134], but were found to be absent from others, such as cattle [190] and dogs [191]. The underlying reasons for the selective expression of α-defensins in mammals is not well understood but clearly, these AMPs arose after mammals diverged from other vertebrates [192]. Primarily founded on the physical proximity of α- and β-defensins on the chromosome as well as similarities between their structures and biological activities, it is generally accepted that the former AMPs have evolved from the latter peptides via gene duplication and subsequent positive diversifying selection [126, 189, 193]. Based on the fact that they had only been identified in therians [194], it had been suggested that α-defensins were at least 130 million years old, arising before the divergence of placental mammals and marsupials [136]. However, the recent identification of α-defensins in the duck-billed platypus pushed their emergence back to before the divergence of monotremes and therians [128], which occurred around 210 million years ago [195]. It is generally believed that mutated α-defensin genes gave rise to θ-defensin genes [138], and these peptides appear to be only expressed in the rhesus macaque monkey [196] and baboons [197, 198]. In humans and their closest primate relatives, including chimpanzees, bonobos, and gorillas, θ-defensin genes exist as pseudogenes and although they are transcribed, a premature termination codon in the signal sequence precludes their translation and to date these peptides have not been reported to occur naturally in human cells [140, 199, 200]. Given that humans evolved from orangutans over 7 million years ago, and that these latter primates have both intact and silenced copies of θ-defensin genes, it has been proposed that the premature stop codon found in some of these genes may have originated at this fork in evolution [201]. Moreover, based on the evolutionary divergence of Hominoidea from Cercopithecoidea, this would suggest that θ-defensins are at least 30 million years old [202, 203]. Interestingly, the pseudogenes of human θ-defensins, or retrocyclins, have been successfully expressed in human cells by using aminoglycosides to bypass the premature stop codon of their mRNA [201, 204] and have been shown to have potent antiviral activity [205], particular against HIV [204, 206, 207]. Given that viruses have evolved in the absence of selective pressures from retrocyclins, it has been proposed that restoration of the endogenous production of these peptides in human cervicovaginal tissues by the application of aminoglycoside-based topical microbicides may help prevent the sexual transmission of HIV [201, 204].

It is clear that AMPs possess an ancient history, and positive selection appears to be the major evolutionary driver of the structural and functional diversity observed for these peptides in the case of both vertebrates [208, 209] and invertebrates [210, 211], engendering AMPs with an ability to combat new or altered microbial pathogens [212, 213]. Indeed, it has been suggested that AMPs and microbial resistance mechanisms have coevolved, leading to a transient host–pathogen balance that has shaped the existing repertoire of these peptides found in nature [57, 170, 172, 173, 214]. For example, plants and humans are susceptible to different pathogens and thus their defensins are exposed to different selective pressures. As plants are generally much more susceptible to infection by fungi compared to other microbes, fungal pathogenicity would select for survivors with antifungal defensins. In contrast, humans are generally much more susceptible to infection by bacteria than fungi and thus bacterial pathogenicity would select for survivors with antibacterial defensins [215]. However, positive selection also appears to be an important factor in retaining duplicated genes and promoting the acquisition of a novel or more specialized function [216], which can provide the host with an evolutionary advantage as observed with the diversification of defensins [217–219]. For example, some plant defensins have no antifungal action, and, rather, inhibit the activity and synthesis of α-amylase, which is a digestive enzyme found in the gut of insects. It has been proposed that that the inhibition of α-amylase activity renders plant material indigestible and thus helps plants to resist feeding insects [141, 220]. In the case of vertebrates, it has been shown that β-defensins have evolved via gene duplication and diversification to become the major components of venom in the duck-billed platypus, believed to be the only venomous monotreme [127, 221]. Interestingly, some species of snakes and lizards produce venoms that include peptides derived from the duplication of β-defensins, and share many features with platypus venoms via convergent evolution [128, 222]. It is also interesting to note that the expression of platypus genes encoding venom defensins was detected in non-venom tissues, suggesting that these peptides may play other biological roles [223], and consistent with this suggestion, non-antimicrobial biological activities have been demonstrated for β-defensins in the case of a number of vertebrates [125, 137, 139, 224, 225]. This functional promiscuity of β-defensins represents an evolutionary advantage for mammals [181], which has been demonstrated for other AMPs [75, 226–231] and is now discussed below using defensins as a major example.

1.3
AMPs: Multifunctional Molecules

As described above, mammalian defensins were first isolated due to their antimicrobial properties [138, 232]; in other cases, however, AMPs were initially recognized for other functions and shown to possess antimicrobial activity at a later date [55]. For example, lactoferrin was originally recognized primarily for its ability to bind iron and later studies demonstrated that the protein was able to kill bacteria via iron-independent mechanisms of membrane interaction that were typical of AMPs [233]. The idea that peptides could serve both as AMPs and modulators of the immune system came from studies on α-melanocyte stimulating hormone (α-MSH). This peptide was well recognized for its endogenous melanogenic properties and ability to control inflammation via immunomodulation, promoting the generation of anti-inflammatory cytokines, and downregulating proinflammatory cytokines [234–237]. However, the widespread distribution of α-MSH in many barrier cells such as keratinocytes, fibroblasts, and melanocytes, and in various immune cells, including neutrophils, monocytes, and macrophages, suggested the potential for a wider role in host defense [238–240]. Confirming this suggestion, α-MSH is now generally included within the AMP's classification based on its potent broad-spectrum antimicrobial activity [241, 242], which includes membranolytic action against bacteria [243–245]. The first hints that established AMPs such as defensins may possess functional promiscuity came from comparative studies between these peptides and chemokines, which play a role in the induction of leukocyte trafficking. These two groups of peptides were originally thought to have distinct roles in the innate immune response, but it was found that their expression levels could be upregulated in response to microbial challenge; further studies showed that many chemokines were able to exert direct antimicrobial activity via mechanisms similar to AMPs while some defensins were capable of activating specific chemokine receptors and exerting chemoattractant activity [246, 247]. Such studies led to the realization that AMPs are also multifunctional peptides that serve a variety of biological roles, including immunomodulatory functions in host defense that can supplement their direct antimicrobial activity and lead to the resolution of infection [231, 248].

1.3.1
Defensins as Effectors of Immunity

The human β-defensins (HBDs) 1–4 are the most studied mammalian defensins and are primarily expressed in various epithelial tissues, including airway epithelia, urogenital tissues, nasolacrimal duct, mammary gland, testes, and epididymis along with some immune cells, such as monocytes and macrophages [232, 249–258]. The best known role of HBDs and other mammalian β-defensins as effectors of innate immunity is direct action against Gram-positive and Gram-negative bacteria, fungi, viruses, and parasites, which has been extensively reviewed elsewhere (Chapter 2) [38, 125, 139, 232, 251, 257]. However, in addition to their direct antimicrobial activities, HBDs play a critical role in regulating inflammation, which is essentially a protective response by the host's immune system to eliminate injurious stimuli, such as infection by pathogenic microbes, and to initiate healing at the inflamed tissue site [137, 139]. As a contribution to regulating inflammation, HBDs function as chemoattractants for a variety of immune cells [224, 232] and in a landmark study, Yang *et al.* demonstrated the recruitment of immature dendritic cells and CD4$^+$ memory T cells by a concentration gradient of HBD1 and 2 [259]. The chemotactic activity of HBD3 towards immature dendritic cells has also been demonstrated, along with that of several murine defensins, which have been shown to chemoattract immature murine dendritic cells [224]. In combination, these results clearly suggested that HBDs and other defensins are able to serve as modulators of the adaptive immune response to infection with their initiation into this process being marked by the recruitment of immature dendritic cells to sites of microbial entry into the host. This recruitment leads to the uptake, processing, and presentation of microbial antigens by mature dendritic cells with the subsequent induction of the antigen-specific immune system [257,

260, 261]. HBDs have also been show to exert chemotactic activity towards phagocytes, which include monocytes and macrophages, and form part of the front-line effector cells involved in innate defense against invading pathogenic microbes [262–266]. For example, it has been shown that HBD1 is chemotactic for monocytes while HBD2–4 are variously chemotactic for monocytes, macrophages, and mast cells [139, 267], which in the latter case can indirectly promote phagocyte recruitment [224, 257]. Once recruited to the site of infection, phagocytes become activated and work in conjunction with a range of effector molecules to internalize and kill infecting microbes. Moreover, the processing of microbes by both immature and mature dendritic cells is phagocytic, and hence contributes to innate immunity by directly killing pathogens. In addition, dendritic cells, particularly upon activation, produce numerous mediators, including cytokines, chemokines and AMPs, that participate in innate immunity [224, 257, 261, 268]. Taken together, these results indicated that HBDs can serve as modulators of both the innate and adaptive immune response to infection [125, 139, 232]. HBDs are able to function as proinflammatory mediators of the immune response by indirect mechanisms as shown by a recent study, which demonstrated that these peptides are able to suppress the apoptosis of neutrophils [269] and prolonging the lifespan of these cells is an inflammatory event that contributes to host defense against invading microbes. It was found that HBDs, particularly HBD3, exerted their suppressive effect on neutrophil apoptosis through binding to CCR6, which is a chemokine receptor. The resulting activation of this receptor led to a series of antiapoptotic events, including the upregulation of antiapoptotic proteins and the downregulation of proapoptotic proteins [269]. However, in many cases HBDs directly promote a proinflammatory immune response by binding to various cell receptors [137]. For example, the ability of HBDs 2 and 3, and to a lesser extent HBD1, to chemoattract immature dendritic cells and CD4$^+$ memory T cells is facilitated by interaction with CCR6 [259, 270], which is a G-protein-coupled receptor (GPCR) [271, 272]. In the case of HBD3, a number of studies have investigated the structure–function relationships underpinning these interactions, and found that both the N-terminal tertiary structure and specific residues proximal to this terminus were important to the receptor binding and chemotactic activity of the peptide [273–275]. These data were strongly supported by the results of structure–function studies on the murine ortholog of HBD3, DEFB14 [274–276], which also chemoattracts cells that express CCR6 along with macrophages from both mice and humans [275, 276]. However, the chemotactic activity of DEFB8, a murine β-defensin, towards immature dendritic cells and CD4$^+$ T cells [277] along with that of HBD3 and 4 towards macrophages was found to be independent of CCR6, suggesting the use of different receptors [273]. In the case of HBD3 this appears to be the chemokine receptor CCR2 [278, 279], which is also a GPCR [280, 281]. The mechanisms used by β-defensins to mediate chemotaxis are poorly understood [139] and are generally beyond the scope of this chapter; however, currently, three alternative models have been proposed [65]. According to the "alternate ligand model," β-defensins directly bind to a specific chemokine receptor, resulting in the initiation of receptor signaling. This model appears to include chemotaxis mediated by the binding of HBDs to CCR6 where these peptides serve as alternate ligands to the chemokine macrophage inflammatory protein-3α (MIP-3α), which is the natural agonist of this receptor. The ability of HBDs to bind CCR6 appeared to be based on similarities between the three-dimensional structures of HBDs and MIP-3α, including the presence of topological motifs that could mediate receptor recognition and the surface distribution of cationic residues on these molecules [232, 246]. The "membrane disruption model" proposes that β-defensins modify the membrane microdomain associated with the chemokine receptor, which indirectly changes its function to either signal without a ligand or to become unresponsive to binding by its specific ligand. In contrast, the "trans-activation model" suggests that β-defensins stimulate the release of a membrane-bound growth factor, which then binds and activates its high-affinity receptor (Figure 1.3) [65].

In addition to binding chemokine receptors, β-defensins have also been shown to interact with antigen-presenting cells (APCs) via Toll-like receptors (TLRs),

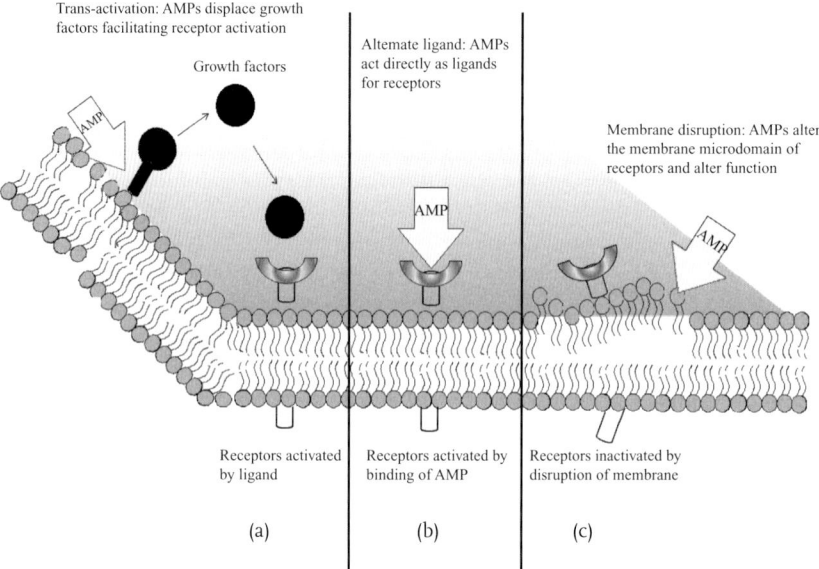

Figure 1.3 Models for the interaction of β-defensins with cell surface receptors: putative models for the interaction of defensins with chemokine receptors (adapted from [65]). According to the "trans-activation model" (a), AMPs stimulate the release of a membrane-bound growth factor, which then binds to its high-affinity receptor with activation resulting. In the "alternate ligand model" (b), AMPs bind directly to the receptor, which results in the initiation of signaling. The "membrane disruption model" (c) proposes that AMPs modify the membrane microdomain associated with the receptor, which indirectly leads to a change in receptor function. This functional change allows the receptor to either signal without a ligand or become insensitive to binding by its specific ligand.

which are integral membrane glycoproteins that recognize microbial components [282, 283]. For example, it has been reported that the activation of APCs by HBD3 is mediated by binding of the peptide to TLR1 and TLR2 and the subsequent activation of the MyD88 (myeloid differentiation primary response gene 88)-dependent signaling pathway [284]. MyD88 is an adapter molecule that is recruited by activated TLRs in order to propagate a signal [283]. Studies on MBD2, which is a murine ortholog of HBD2, showed that interactions between the peptide and TLR4 mediated the upregulation of costimulatory proteins on immature dendritic cells, including CD40, CD80, and CD86 [285], which play a role in the maturation of these cells [286]. MBD2 was found to stimulate the maturation of immature dendritic cells, which led to an adaptive immune response, including the production of proinflammatory cytokines [285, 287]. Most recently, the induction of proinflammatory cytokines by HBD3 was demonstrated in human macrophages by the enhanced expression of a variety of gene transcripts, including those for tumor necrosis factor-α, interleukin (IL)-1α, IL-6, and IL-8 [279]. In human monocytes, the increased expression of proinflammatory cytokines was detected at the protein level for IL-1α, IL-6, and IL-8, which appeared to be induced by the binding of HBD3 to TLR1 and TLR2 [288]. A number of studies have shown that HBD2–4 can stimulate human keratinocytes to increase the production of a variety of proinflammatory mediators, including IL-6, IL-8, IL-10, IL-20, and MIP-3α [289–291]. In some cases, there was evidence to suggest that the induction of these mediators was dependent on the interaction of HBDs with a GPCR [289]. The use of a GPCR was also implicated in the ability of HBD2–4 to enhance the expression of the pruritogenic cytokine, IL-31, in a variety of human mast cells, which constituted a novel source of this cytokine. These studies also showed that the expression of IL-31 was elevated in psoriatic skin mast cells, which led to the suggestion that the HBD-stimulated production of pruritogenic factors by mast cells provided a novel mechanism by, which human AMPs may contribute to inflammatory reactions and suggested a role for these HBDs in the pathogenesis of psoriasis and other skin disorders [292].

Collectively, the studies described above demonstrate that HBDs and other β-defensins are able to bind a variety of cell surface receptors and enhance the adaptive and innate immune response. However, studies over recent years have suggested that these peptides are also able to attenuate the immune response by suppressing a proinflammatory response [137, 293]. The ability of HBDs and other β-defensins to exert an immunosuppressive effect on proinflammatory responses induced by bacterial and viral components has been suggested by investigations both *in vitro* [137, 294–297] and *in vivo* [293, 298, 299]. In general, the mechanisms underpinning these immunosuppressive effects are unclear but recent work has suggested that HBDs and other β-defensins may combine both pro- and anti-inflammatory capacities with the balance of these effects determined by the levels of expression of the HBD peptides. For example, HBDs may be produced at high levels at the site of microbial infection, resulting in a proinflammatory response and the chemotactic recruitment of immune cells such as macrophages, as described above. However, as the microbial threat is nullified and the levels of

HBDs present at the site of infection decreases, the ability of these peptides to act as proinflammatory suppressors contributes to the resolution of infection [137].

1.3.2
Defensins and Wound Healing

A number of studies showed that HBD1–4 was expressed in skin substitutes prepared from keratinocyte cultures and split skin grafts from healthy and burned donors [300–302]. It has also been shown that a number of these HBDs stimulate the migration and proliferation of epidermal keratinocytes, which led to the suggestion that these peptides may promote cutaneous wound healing [289]. Consistent with this suggestion, the expression of HBD2 appeared to be upregulated in acute and chronic wounds while the presence of the peptide was not detected in healthy skin [303, 304]. It has been shown that the expression of HBD2 is significantly induced by epidermal growth factor (EGF) when costimulated with the cytokine, IL-1α [305]. It has previously been shown that this defensin is upregulated by IL-1α in the human corneal epithelium in response to ocular surface injury and it was suggested that the presence of the peptide in the regenerating corneal epithelium may indicate that it plays a direct role in the wound-healing process [306]. HBD2 also appears to play a role in the restitutive migration of epithelial cells and wound healing of the mucosal barrier in the intestine [307], while other studies have shown that, similarly to vascular endothelial growth factor, the peptide is able to promote the migration and wound healing of endothelial cells, manifested by the formation of capillary-like tubes with human umbilical vein endothelial cells [308]. More recent studies have shown that a high-glucose environment reduces HBD2 expression in keratinocytes via the downregulation of STAT protein signaling, which led to the observation that reduced expression of the peptide would have severe negative effects on the wound healing process under diabetic conditions, including impaired keratinocyte migration and suboptimal neovascular formation [309]. HBD3 is highly expressed in keratinocytes, especially at wound sites in response to the growth factors such as EGF, transforming growth factor-α, and insulin growth factor-1 [310–312], and it has been shown that the peptide promotes the proliferation and migration of keratinocytes through phosphorylation of the EGF receptor and STAT proteins [289]. Consistent with these findings, studies on murine models showed that the expression of rat β-defensin 3 (RBD3) was enhanced at both the mRNA and protein levels after skin wounding. However, this study also showed that when the skin of diabetic rats was wounded, RBD3 induction was negligible and similar observations were made for HBD3 expression in human keratinocytes under various glucose-treatment conditions, indicating that both defensins were dysfunctional under hyperglycemic conditions [313]. Reinforcing the importance of HBD3 to the process of wound healing, several studies demonstrated that gene transfer of the peptide to infected excisional wounds on the backs of diabetic pigs enhanced wound closure by around 25% compared to controls [314, 315].

1.3.3
Defensins and Canine Coat Color

CBD103 is a canine β-defensin with potent antibacterial activity that is widely expressed in the skin of dogs, thereby participating in the innate defense of the host [191, 316–318]. However, in addition to its immune-related functions, a number of recent studies have suggested that CBD103 may play a role in determining the color of some dog coats [137, 319]. A number of genes and their protein products were known to mediate the pigmentation of dog coats although interactions between the melanocortin 1 receptor (Mc1r) and its antagonist, the agouti signal peptide (ASP) appeared to be the primary determinants of this process [320]. Essentially, constitutive signaling by Mc1r located on the surface of melanocytes controls the production of eumelanin, or black pigment, and the binding of ASP to the receptor antagonizes this signaling, resulting in the production of pheomelanin, or red/yellow pigment [319, 321]. Differential signaling and genetic variation in both ASP and Mc1r was able to explain much of the variation found in canine coat color [319], but it had long been known that these interactions could not account for the dominant black coat color observed in breeds such as Labrador and Golden retrievers [322]. In response, recent studies showed that this dominant black phenotype was associated with an allele located at the K locus of the canine genome [323, 324], which encoded a variant of CBD103 that lacked an N-terminal glycine residue [325]. This mutant defensin was found to act as a neutral antagonist to Mc1r, thereby preventing ASP from associating with the receptor and allowing unrestricted signaling to synthesize black pigment [326]. Interestingly, through mating with domestic dogs, this dominant mutation has been identified in North American wolves and occurs at high frequency in those of these animals that inhabit forested locations where dark color would provide an advantage, demonstrating a molecular signature of positive selection [327]. Most recently, evidence has been presented suggesting that orthologs of CBD103 may be linked to the determination of coat color phenotypes in cattle [328] and function as novel Mc1r agonists in the regulation of melanocyte responses in humans [326, 329].

1.4
Discussion

In this chapter, the history of AMPs has been described, and the discovery of these peptides provided an explanation as to why plants and insects, which lack an adaptive immune systems, are able to resist infections. This chapter has shown that the starting point for the history of AMPs is popularly taken to be the 1980s based on the first descriptions of cecropins in 1981 [32] and magainins in 1987 [33]. However, there is some debate as to the validity of this assertion, which appears to be not without foundation, given that in 1962 AMPs were reported in the European frog, *B. variegate* [23], predating the first reports of magainins by around 25 years. However, while accreditation for the discovery of AMPs is an academic matter, the subsequent explosive increase in their rate of identification and characterization (Table 1.1) has provided some very tangible benefits in relation to the understanding and utilization of these peptides. The most obvious of these benefits is the promise shown by these peptides to act as lead molecules for development as novel antimicrobial agents and as anticancer agents (Chapter 6). However, the availability of databases of AMPs with detailed descriptions of their structural and antimicrobial properties has provided other benefits, such as the construction of models to describe their membrane interactions, which is a key factor in the ability of these peptides to kill microbes (Chapters 5). Such understanding can also provide insight into other lipid–peptide interactions within biological systems. The availability of databases of AMPs with structural information has also contributed to the development of theoretical techniques to analyze the properties of AMPs, including the prediction of their membrane interactive potential (Chapter 4) along with phylogenetic analysis of their evolutionary relationships as discussed above. This chapter has shown that AMPs have an evolutionarily ancient history with pre-prodermaseptins from frogs of the Ranidae and Hylidae families, originating from an ancestral gene over 200 million years old. Moreover, these AMPs have evolved to give modern frogs profiles of these peptides that are sufficiently characteristic to aid the taxonomic classification of these creatures and link their biodiversity to their recent evolutionary history. In the case of vertebrate β-defensins, these peptides appear to have evolved from the C-terminal domain of ancestral invertebrate defensins with an evolutionary history that can be traced back to bilateran organisms that lived over half a billion years ago (Figure 1.2).

An emerging theme from the voluminous literature on AMPs is the multifunctional nature of these peptides. In the case of β-defensins, these peptides contribute to host defense by direct antimicrobial action and by their ability to modulate both the innate and adaptive immune systems. These peptides have been shown to promote the proinflammatory response via indirect mechanisms such as inhibiting the apoptosis of immune cells and directly through their ability to bind to immune cell receptors and induce the chemoattraction of these cells to sites of microbial infection, thereby facilitating pathogen clearance. There is also increasing evidence that β-defensins are able to suppress the proinflammatory response, thereby contributing to the resolution of infection. Increasingly, other biological functions for β-defensins are being reported and here we have briefly described several of these functions, including the ability of these peptides to influence canine coat color and to contribute to wound healing. However, beyond the scope of this chapter, β-defensins have also been reported to play roles in mammalian fertility and development along with conditions ranging from psoriasis and atopic dermatitis to cancer and Crohn's disease, which have been extensively reviewed elsewhere [19, 125, 137, 139, 251, 330]. Taking this chapter as a whole, it is clear that AMPs are ancient molecules; however, given their evolutionary significance and functional promiscuity, they would appear to have a promising future in areas ranging from phylogeny to therapeutics.

References

1. Dubos, R.J. (1939) Studies on bactericidal agent extracted from a soil *Bacillus*: I. Preparation of the agent. Its activity *in vitro*. *The Journal of Experimental Medicine*, **70**, 1–10.
2. Dubos, R.J. (1939) Studies on bactericidal agent extracted from a soil *Bacillus*: II. Protective effect of the vactericidal agent against experimental *Pneumococcus* infections in mice. *The Journal of Experimental Medicine*, **70**, 11–17.
3. Gause, G.F. and Brazhnikova, M.G. (1944) Gramicidin S and its use in the treatment of infected wounds. *Nature*, **154**, 703–703.
4. Van Epps, H.L. (2006) Rene Dubos: unearthing antibiotics. *The Journal of Experimental Medicine*, **203**, 259–259.
5. Stec, B. (2006) Plant thionins – the structural perspective. *Cellular and Molecular Life Sciences*, **63**, 1370–1385.
6. Jago, W. and Jago, W. (1926) Toxic action of wheat flour to brewer's yeast, in *Industrial Fermentations* (ed. W. Allen), The Chemical Catering Company, New York, pp. 128–167.
7. Fleming, A. (1922) On a remarkable bacteriolytic element found in tissues and secretions. *Proceedings of the Royal Society of London B*, **93**, 306–317.
8. Benkerroum, N. (2008) Antimicrobial activity of lysozyme with special relevance to milk. *African Journal of Biotechnology*, **7**, 4856–4867.
9. Vocadlo, D.J., Davies, G.J., Laine, R., and Withers, S.G. (2001) Catalysis by hen egg-white lysozyme proceeds via a covalent intermediate. *Nature*, **412**, 835–838.
10. Fleming, A. (2001) On the antibacterial action of cultures of a penicillium, with special reference to their use in the isolation of *B. influenzae*. *Bulletin of the World Health Organization*, **79**, 780–790.
11. Brown, K. (2004) The history of penicillin from discovery to the drive to production. *Pharmaceutical Historian*, **34**, 37–43.
12. Zaffiri, L., Gardner, J., and Toledo-Pereyra, L.H. (2012) History of antibiotics. From Salvarsan to Cephalosporins. *Journal of Investigative Surgery*, **25**, 67–77.
13. Bentley, R. (2009) Different roads to discovery; Prontosil (hence sulfa drugs) and penicillin (hence beta-lactams). *Journal of Industrial Microbiology & Biotechnology*, **36**, 775–786.
14. Fernandez de Caleya, R., Gonzalez-Pascual, B., Garcia-Olmedo, F., and Carbonero, P. (1972) Susceptibility of phytopathogenic bacteria to wheat purothionins *in vitro*. *Applied Microbiology*, **23**, 998–1000.
15. Ohtani, S., Okada, T., Yoshizumi, H., and Kagamiyama, H. (1977) Complete primary structures of 2 subunits of purothionin-A, a lethal protein for brewers yeast from wheat flour. *Journal of Biochemistry*, **82**, 753–767.
16. Mak, A.S. and Jones, B.L. (1976) Amino acid sequence of wheat beta-purothionin. *Canadian Journal of Biochemistry*, **54**, 835–842.
17. Davies, J. (2006) Where have all the antibiotics gone? *The Canadian Journal of Infectious Diseases & Medical Microbiology*, **17**, 287–290.
18. Katz, M.L., Mueller, L.V., Polyakov, M., and Weinstock, S.F. (2006) Where have all the antibiotic patents gone? *Nature Biotechnology*, **24**, 1529–1531.
19. Nakatsuji, T. and Gallo, R.L. (2012) Antimicrobial peptides: old molecules with new ideas. *The Journal of Investigative Dermatology*, **132**, 887–895.
20. Hirsch, J.G. (1956) Phagocytin: a bactericidal substance from polymorphonuclear leucocytes. *The Journal of Experimental Medicine*, **103**, 589–611.
21. Zeya, H.I. and Spitznagel, J.K. (1966) Cationic proteins of polymorphonuclear leukocyte lysosomes. II. Composition, properties, and mechanism of antibacterial action. *Journal of Bacteriology*, **91**, 755–762.
22. Bagnicka, E., Jozwik, A., Strzalkowska, N., Krzyzewski, J., and Zwierzchowski, L.

(2011) Antimicrobial peptides – outline of the history of studies and mode of action. *Medycyna Weterynaryjna*, **67**, 444–448.

23 Kiss, G. and Michl, H. (1962) Uber das Giftsekret der Gelbbauchunke, *Bombina variegata* L. *Toxicon*, **1**, 33–34.

24 Groves, M.L., Peterson, R.F., and Kiddy, C.A. (1965) Polymorphism in the red protein isolated from milk of individual cows. *Nature*, **207**, 1007–1008.

25 Stephens, J.M. and Marshall, J.H. (1962) Some properties of an immune factor isolated from the blood of actively immunised wax moth larvae. *Canadian Journal of Microbiology*, **8**, 719–725.

26 Levy, O. (2004) Antimicrobial proteins and peptides: anti-infective molecules of mammalian leukocytes. *Journal of Leukocyte Biology*, **76**, 909–925.

27 Selsted, M.E., Brown, D.M., Delange, R.J., Harwig, S.S.L., and Lehrer, R.I. (1985) Primary structures of 6 antimicrobial peptides of rabbit peritoneal neutrophils. *Journal of Biological Chemistry*, **260**, 4579–4584.

28 Selsted, M.E., Szklarek, D., and Lehrer, R.I. (1984) Purification and antibacterial activity of antimicrobial peptides of rabbit granulocytes. *Infection and Immunity*, **45**, 150–154.

29 Selsted, M.E., Brown, D.M., Delange, R.J., and Lehrer, R.I. (1983) Primary structures of MCP-1 and MCP-2, natural peptide antibiotics of rabbit lung macrophages. *Journal of Biological Chemistry*, **258**, 4485–4489.

30 Ganz, T., Selsted, M.E., Szklarek, D., Harwig, S.S.L., Daher, K., Bainton, D.F., and Lehrer, R.I. (1985) Defensins – natural peptide antibiotics of human neutrophils. *Journal of Clinical Investigation*, **76**, 1427–1435.

31 Hultmark, D., Steiner, H., Rasmuson, T., and Boman, H.G. (1980) Insect immunity – purification and properties of 3 inducible bactericidal proteins from hemolymph of immunised pupae of *Hyalophora cecropia*. *European Journal of Biochemistry*, **106**, 7–16.

32 Steiner, H., Hultmark, D., Engstrom, A., Bennich, H., and Boman, H.G. (1981) Sequence and specificity of two anti-bacterial proteins involved in insect immunity. *Nature*, **292**, 246–248.

33 Zasloff, M. (1987) Magainins, a class of antimicrobial peptides from *Xenopus* skin: isolation, characterization of two active forms, and partial cDNA sequence of a precursor. *Proceedings of the National Academy of Sciences of the United States of America*, **84**, 5449–5453.

34 Selsted, M.E., Tang, Y.Q., Morris, W.L., McGuire, P.A., Novotny, M.J., Smith, W., Henschen, A.H., and Cullor, J.S. (1993) Purification, primary structures, and antibacterial activities of beta-defensins, a new family of antimicrobial peptides from bovine neutrophils. *Journal of Biological Chemistry*, **268**, 6641–6648.

35 Tang, Y.Q., Yuan, J., Osapay, G., Osapay, K., Tran, D., Miller, C.J., Ouellette, A.J., and Selsted, M.E. (1999) A cyclic antimicrobial peptide produced in primate leukocytes by the ligation of two truncated alpha-defensins. *Science*, **286**, 498–502.

36 Brogden, K.A., Ackermann, M., and Huttner, K.M. (1997) Small, anionic, and charge-neutralizing propeptide fragments of zymogens are antimicrobial. *Antimicrobial Agents and Chemotherapy*, **41**, 1615–1617.

37 Brogden, K.A., Ackermann, M., McCray, P.B., and Tack, B.F. (2003) Antimicrobial peptides in animals and their role in host defences. *International Journal of Antimicrobial Agents*, **22**, 465–478.

38 Ganz, T. (2003) The role of antimicrobial peptides in innate immunity. *Integrative and Comparative Biology*, **43**, 300–304.

39 Lemaitre, B., Nicolas, E., Michaut, L., Reichhart, J.-M., and Hoffmann, J.A. (1996) The Dorsoventral Regulatory Gene Cassette spätzle/Toll/cactus controls the potent antifungal response in *Drosophila* adults. *Cell*, **86**, 973–983.

40 Duran, N., Marcato, P.D., Duran, M., Yadav, A., Gade, A., and Rai, M. (2011) Mechanistic aspects in the biogenic synthesis of extracellular metal nanoparticles by peptides, bacteria, fungi, and plants. *Applied Microbiology and Biotechnology*, **90**, 1609–1624.

41 Barbosa Pelegrini, P., Del Sarto, R.P., Silva, O.N., Franco, O.L., and Grossi-de-Sa, M.F. (2011) Antibacterial peptides from plants: what they are and how they probably work. *Biochemistry*, **2011**, 250349.

42 Bulet, P. and Stocklin, R. (2005) Insect antimicrobial peptides: structures, properties and gene regulation. *Protein and Peptide Letters*, **12**, 3–11.

43 Ravi, C., Jeyashree, A., and Renuka Devi, K. (2011) Antimicrobial peptides from insects: an overview. *Research in Biotechnology*, **2**, 1–7.

44 Ntwasa, M., Goto, A., and Kurata, S. (2012) Coleopteran antimicrobial peptides: prospects for clinical applications. *International Journal of Microbiology*, **2012**, 101989.

45 Lee, J.K., Jeon, J.-K., Kim, S.-K., and Byun, H.-G. (2012) Characterization of bioactive peptides obtained from marine invertebrates. *Advances in Food and Nutrition Research*, **65**, 47–72.

46 Sperstad, S.V., Haug, T., Blencke, H.-M., Styrvold, O.B., Li, C., and Stensvag, K. (2011) Antimicrobial peptides from marine invertebrates: challenges and perspectives in marine antimicrobial peptide discovery. *Biotechnology Advances*, **29**, 519–530.

47 Hancock, R.E.W., Brown, K.L., and Mookherjee, N. (2006) Host defence peptides from invertebrates–emerging antimicrobial strategies. *Immunobiology*, **211**, 315–322.

48 Bulet, P., Stocklin, R., and Menin, L. (2004) Anti-microbial peptides: from invertebrates to vertebrates. *Immunological Reviews*, **198**, 169–184.

49 Amiche, M. and Galanth, C. (2011) Dermaseptins as models for the elucidation of membrane-acting helical amphipathic antimicrobial peptides. *Current Pharmaceutical Biotechnology*, **12**, 1184–1193.

50 Thomas, P., Kumar, T.V.V., Reshmy, V., Kumar, K.S., and George, S. (2012) A mini review on the antimicrobial peptides isolated from the genus Hylarana (Amphibia: Anura) with a proposed nomenclature for amphibian skin peptides. *Molecular Biology Reports*, **39**, 6943–6947.

51 Fernandez, D.I., Gehman, J.D., and Separovic, F. (2009) Membrane interactions of antimicrobial peptides from Australian frogs. *Biochimica et Biophysica Acta – Biomembranes*, **1788**, 1630–1638.

52 Nicolas, P. and El Amri, C. (2009) The dermaseptin superfamily: a gene-based combinatorial library of antimicrobial peptides. *Biochimica et Biophysica Acta – Biomembranes*, **1788**, 1537–1550.

53 Zasloff, M. (2002) Antimicrobial peptides of multicellular organisms. *Nature*, **415**, 389–395.

54 Bevins, C.L. and Salzman, N.H. (2011) Paneth cells, antimicrobial peptides and maintenance of intestinal homeostasis. *Nature Reviews Microbiology*, **9**, 356–368.

55 Wiesner, J. and Vilcinskas, A. (2010) Antimicrobial peptides: the ancient arm of the human immune system. *Virulence*, **1**, 440–464.

56 Cederlund, A., Gudmundsson, G.H., and Agerberth, B. (2011) Antimicrobial peptides important in innate immunity. *FEBS Journal*, **278**, 3942–3951.

57 Pasupuleti, M., Schmidtchen, A., and Malmsten, M. (2012) Antimicrobial peptides: key components of the innate immune system. *Critical Reviews in Biotechnology*, **32**, 143–171.

58 Cox, D., Kerrigan, S.W., and Watson, S.P. (2011) Platelets and the innate immune system: mechanisms of bacterial-induced platelet activation. *Journal of Thrombosis and Haemostasis*, **9**, 1097–1107.

59 Malik, A. and Batra, J.K. (2012) Antimicrobial activity of human eosinophil granule proteins: involvement in host defence against pathogens. *Critical Reviews in Microbiology*, **38**, 168–181.

60 Risso, A. (2000) Leukocyte antimicrobial peptides: multifunctional effector molecules of innate immunity. *Journal of Leukocyte Biology*, **68**, 785–792.

61 Hancock, R.E.W. and Scott, M.G. (2000) The role of antimicrobial peptides in animal defenses. *Proceedings of the National Academy of Sciences of the United States of America*, **97**, 8856–8861.

62 Scott, M.G. and Hancock, R.E.W. (2000) Cationic antimicrobial peptides and their multifunctional role in the immune system. *Critical Reviews in Immunology*, **20**, 407–431.

63 Schittek, B., Hipfel, R., Sauer, B., Bauer, J., Kalbacher, H., Stevanovic, S., Schirle, M., Schroeder, K., Blin, N., Meier, F., Rassner, G., and Garbe, C. (2001) Dermcidin: a novel human antibiotic peptide secreted by sweat glands. *Nature Immunology*, **2**, 1133–1137.

64 Rieg, S., Seeber, S., Steffen, H., Humeny, A., Kalbacher, H., Stevanovic, S., Kimura, A., Garbe, C., and Schittek, B. (2006) Generation of multiple stable dermcidin-derived antimicrobial peptides in sweat of different body sites. *Journal of Investigative Dermatology*, **126**, 354–365.

65 Lai, Y. and Gallo, R.L. (2009) AMPed up immunity: how antimicrobial peptides have multiple roles in immune defense. *Trends in Immunology*, **30**, 131–141.

66 Cassone, M. and Otvos, L., Jr (2010) Synergy among antibacterial peptides and between peptides and small-molecule antibiotics. *Expert Review of Anti-Infective Therapy*, **8**, 703–716.

67 Westerhoff, H.V., Zasloff, M., Rosner, J.L., Hendler, R.W., Dewaal, A., Gomes, A.V., Jongsma, A.P.M., Riethorst, A., and Juretic, D. (1995) Functional synergism of the magainins PGLA and Magainin-2 in *Escherichia coli*, tumor cells and liposomes. *European Journal of Biochemistry*, **228**, 257–264.

68 Matsuzaki, K., Mitani, Y., Akada, K., Murase, O., Yoneyama, S., Zasloff, M., and Miyajima, K. (1998) Mechanism of synergism between antimicrobial peptides magainin 2 and PGLa. *Biochemistry*, **37**, 15144–15153.

69 Yan, H. and Hancock, R.E.W. (2001) Synergistic interactions between mammalian antimicrobial defense peptides. *Antimicrobial Agents and Chemotherapy*, **45**, 1558–1560.

70 Liu, R., Liu, H., Ma, Y., Wu, J., Yang, H., Ye, H., and Lai, R. (2011) There are abundant antimicrobial peptides in brains of two kinds of *Bombina* toads. *Journal of Proteome Research*, **10**, 1806–1815.

71 Gruber, C.W., Cemazar, M., Anderson, M.A., and Craik, D.J. (2007) Insecticidal plant cyclotides and related cystine knot toxins. *Toxicon*, **49**, 561–575.

72 Gruber, C.W., Elliott, A.G., Ireland, D.C., Delprete, P.G., Dessein, S., Goransson, U., Trabi, M., Wang, C.K., Kinghorn, A.B., Robbrecht, E., and Craik, D.J. (2008) Distribution and evolution of circular miniproteins in flowering plants. *Plant Cell*, **20**, 2471–2483.

73 Voskoboinik, I., Dunstone, M.A., Baran, K., Whisstock, J.C., and Trapani, J.A. (2010) Perforin: structure, function, and role in human immunopathology. *Immunological Reviews*, **235**, 35–54.

74 Kondos, S.C., Hatfaludi, T., Voskoboinik, I., Trapani, J.A., Law, R.H.P., Whisstock, J.C., and Dunstone, M.A. (2010) The structure and function of mammalian membrane-attack complex/perforin-like proteins. *Tissue Antigens*, **76**, 341–351.

75 Brock, J.H. (2012) Lactoferrin–50years on. *Biochemistry and Cell Biology*, **90**, 245–251.

76 Legrand, D. (2012) Lactoferrin, a key molecule in immune and inflammatory processes. *Biochemistry and Cell Biology*, **90**, 252–268.

77 Fabian, T.K., Hermann, P., Beck, A., Fejerdy, P., and Fabian, G. (2012) Salivary defense proteins: their network and role in innate and acquired oral immunity. *International Journal of Molecular Sciences*, **13**, 4295–4320.

78 Wade, D. and Englund, J. (2002) Synthetic antibiotic peptides database. *Protein and Peptide Letters*, **9**, 53–57.

79 Whitmore, L. and Wallace, B.A. (2004) The Peptaibol Database: a database for sequences and structures of naturally occurring peptaibols. *Nucleic Acids Research*, **32**, D593–D594.

80 Mulvenna, J.P., Wang, C., and Craik, D.J. (2006) CyBase: a database of cyclic protein sequence and structure. *Nucleic Acids Research*, **34**, D192–D194.

81 Gueguen, Y., Garnier, J., Robert, L., Lefranc, M.P., Mougenot, I., de Lorgeril, J., Janech, M., Gross, P.S., Warr, G.W., Cuthbertson, B., Barracco, M.A., Bulet, P., Aumelas, A., Yang, Y.S., Bo, D.,

Xiang, J.H., Tassanakajon, A., Piquemal, D., and Bachere, E. (2006) PenBase, the shrimp antimicrobial peptide penaeidin database: sequence-based classification and recommended nomenclature. *Developmental and Comparative Immunology*, **30**, 283–288.

82 Hammami, R., Zouhir, A., Ben Hamida, J., and Fliss, I. (2007) BACTIBASE: a new web-accessible database for bacteriocin characterization. *BMC Microbiology*, **7**, 89.

83 Seebah, S., Suresh, A., Zhuo, S., Choong, Y.H., Chua, H., Chuon, D., Beuerman, R., and Verma, C. (2007) Defensins knowledgebase: a manually curated database and information source focused on the defensins family of antimicrobial peptides. *Nucleic Acids Research*, **35**, D265–D268.

84 Fjell, C.D., Hancock, R.E.W., and Cherkasov, A. (2007) AMPer: a database and an automated discovery tool for antimicrobial peptides. *Bioinformatics*, **23**, 1148–1155.

85 Hammami, R., Ben Hamida, J., Vergoten, G., and Fliss, I. (2009) PhytAMP: a database dedicated to antimicrobial plant peptides. *Nucleic Acids Research*, **37**, D963–D968.

86 Li, Y. and Chen, Z. (2008) RAPD: a database of recombinantly-produced antimicrobial peptides. *Fems Microbiology Letters*, **289**, 126–129.

87 Wang, G., Li, X., and Wang, Z. (2009) APD2: the updated antimicrobial peptide database and its application in peptide design. *Nucleic Acids Research*, **37**, D933–D937.

88 Thomas, S., Karnik, S., Barai, R.S., Jayaraman, V.K., and Idicula-Thomas, S. (2010) CAMP: a useful resource for research on antimicrobial peptides. *Nucleic Acids Research*, **38**, D774–D780.

89 Piotto, S.P., Sessa, L., Concilio, S., and Iannelli, P. (2012) YADAMP: yet another database of antimicrobial peptides. *International Journal of Antimicrobial Agents*, **39**, 346–351.

90 Sundararajan, V.S., Gabere, M.N., Pretorius, A., Adam, S., Christoffels, A., Lehvaeslaiho, M., Archer, J.A.C., and Bajic, V.B. (2012) DAMPD: a manually curated antimicrobial peptide database. *Nucleic Acids Research*, **40**, D1108–D1112.

91 Novković, M., Simunić, J., Bojović, V., Tossi, A., and Juretić, D. (2012) DADP: the database of anuran defense peptides. *Bioinformatics*, **28**, 1406–1407.

92 Conlon, J.M., Iwamuro, S., and King, J.D. (2009) Dermal cytolytic peptides and the system of innate immunity in anurans. *Annals of the New York Academy of Sciences*, **1163**, 75–82.

93 Conlon, J. (2011) The contribution of skin antimicrobial peptides to the system of innate immunity in anurans. *Cell and Tissue Research*, **343**, 201–212.

94 Conlon, J. (2011) Structural diversity and species distribution of host-defense peptides in frog skin secretions. *Cellular and Molecular Life Sciences*, **68**, 2303–2315.

95 Apponyi, M.A., Pukala, T.L., Brinkworth, C.S., Maselli, V.M., Bowie, J.H., Tyler, M.J., Booker, G.W., Wallace, J.C., Carver, J.A., Separovic, F., Doyle, J., and Llewellyn, L. (2004) Host-defence peptides of Australian anurans: structure, mechanism of action and evolutionary significance. *Peptides*, **25**, 1035–1054.

96 Pukala, T.L., Bowie, J.H., Maselli, V.M., Musgrave, I.F., and Tyler, M.J. (2006) Host-defence peptides from the glandular secretions of amphibians: structure and activity. *Natural Product Reports*, **23**, 368–393.

97 Belanger, R.M. and Corkum, L.D. (2009) Review of aquatic sex pheromones and chemical communication in anurans. *Journal of Herpetology*, **43**, 184–191.

98 Benson, D.A., Karsch-Mizrachi, I., Clark, K., Lipman, D.J., Ostell, J., and Sayers, E.W. (2012) GenBank. *Nucleic Acids Research*, **40**, D48–D53.

99 Mcwilliam, H., Valentin, F., Goujon, M., Li, W., Narayanasamy, M., Martin, J., Miyar, T., and Lopez, R. (2009) Web services at the European Bioinformatics Institute–2009. *Nucleic Acids Research*, **37**, W6–W10.

100 Kulikova, T., Akhtar, R., Aldebert, P., Althorpe, N., Andersson, M., Baldwin, A., Bates, K., Bhattacharyya, S., Bower, L., Browne, P., Castro, M., Cochrane, G., Duggan, K., Eberhardt, R., Faruque, N.,

Hoad, G., Kanz, C., Lee, C., Leinonen, R., Lin, Q., Lombard, V., Lopez, R., Lorenc, D., McWilliam, H., Mukherjee, G., Nardone, F., Pilar, M., Pastor, G., Plaister, S., Sobhany, S., Stoehr, P., Vaughan, R., Wu, D., Zhu, W., and Apweiler, R. (2007) EMBL Nucleotide Sequence Database in 2006. *Nucleic Acids Research*, **35**, D16–D20.

101 Nicolas, P., Vanhoye, D., and Amiche, M. (2003) Molecular strategies in biological evolution of antimicrobial peptides. *Peptides*, **24**, 1669–1680.

102 Vanhoye, D., Bruston, F., Nicolas, P., and Amiche, M. (2003) Antimicrobial peptides from hylid and ranin frogs originated from a 150-million-year-old ancestral precursor with a conserved signal peptide but a hypermutable antimicrobial domain. *European Journal of Biochemistry*, **270**, 2068–2081.

103 Chen, T.B., Scott, C., Tang, L.J., Zhou, M., and Shaw, C. (2005) The structural organization of aurein precursor cDNAs from the skin secretion of the Australian green and golden bell frog, *Litoria aurea*. *Regulatory Peptides*, **128**, 75–83.

104 Jackway, R.J., Pukala, T.L., Maselli, V.M., Musgrave, I.F., Bowie, J.H., Liu, Y., Surinya-Johnson, K.H., Donnellan, S.C., Doyle, J.R., Llewellyn, L.E., and Tyler, M.J. (2008) Disulfide-containing peptides from the glandular skin secretions of froglets of the genus *Crinia*: structure, activity and evolutionary trends. *Regulatory Peptides*, **151**, 80–87.

105 Jackway, R.J., Pukala, T.L., Donnellan, S.C., Sherman, P.J., Tyler, M.J., and Bowie, J.H. (2011) Skin peptide and cDNA profiling of Australian anurans: genus and species identification and evolutionary trends. *Peptides*, **32**, 161–172.

106 Duda, T.F., Vanhoye, D., and Nicolas, P. (2002) Roles of diversifying selection and coordinated evolution in the evolution of amphibian antimicrobial peptides. *Molecular Biology and Evolution*, **19**, 858–864.

107 Tennessen, J.A. and Blouin, M.S. (2007) Selection for antimicrobial peptide diversity in frogs leads to gene duplication and low allelic variation. *Journal of Molecular Evolution*, **65**, 605–615.

108 Savage, J.M. (1973) *The Geographic Distribution of Frog Patterns and Predictions*. University of Missouri Press, Columbia, MO.

109 Smith, A.G. and Hallam, A. (1970) The fit of the southern continents. *Nature*, **225**, 139–144.

110 San Mauro, D. (2010) A multilocus timescale for the origin of extant amphibians. *Molecular Phylogenetics and Evolution*, **56**, 554–561.

111 Kumar, S. and Hedges, S.B. (1998) A molecular timescale for vertebrate evolution. *Nature*, **392**, 917–920.

112 Amiche, M., Seon, A.A., Pierre, T.N., and Nicolas, P. (1999) The dermaseptin precursors: a protein family with a common preproregion and a variable C-terminal antimicrobial domain. *FEBS Letters*, **456**, 352–356.

113 Koenig, E. and Bininda-Emonds, O.R.P. (2011) Evidence for convergent evolution in the antimicrobial peptide system in anuran amphibians. *Peptides*, **32**, 20–25.

114 Janion, C. (2008) Inducible SOS response system of DNA repair and mutagenesis in *Escherichia coli*. *International Journal of Biological Sciences*, **4**, 338–344.

115 Conticello, S.G., Gilad, Y., Avidan, N., Ben-Asher, E., Levy, Z., and Fainzilber, M. (2001) Mechanisms for evolving hypervariability: the case of conopeptides. *Molecular Biology and Evolution*, **18**, 120–131.

116 Rattray, A.J. and Strathern, J.N. (2003) Error-prone DNA polymerases: when making a mistake is the only way to get ahead. *Annual Review of Genetics*, **37**, 31–66.

117 Wong, J.H., Xia, L.X., and Ng, T.B. (2007) A review of defensins of diverse origins. *Current Protein & Peptide Science*, **8**, 446–459.

118 Yamauchi, H., Maehara, N., Takanashi, T., and Nakashima, T. (2010) Defensins as host defense molecules against microbes: the characteristics of the defensins from arthropods, mollusks and fungi. *Bulletin of the Forestry and Forest Products Research Institute*, **9**, 1–18.

119 Wilmes, M., Cammue, B.P.A., Sahl, H.G., and Thevissen, K. (2011) Antibiotic activities of host defense peptides: more to it than lipid bilayer perturbation. *Natural Product Reports*, **28**, 1350–1358.

120 Aerts, A.M., Francois, I., Cammue, B.P.A., and Thevissen, K. (2008) The mode of antifungal action of plant, insect and human defensins. *Cellular and Molecular Life Sciences*, **65**, 2069–2079.

121 Zou, J., Mercier, C., Koussounadis, A., and Secombes, C. (2007) Discovery of multiple beta-defensin like homologues in teleost fish. *Molecular Immunology*, **44**, 638–647.

122 Jin, J.-Y., Zhou, L., Wang, Y., Li, Z., Zhao, J.-G., Zhang, Q.-Y., and Gui, J.-F. (2010) Antibacterial and antiviral roles of a fish beta-defensin expressed both in pituitary and testis. *PLoS ONE*, **5**, e12883.

123 Alibardi, L., Celeghin, A., and Valle, L.D. (2012) Wounding in lizards results in the release of beta-defensins at the wound site and formation of an antimicrobial barrier. *Developmental and Comparative Immunology*, **36**, 557–565.

124 Dalla Valle, L., Benato, F., Maistro, S., Quinzani, S., and Alibardi, L. (2012) Bioinformatic and molecular characterization of beta-defensins-like peptides isolated from the green lizard *Anolis carolinensis*. *Developmental and Comparative Immunology*, **36**, 222–229.

125 Arnett, E. and Seveau, S. (2011) The multifaceted activities of mammalian defensins. *Current Pharmaceutical Design*, **17**, 4254–4269.

126 Amid, C., Rehaume, L.M., Brown, K.L., Gilbert, J.G.R., Dougan, G., Hancock, R.E.W., and Harrow, J.L. (2009) Manual annotation and analysis of the defensin gene cluster in the C57BL/6J mouse reference genome. *BMC Genomics*, **10**, 606.

127 Ligabue-Braun, R., Verli, H., and Carlini, C.R. (2012) Venomous mammals: a review. *Toxicon*, **59**, 680–695.

128 Whittington, C.M., Papenfuss, A.T., Bansal, P., Torres, A.M., Wong, E.S.W., Deakin, J.E., Graves, T., Alsop, A., Schatzkamer, K., Kremitzki, C., Ponting, C.P., Temple-Smith, P., Warren, W.C., Kuchel, P.W., and Belov, K. (2008) Defensins and the convergent evolution of platypus and reptile venom genes. *Genome Research*, **18**, 986–994.

129 Herve-Grepinet, V., Rehault-Godbert, S., Labas, V., Magallon, T., Derache, C., Lavergne, M., Gautron, J., Lalmanach, A.-C., and Nys, Y. (2010) Purification and characterization of avian beta-defensin 11, an antimicrobial peptide of the hen egg. *Antimicrobial Agents and Chemotherapy*, **54**, 4401–4408.

130 Rehault-Godbert, S., Herve-Grepinet, V., Gautron, J., Cabau, C., Nys, Y., and Hincke, M. (2011) Molecules involved in chemical defence of the chicken egg, in *Improving the Safety and Quality of Eggs and Egg Products, Vol 1: Egg Chemistry, Production and Consumption* (eds Y. Nys, M. Bain, and F. Immerseel), Woodhead Publishing, Sawston, pp. 183–208.

131 Abdel Mageed, A.M., Isobe, N., and Yoshimura, Y. (2009) Immunolocalization of avian beta-defensins in the hen oviduct and their changes in the uterus during eggshell formation. *Reproduction*, **138**, 971–978.

132 van Dijk, A., Veldhuizen, E.J.A., and Haagsman, H.P. (2008) Avian defensins. *Veterinary Immunology and Immunopathology*, **124**, 1–18.

133 Morris, K., Wong, E.S.W., and Belov, K. (2010) Use of genomic information to gain insights into immune function in marsupials: a review of divergent immune genes, in *Marsupial Genetics and Genomics* (eds J.E. Deakin, P.D. Waters, and J.A. Marshall Graves), Springer, Dordrecht, pp. 381–400.

134 Bruhn, O., Groetzinger, J., Cascorbi, I., and Jung, S. (2011) Antimicrobial peptides and proteins of the horse – insights into a well-armed organism. *Veterinary Research*, **42**, 98.

135 Sang, Y. and Blecha, F. (2009) Porcine host defense peptides: expanding repertoire and functions. *Developmental & Comparative Immunology*, **33**, 334–343.

136 Lynn, D.J. and Bradley, D.G. (2007) Discovery of α-defensins in basal

137 Semple, F. and Dorin, J.R. (2012) β-Defensins: multifunctional Modulators of infection, inflammation and more? *Journal of Innate Immunity*, **4**, 337–348.

138 Lehrer, R.I. and Lu, W. (2012) α-Defensins in human innate immunity. *Immunological Reviews*, **245**, 84–112.

139 Hazlett, L. and Wu, M. (2011) Defensins in innate immunity. *Cell and Tissue Research*, **343**, 175–188.

140 Nguyen, T.X., Cole, A.M., and Lehrer, R.I. (2003) Evolution of primate theta-defensins: a serpentine path to a sweet tooth. *Peptides*, **24**, 1647–1654.

141 Carvalho, A.D. and Gomes, V.M. (2011) Plant defensins and defensin-like peptides–biological activities and biotechnological applications. *Current Pharmaceutical Design*, **17**, 4270–4293.

142 Kaur, J., Sagaram, U.S., and Shah, D. (2011) Can plant defensins be used to engineer durable commercially useful fungal resistance in crop plants? *Fungal Biology Reviews*, **25**, 128–135.

143 Lay, F.T. and Anderson, M.A. (2005) Defensins–components of the innate immune system in plants. *Current Protein & Peptide Science*, **6**, 85–101.

144 Yang, Y., Teng, D., Zhang, J., Tian, Z., Wang, S., and Wang, J. (2011) Characterization of recombinant plectasin: solubility, antimicrobial activity and factors that affect its activity. *Process Biochemistry*, **46**, 1050–1055.

145 Zhu, S., Gao, B., Harvey, P.J., and Craik, D.J. (2012) Dermatophytic defensin with antiinfective potential. *Proceedings of the National Academy of Sciences of the United States of America*, **109**, 8495–8500.

146 Thevissen, K., Kristensen, H.-H., Thomma, B.P.H.J., Cammue, B.P.A., and Francois, I.E.J.A. (2007) Therapeutic potential of antifungal plant and insect defensins. *Drug Discovery Today*, **12**, 966–971.

147 Zhang, Z.T. and Zhu, S.Y. (2009) Drosomycin, an essential component of antifungal defence in *Drosophila*. *Insect Molecular Biology*, **18**, 549–556.

148 Baumann, T., Kuhn-Nentwig, L., Largiader, C.R., and Nentwig, W. (2010) Expression of defensins in non-infected araneomorph spiders. *Cellular and Molecular Life Sciences*, **67**, 2643–2651.

149 Zhao, H.W., Kong, Y., Wang, H.J., Yan, T.H., Feng, F.F., Bian, J.M., Yang, Y., and Yu, H.N. (2011) A defensin-like antimicrobial peptide from the venoms of spider, *Ornithoctonus hainana*. *Journal of Peptide Science*, **17**, 540–544.

150 Taylor, D. (2006) Innate immunity in ticks: a review. *Journal of the Acarological Society of Japan*, **15**, 109–127.

151 Chrudimska, T., Chrudimsky, T., Golovchenko, M., Rudenko, N., and Grubhoffer, L. (2010) New defensins from hard and soft ticks: similarities, differences, and phylogenetic analyses. *Veterinary Parasitology*, **167**, 298–303.

152 Cociancich, S., Goyffon, M., Bontems, F., Bulet, P., Bouet, F., Menez, A., and Hoffmann, J. (1993) Purification and characterisation of a scorpion defensin, a 4 kDa antibacterial peptide presenting structural similarities with insect defensins and scorpion toxins. *Biochemical and Biophysical Research Communications*, **194**, 17–22.

153 EhretSabatier, L., Loew, D., Goyffon, M., Fehlbaum, P., Hoffmann, J.A., vanDorsselaer, A., and Bulet, P. (1996) Characterization of novel cysteine-rich antimicrobial peptides from scorpion blood. *Journal of Biological Chemistry*, **271**, 29537–29544.

154 Kouno, T., Mizuguchi, M., Aizawa, T., Shinoda, H., Demura, M., Kawabata, S., and Kawano, K. (2009) A novel beta-defensin structure: big defensin changes its N-terminal structure to associate with the target membrane. *Biochemistry*, **48**, 7629–7635.

155 Kawabata, S., Saito, T., Saeki, K., Okino, N., Mizutani, A., Toh, Y., and Iwanaga, S. (1997) cDNA cloning, tissue distribution, and subcellular localization of horseshoe crab big defensin. *Biological Chemistry*, **378**, 289–292.

156 Saito, T., Kawabata, S., Shigenaga, T., Takayenoki, Y., Cho, J.K., Nakajima, H., Hirata, M., and Iwanaga, S. (1995) A novel big defensin identified in

horseshoe crab hemocytes – isolation, amino acid sequence and antibacterial activity. *Journal of Biochemistry*, **117**, 1131–1137.

157 Pisuttharachai, D., Yasuike, M., Aono, H., Yano, Y., Murakami, K., Kondo, H., Aoki, T., and Hirono, I. (2009) Characterization of two isoforms of Japanese spiny lobster *Panulirus japonicus* defensin cDNA. *Developmental and Comparative Immunology*, **33**, 434–438.

158 Adhya, M., Jeung, H.-D., Kang, H.-S., Choi, K.-S., Lee, D.S., and Cho, M. (2012) Cloning and localization of MCdef, a defensin from Manila clams (*Ruditapes philippinarum*). *Comparative Biochemistry and Physiology B-Biochemistry & Molecular Biology*, **161**, 25–31.

159 Zhao, J., Li, C., Chen, A., Li, L., Su, X., and Li, T. (2010) Molecular characterization of a novel big defensin from clam *Venerupis philippinarum*. *PloS ONE*, **5**, e13480.

160 Gerdol, M., De Moro, G., Manfrin, C., Venier, P., and Pallavicini, A. (2012) Big defensins and mytimacins, new AMP families of the Mediterranean mussel *Mytilus galloprovincialis*. *Developmental and Comparative Immunology*, **36**, 390–399.

161 Peng, K., Wang, J.-H., Sheng, J.Q., Zeng, L.-G., and Hong, Y.J. (2012) Molecular characterization and immune analysis of a defensin from freshwater pearl mussel, *Hyriopsis schlegelii*. *Aquaculture*, **334**, 45–50.

162 Ren, Q., Li, M., Zhang, C.-Y., and Chen, K.-P. (2011) Six defensins from the triangle-shell pearl mussel *Hyriopsis cumingii*. *Fish & Shellfish Immunology*, **31**, 1232–1238.

163 Li, C.-H., Zhao, J.-M., and Song, L.-S. (2009) A review of advances in research on marine molluscan antimicrobial peptides and their potential application in aquaculture. *Molluscan Research*, **29**, 17–26.

164 Mygind, P.H., Fischer, R.L., Schnorr, K.M., Hansen, M.T., Sonksen, C.P., Ludvigsen, S., Raventos, D., Buskov, S., Christensen, B., De Maria, L., Taboureau, O., Yaver, D., Elvig-Jorgensen, S.G., Sorensen, M.V., Christensen, B.E., Kjaerulff, S., Frimodt-Moller, N., Lehrer, R.I., Zasloff, M., and Kristensen, H.-H. (2005) Plectasin is a peptide antibiotic with therapeutic potential from a saprophytic fungus. *Nature*, **437**, 975–980.

165 Dimarcq, J.-L., Bulet, P., Hetru, C., and Hoffmann, J. (1998) Cysteine-rich antimicrobial peptides in invertebrates. *Peptide Science*, **47**, 465–477.

166 Zhu, S. (2008) Discovery of six families of fungal defensin-like peptides provides insights into origin and evolution of the CSαβ defensins. *Molecular Immunology*, **45**, 828–838.

167 Dassanayake, R.S., Gunawardene, Y., and Tobe, S.S. (2007) Evolutionary selective trends of insect/mosquito antimicrobial defensin peptides containing cysteine-stabilized alpha/beta motifs. *Peptides*, **28**, 62–75.

168 Semple, C.A., Gautier, P., Taylor, K., and Dorin, J.R. (2006) The changing of the guard: molecular diversity and rapid evolution of beta-defensins. *Molecular Diversity*, **10**, 575–584.

169 Gachomo, E.W., Jimenez-Lopez, J.C., Kayode, A.P.P., Baba-Moussa, L., and Kotchoni, S.O. (2012) Structural characterization of plant defensin protein superfamily. *Molecular Biology Reports*, **39**, 4461–4469.

170 Yount, N.Y. and Yeaman, M.R. (2004) Multidimensional signatures in antimicrobial peptides. *Proceedings of the National Academy of Sciences of the United States of America*, **101**, 7363–7368.

171 Sagaram, U.S., Pandurangi, R., Kaur, J., Smith, T.J., and Shah, D.M. (2011) Structure–activity determinants in antifungal plant defensins MsDef1 and MtDef4 with different modes of action against *Fusarium graminearum*. *PLoS ONE*, **6**, e18550.

172 Yount, N.Y. and Yeaman, M.R. (2006) Structural congruence among membrane-active host defense polypeptides of diverse phylogeny. *Biochimica et Biophysica Acta – Biomembranes*, **1758**, 1373–1386.

173 Yeaman, M.R. and Yount, N.Y. (2007) Unifying themes in host defence effector polypeptides. *Nature Reviews Microbiology*, **5**, 727–740.

174 Gao, B., del Carmen Rodriguez, M., Lanz-Mendoza, H., and Zhu, S. (2009) AdDLP, a bacterial defensin-like peptide, exhibits anti-*Plasmodium* activity. *Biochemical and Biophysical Research Communications*, **387**, 393–398.

175 Zhu, S. (2007) Evidence for myxobacterial origin of eukaryotic defensins. *Immunogenetics*, **59**, 949–954.

176 Lopez-Garcia, P. and Moreira, D. (2006) Selective forces for the origin of the eukaryotic nucleus. *Bioessays*, **28**, 525–533.

177 Moreira, D. and Lopez-Garcia, P. (1998) Symbiosis between methanogenic archaea and delta-proteobacteria as the origin of eukaryotes: the syntrophic hypothesis. *Journal of Molecular Evolution*, **47**, 517–530.

178 Holland, L.Z., Albalat, R., Azumi, K., Benito-Gutierrez, E., Blow, M.J., Bronner-Fraser, M., Brunet, F., Butts, T., Candiani, S., Dishaw, L.J., Ferrier, D.E.K., Garcia-Fernandez, J., Gibson-Brown, J.J., Gissi, C., Godzik, A., Hallbook, F., Hirose, D., Hosomichi, K., Ikuta, T., Inoko, H., Kasahara, M., Kasamatsu, J., Kawashima, T., Kimura, A., Kobayashi, M., Kozmik, Z., Kubokawa, K., Laudet, V., Litman, G.W., McHardy, A.C., Meulemans, D., Nonaka, M., Olinski, R.P., Pancer, Z., Pennacchio, L.A., Pestarino, M., Rast, J.P., Rigoutsos, I., Robinson-Rechavi, M., Roch, G., Saiga, H., Sasakura, Y., Satake, M., Satou, Y., Schubert, M., Sherwood, N., Shiina, T., Takatori, N., Tello, J., Vopalensky, P., Wada, S., Xu, A., Ye, Y., Yoshida, K., Yoshizaki, F., Yu, J.-K., Zhang, Q., Zmasek, C.M., de Jong, P.J., Osoegawa, K., Putnam, N.H., Rokhsar, D.S., Satoh, N., and Holland, P.W.H. (2008) The amphioxus genome illuminates vertebrate origins and cephalochordate biology. *Genome Research*, **18**, 1100–1111.

179 Erwin, D.H. and Davidson, E.H. (2002) The last common bilaterian ancestor. *Development*, **129**, 3021–3032.

180 Panopoulou, G. and Poustka, A.J. (2005) Timing and mechanism of ancient vertebrate genome duplications – the adventure of a hypothesis. *Trends in Genetics*, **21**, 559–567.

181 Zhu, S. and Gao, B. (2012) Evolutionary origin of β-defensins. *Developmental & Comparative Immunology*, doi: 10.1016/j.dci.2012.02.011.

182 Wei, Y.X., Guo, D.S., Li, R.G., Chen, H.W., and Chen, P.X. (2003) Purification of a big defensin from *Ruditapes philippinesis* and its antibacterial activity. *Shengwu Huaxue yu Shengwu Wuli Xuebao*, **35**, 1145–1148.

183 Zhao, J., Song, L., Li, C., Ni, D., Wu, L., Zhu, L., Wang, H., and Xu, W. (2007) Molecular cloning, expression of a big defensin gene from bay scallop *Argopecten irradians* and the antimicrobial activity of its recombinant protein. *Molecular Immunology*, **44**, 360–368.

184 Rosa, R.D., Santini, A., Fievet, J., Bulet, P., Destoumieux-Garzon, D., and Bachere, E. (2011) Big defensins, a diverse family of antimicrobial peptides that follows different patterns of expression in hemocytes of the oyster *Crassostrea gigas*. *PLoS ONE*, **6**, e25594.

185 Teng, L., Gao, B., and Zhang, S. (2012) The first chordate big defensin: identification, expression and bioactivity. *Fish & Shellfish Immunology*, **32**, 572–577.

186 Amemiya, C.T., Prohaska, S.J., Hill-Force, A., Cook, A., Wasserscheid, J., Ferrier, D.E.K., Pascual-Anaya, J., Garcia-Fernandez, J., Dewar, K., and Stadler, P.F. (2008) The amphioxus Hox cluster: characterization, comparative genomics, and evolution. *Journal of Experimental Zoology B*, **310B**, 465–477.

187 Garciafernandez, J. and Holland, P.W.H. (1994) Archetypal organisation of the amphioxus hox gene cluster. *Nature*, **370**, 563–566.

188 Froy, O. and Gurevitz, M. (2003) Arthropod and mollusk defensins – evolution by exon-shuffling. *Trends in Genetics*, **19**, 684–687.

189 Patil, A., Hughes, A.L., and Zhang, G. (2004) Rapid evolution and

diversification of mammalian α-defensins as revealed by comparative analysis of rodent and primate genes. *Physiological Genomics*, **20**, 1–11.

190 Fjell, C.D., Jenssen, H., Fries, P., Aich, P., Griebel, P., Hilpert, K., Hancock, R.E.W., and Cherkasov, A. (2008) Identification of novel host defense peptides and the absence of α-defensins in the bovine genome. *Proteins*, **73**, 420–430.

191 Leonard, B.C., Marks, S.L., Outerbridge, C.A., Affolter, V.K., Kananurak, A., Young, A., Moore, P.F., Bannasch, D.L., and Bevins, C.L. (2012) Activity, expression and genetic variation of canine beta-defensin 103: a multifunctional antimicrobial peptide in the skin of domestic dogs. *Journal of Innate Immunity*, **4**, 248–259.

192 Xiao, Y., Hughes, A., Ando, J., Matsuda, Y., Cheng, J.-F., Skinner-Noble, D., and Zhang, G. (2004) A genome-wide screen identifies a single beta-defensin gene cluster in the chicken: implications for the origin and evolution of mammalian defensins. *BMC Genomics*, **5**, 56.

193 Lynn, D.J., Lloyd, A.T., Fares, M.A., and O'Farrelly, C. (2004) Evidence of positively selected sites in mammalian α-defensins. *Molecular Biology and Evolution*, **21**, 819–827.

194 Belov, K., Sanderson, C.E., Deakin, J.E., Wong, E.S.W., Assange, D., McColl, K.A., Gout, A., de Bono, B., Barrow, A.D., Speed, T.P., Trowsdale, J., and Papenfuss, A.T. (2007) Characterization of the opossum immune genome provides insights into the evolution of the mammalian immune system. *Genome Research*, **17**, 982–991.

195 Woodburne, M.O., Rich, T.H., and Springer, M.S. (2003) The evolution of tribosphery and the antiquity of mammalian clades. *Molecular Phylogenetics and Evolution*, **28**, 360–385.

196 Tran, D., Tran, P., Roberts, K., Osapay, G., Schaal, J., Ouellette, A., and Selsted, M.E. (2008) Microbicidal properties and cytocidal selectivity of rhesus macaque theta defensins. *Antimicrobial Agents and Chemotherapy*, **52**, 944–953.

197 Garcia, A.E., Osapay, G., Tran, P.A., Yuan, J., and Selsted, M.E. (2008) Isolation, synthesis, and antimicrobial activities of naturally occurring theta-defensin isoforms from baboon leukocytes. *Infection and Immunity*, **76**, 5883–5891.

198 Stegemann, C., Tsvetkova, E.V., Aleshina, G.M., Lehrer, R.I., Kokryakov, V.N., and Hoffmann, R. (2010) *De novo* sequencing of two new cyclic theta-defensins from baboon (*Papio hamadryas*) leukocytes by matrix-assisted laser desorption/ionization mass spectrometry. *Rapid Communications in Mass Spectrometry*, **24**, 599–604.

199 Selsted, M.E. (2004) Theta-defensins: cyclic antimicrobial peptides produced by binary ligation of truncated alpha-defensins. *Current Protein & Peptide Science*, **5**, 365–371.

200 Yang, C., Boone, L., Nguyen, T.X., Rudolph, D., Limpakarnjanarat, K., Mastro, T.D., Tappero, J., Cole, A.M., and Lal, R.B. (2005) Theta-defensin pseudogenes in HIV-1-exposed, persistently seronegative female sex-workers from Thailand. *Infection Genetics and Evolution*, **5**, 11–15.

201 Venkataraman, N., Cole, A.L., Ruchala, P., Waring, A.J., Lehrer, R.I., Stuchlik, O., Pohl, J., and Cole, A.M. (2009) Reawakening retrocyclins: ancestral human defensins active against HIV-1. *PLoS Biology*, **7**, e1000095.

202 Steiper, M.E. and Young, N.M. (2006) Primate molecular divergence dates. *Molecular Phylogenetics and Evolution*, **41**, 384–394.

203 Wilkinson, R.D., Steiper, M.E., Soligo, C., Martin, R.D., Yang, Z., and Tavare, S. (2011) Dating primate divergences through an integrated analysis of palaeontological and molecular data. *Systematic Biology*, **60**, 16–31.

204 Penberthy, W.T., Chari, S., Cole, A.L., and Cole, A.M. (2011) Retrocyclins and their activity against HIV-1. *Cellular and Molecular Life Sciences*, **68**, 2231–2242.

205 Ding, J., Chou, Y.-Y., and Chang, T.L. (2009) Defensins in viral infections. *Journal of Innate Immunity*, **1**, 413–420.

206 Ruchala, P., Cho, S., Cole, A.L., Carpenter, C., Jung, C.L., Luong, H., Micewicz, E.D., Waring, A.J.,

Cole, A.M., Herold, B.C., and Lehrer, R.I. (2011) Simplified theta-defensins: search for new antivirals. *International Journal of Peptide Research and Therapeutics*, **17**, 325–336.

207 Doss, M., Ruchala, P., Tecle, T., Gantz, D., Verma, A., Hartshorn, A., Crouch, E.C., Luong, H., Micewicz, E.D., Lehrer, R.I., and Hartshorn, K.L. (2012) Hapivirins and diprovirins: novel theta-defensin analogs with potent activity against influenza A virus. *Journal of Immunology*, **188**, 2759–2768.

208 Das, S., Nikolaidis, N., Goto, H., McCallister, C., Li, J., Hirano, M., and Cooper, M.D. (2010) Comparative genomics and evolution of the alpha-defensin multigene family in primates. *Molecular Biology and Evolution*, **27**, 2333–2343.

209 Barreiro, L.B. and Quintana-Murci, L. (2010) From evolutionary genetics to human immunology: how selection shapes host defence genes. *Nature Reviews Genetics*, **11**, 17–30.

210 Ghosh, J., Lun, C.M., Majeske, A.J., Sacchi, S., Schrankel, C.S., and Smith, L.C. (2011) Invertebrate immune diversity. *Developmental and Comparative Immunology*, **35**, 959–974.

211 Viljakainen, L., Evans, J.D., Hasselmann, M., Rueppell, O., Tingek, S., and Pamilo, P. (2009) Rapid evolution of immune proteins in social insects. *Molecular Biology and Evolution*, **26**, 1791–1801.

212 Tennessen, J.A. (2005) Molecular evolution of animal antimicrobial peptides: widespread moderate positive selection. *Journal of Evolutionary Biology*, **18**, 1387–1394.

213 Tennessen, J.A. (2008) Positive selection drives a correlation between non-synonymous/synonymous divergence and functional divergence. *Bioinformatics*, **24**, 1421–1425.

214 Peschel, A. and Sahl, H.-G. (2006) The co-evolution of host cationic antimicrobial peptides and microbial resistance. *Nature Reviews Microbiology*, **4**, 529–536.

215 Stotz, H.U., Thomson, J.G., and Wang, Y. (2009) Plant defensins: defense, development and application. *Plant Signaling & Behavior*, **4**, 1010–1012.

216 Long, M., Betran, E., Thornton, K., and Wang, W. (2003) The origin of new genes: glimpses from the young and old. *Nature Reviews Genetics*, **4**, 865–875.

217 Hughes, A.L. (2001) Defensins: evolution, in eLS, Wiley, New York, doi: 10.1002/9780470015902.a0006136.pub2.

218 Hughes, A.L. (1999) Evolutionary diversification of the mammalian defensins. *Cellular and Molecular Life Sciences*, **56**, 94–103.

219 Hughes, A.L. and Yeager, M. (1997) Coordinated amino acid changes in the evolution of mammalian defensins. *Journal of Molecular Evolution*, **44**, 675–682.

220 Carvalho, A.D. and Gomes, V.M. (2009) Plant defensins – prospects for the biological functions and biotechnological properties. *Peptides*, **30**, 1007–1020.

221 Whittington, C.M., Papenfuss, A.T., Locke, D.P., Mardis, E.R., Wilson, R.K., Abubucker, S., Mitreva, M., Wong, E.S.W., Hsu, A.L., Kuchel, P.W., Belov, K., and Warren, W. (2010) Novel venom gene discovery in the platypus. *Genome Biology*, **11**, R95.

222 Warren, W.C., Hillier, L.W., Graves, J.A.M., Birney, E., Ponting, C.P., Grutzner, F., Belov, K., Miller, W., Clarke, L., Chinwalla, A.T., Yang, S.-P., Heger, A., Locke, D.P., Miethke, P., Waters, P.D., Veyrunes, F., Fulton, L., Fulton, B., Graves, T., Wallis, J., Puente, X.S., Lopez-Otin, C., Ordonez, G.R., Eichler, E.E., Chen, L., Cheng, Z., Deakin, J.E., Alsop, A., Thompson, K., Kirby, P., Papenfuss, A.T., Wakefield, M.J., Olender, T., Lancet, D., Huttley, G.A., Smit, A.F.A., Pask, A., Temple-Smith, P., Batzer, M.A., Walker, J.A., Konkel, M.K., Harris, R.S., Whittington, C.M., Wong, E.S.W., Gemmell, N.J., Buschiazzo, E., Jentzsch, I.M.V., Merkel, A., Schmitz, J., Zemann, A., Churakov, G., Kriegs, J.O., Brosius, J., Murchison, E.P., Sachidanandam, R., Smith, C., Hannon, G.J., Tsend-Ayush, E., McMillan, D., Attenborough, R., Rens, W., Ferguson-Smith, M., Lefevre, C.M., Sharp, J.A., Nicholas, K.R., Ray, D.A.,

Kube, M., Reinhardt, R., Pringle, T.H., Taylor, J., Jones, R.C., Nixon, B., Dacheux, J.-L., Niwa, H., Sekita, Y., Huang, X., Stark, A., Kheradpour, P., Kellis, M., Flicek, P., Chen, Y., Webber, C., Hardison, R., Nelson, J., Hallsworth-Pepin, K., Delehaunty, K., Markovic, C., Minx, P., Feng, Y., Kremitzki, C., Mitreva, M., Glasscock, J., Wylie, T., Wohldmann, P., Thiru, P., Nhan, M.N., Pohl, C.S., Smith, S.M., Hou, S., Renfree, M.B., *et al.* (2008) Genome analysis of the platypus reveals unique signatures of evolution. *Nature*, **453**, 175-U1.

223 Whittington, C.M. and Belov, K. (2009) Platypus venom genes expressed in non-venom tissues. *Australian Journal of Zoology*, **57**, 199–202.

224 Yang, D., Biragyn, A., Kwak, L.W., and Oppenheim, J.J. (2002) Mammalian defensins in immunity: more than just microbicidal. *Trends in Immunology*, **23**, 291–296.

225 Yang, D., Chertov, O., and Oppenheim, J.J. (2001) The role of mammalian antimicrobial peptides and proteins in awakening of innate host defenses and adaptive immunity. *Cellular and Molecular Life Sciences*, **58**, 978–989.

226 Kai-Larsen, Y. and Agerberth, B. (2008) The role of the multifunctional peptide LL-37 in host defense. *Frontiers in Bioscience*, **13**, 3760–3767.

227 Kindrachuk, J. and Napper, S. (2010) Structure–activity relationships of multifunctional host defence peptides. *Mini-Reviews in Medicinal Chemistry*, **10**, 596–614.

228 Metz-Boutigue, M.H., Shooshtarizadeh, P., Prevost, G., Haikel, Y., and Chich, J.F. (2010) Antimicrobial peptides present in mammalian skin and gut are multifunctional defence molecules. *Current Pharmaceutical Design*, **16**, 1024–1039.

229 Zaiou, M. (2007) Multifunctional antimicrobial peptides: therapeutic targets in several human diseases. *Journal of Molecular Medicine*, **85**, 317–329.

230 Bernard, J.J. and Gallo, R.L. (2011) Protecting the boundary: the sentinel role of host defense peptides in the skin. *Cellular and Molecular Life Sciences*, **68**, 2189–2199.

231 Steinstraesser, L., Kraneburg, U., Jacobsen, F., and Al-Benna, S. (2011) Host defense peptides and their antimicrobial-immunomodulatory duality. *Immunobiology*, **216**, 322–333.

232 Pazgier, M., Hoover, D.M., Yang, D., Lu, W., and Lubkowski, J. (2006) Human beta-defensins. *Cellular and Molecular Life Sciences*, **63**, 1294–1313.

233 Jenssen, H. and Hancock, R.E.W. (2009) Antimicrobial properties of lactoferrin. *Biochimie*, **91**, 19–29.

234 Catania, A., Lonati, C., Sordi, A., Carlin, A., Leonardi, P., and Gatti, S. (2010) The melanocortin system in control of inflammation. *TheScientificWorldJournal*, **10**, 1840–1853.

235 Eves, P.C. and Haycock, J.W. (2010) Melanocortin signalling mechanisms, in *Melanocortins: Multiple Actions and Therapeutic Potential* (ed. A. Catania), Springer, Berlin, pp. 19–28.

236 Catania, A., Gatti, S., Colombo, G., and Lipton, J.M. (2003) Alpha-melanocyte stimulating hormone in modulation of inflammatory reactions. *Pediatric Endocrinology Reviews*, **1**, 101–108.

237 Rousseau, K., Kauser, S., Pritchard, L.E., Warhurst, A., Oliver, R.L., Slominski, A., Wei, E.T., Thody, A.J., Tobin, D.J., and White, A. (2007) Proopiomelanocortin (POMC), the ACTH/melanocortin precursor, is secreted by human epidermal keratinocytes and melanocytes and stimulates melanogenesis. *FASEB Journal*, **21**, 1844–1856.

238 Catania, A., Colombo, G., Rossi, C., Carlin, A., Sordi, A., Lonati, C., Turcatti, F., Leonardi, P., Grieco, P., and Gatti, S. (2006) Antimicrobial properties of alpha-MSH and related synthetic melanocortins. *TheScientificWorldJournal*, **6**, 1241–1246.

239 Cutuli, M., Cristiani, S., Lipton, J.M., and Catania, A. (2000) Antimicrobial effects of alpha-MSH peptides. *Journal of Leukocyte Biology*, **67**, 233–239.

240 Donnarumma, G., Paoletti, I., Buommino, E., Tufano, M.A., and Baroni, A. (2004) Alpha-MSH reduces the internalization of *Staphylococcus*

aureus and down-regulates HSP 70, integrins and cytokine expression in human keratinocyte cell lines. *Experimental Dermatology*, **13**, 748–754.

241 Brogden, K.A., Guthmiller, J.M., Salzet, M., and Zasloff, M. (2005) The nervous system and innate immunity: the neuropeptide connection. *Nature Immunology*, **6**, 558–564.

242 Catania, A., Cutuli, M., Garofalo, L., Carlin, A., Airaghi, L., Barcellini, W., and Lipton, J.M. (2000) The neuropeptide α-MSH in host defense. *Annals of the New York Academy of Sciences*, **917**, 227–231.

243 Madhuri, Shireen, T., Venugopal, S.K., Ghosh, D., Gadepalli, R., Dhawan, B., and Mukhopadhyay, K. (2009) In vitro antimicrobial activity of alpha-melanocyte stimulating hormone against major human pathogen *Staphylococcus aureus*. *Peptides*, **30**, 1627–1635.

244 Singh, M. and Mukhopadhyay, K. (2011) C-terminal amino acids of alpha-melanocyte-stimulating hormone are requisite for its antibacterial activity against *Staphylococcus aureus*. *Antimicrobial Agents and Chemotherapy*, **55**, 1920–1929.

245 Charnley, M., Moir, A.J.G., Douglas, C.W.I., and Haycock, J.W. (2008) Anti-microbial action of melanocortin peptides and identification of a novel X-Pro-D/L-Val sequence in Gram-positive and Gram-negative bacteria. *Peptides*, **29**, 1004–1009.

246 Durr, M. and Peschel, A. (2002) Chemokines meet defensins: the merging concepts of chemoattractants and antimicrobial peptides in host defense. *Infection and Immunity*, **70**, 6515–6517.

247 Charo, I.F. and Ransohoff, R.M. (2006) The many roles of chemokines and chemokine receptors in inflammation. *New England Journal of Medicine*, **354**, 610–621.

248 Mayer, M.L., Easton, D.M., and Hancock, R.E.W. (2010) *Fine Tuning Host Responses in the Face of Infection: Emerging Roles and Clinical Applications of Host Defence Peptides*, CABI, Wallingford.

249 Diamond, G., Laube, D., and ME, K.-P. (2004) Mammalian β-defensins, in *Mucosal Defence in Mammalian Host Defence Peptides* (eds R.E.W. Hancock and D.A. Devine), Cambridge University Press, Cambridge, pp. 111–138.

250 Diamond, G., Beckloff, N., and Ryan, L.K. (2008) Host defense peptides in the oral cavity and the lung: similarities and differences. *Journal of Dental Research*, **87**, 915–927.

251 Diamond, G. and Ryan, L.K. (2011) Beta-defensins: what are they REALLY doing in the oral cavity? *Oral Diseases*, **17**, 628–635.

252 Weichhart, T., Haidinger, M., Horl, W.H., and Saemann, M.D. (2008) Current concepts of molecular defence mechanisms operative during urinary tract infection. *European Journal of Clinical Investigation*, **38**, 29–38.

253 Salzman, N.H., Hung, K.C., Haribhai, D., Chu, H.T., Karlsson-Sjoberg, J., Amir, E., Teggatz, P., Barman, M., Hayward, M., Eastwood, D., Stoel, M., Zhou, Y.J., Sodergren, E., Weinstock, G.M., Bevins, C.L., Williams, C.B., and Bos, N.A. (2010) Enteric defensins are essential regulators of intestinal microbial ecology. *Nature Immunology*, **11**, 76-U1.

254 McDermott, A.M. (2009) The role of antimicrobial peptides at the ocular surface. *Ophthalmic Research*, **41**, 60–75.

255 Selsted, M.E. and Ouellette, A.J. (2005) Mammalian defensins in the antimicrobial immune response. *Nature Immunology*, **6**, 551–557.

256 Paulsen, F.P., Pufe, T., Schaudig, U., Held-Feindt, J., Lehmann, J., Schroder, J.M., and Tillmann, B.N. (2001) Detection of natural peptide antibiotics in human nasolacrimal ducts. *Investigative Ophthalmology & Visual Science*, **42**, 2157–2163.

257 Yang, D., Biragyn, A., Hoover, D.M., Lubkowski, J., and Oppenheim, J.J. (2004) Multiple roles of antimicrobial defensins, cathelicidins, and eosinophil-derived neurotoxin in host defense. *Annual Review of Immunology*, **22**, 181–215.

258 Garreis, F., Schlorf, T., Worlitzsch, D., Steven, P., Brauer, L., Jager, K., and

Paulsen, F.P. (2010) Roles of human beta-defensins in innate immune defense at the ocular surface: arming and alarming corneal and conjunctival epithelial cells. *Histochemistry and Cell Biology*, **134**, 59–73.

259 Yang, D., Chertov, O., Bykovskaia, N., Chen, Q., Buffo, M.J., Shogan, J., Anderson, M., Schroder, J.M., Wang, J.M., Howard, O.M.Z., and Oppenheim, J.J. (1999) Beta-defensins: linking innate and adaptive immunity through dendritic and T cell CCR6. *Science*, **286**, 525–528.

260 Brogden, K.A., Heidari, M., Sacco, R.E., Palmquist, D., Guthmiller, J.M., Johnson, G.K., Jia, H.P., Tack, B.F., and McCray, P.B. (2003) Defensin-induced adaptive immunity in mice and its potential in preventing periodontal disease. *Oral Microbiology and Immunology*, **18**, 95–99.

261 Blanco, P., Palucka, A.K., Pascual, V., and Banchereau, J. (2008) Dendritic cells and cytokines in human inflammatory and autoimmune diseases. *Cytokine & Growth Factor Reviews*, **19**, 41–52.

262 Soruri, A., Grigat, J., Forssmann, U., Riggert, J., and Zwirner, J. (2007) Beta-defensins chemoattract macrophages and mast cells but not lymphocytes and dendritic cells: CCR6 is not involved. *European Journal of Immunology*, **37**, 2474–2486.

263 Schutte, B.C. and McCray, P.B. (2002) β-Defensins in lung host defense. *Annual Review of Physiology*, **64**, 709–748.

264 Niyonsaba, F., Iwabuchi, K., Matsuda, H., Ogawa, H., and Nagaoka, I. (2002) Epithelial cell-derived human β-defensin-2 acts as a chemotaxin for mast cells through a pertussis toxin-sensitive and phospholipase C-dependent pathway. *International Immunology*, **14**, 421–426.

265 García, J.-R., Jaumann, F., Schulz, S., Krause, A., Rodríguez-Jiménez, J., Forssmann, U., Adermann, K., Klüver, E., Vogelmeier, C., Becker, D., Hedrich, R., Forssmann, W.-G., and Bals, R. (2001) Identification of a novel, multifunctional β-defensin (human β-defensin 3) with specific antimicrobial activity. *Cell and Tissue Research*, **306**, 257–264.

266 Niyonsaba, F., Ogawa, H., and Nagaoka, I. (2004) Human beta-defensin-2 functions as a chemotactic agent for tumour necrosis factor-alpha-treated human neutrophils. *Immunology*, **111**, 273–281.

267 McDermott, A.M. (2004) Defensins and other antimicrobial peptides at the ocular surface. *Ocular Surface*, **2**, 229–247.

268 de Jong, H.K., van der Poll, T., and Wiersinga, W.J. (2010) The systemic pro-inflammatory response in sepsis. *Journal of Innate Immunity*, **2**, 422–430.

269 Nagaoka, I., Niyonsaba, F., Tsutsumi-Ishii, Y., Tamura, H., and Hirata, M. (2008) Evaluation of the effect of human beta-defensins on neutrophil apoptosis. *International Immunology*, **20**, 543–553.

270 Rohrl, J., Yang, D., Oppenheim, J.J., and Hehlgans, T. (2010) Specific binding and chemotactic activity of mBD4 and its functional orthologue hBD2 to CCR6-expressing Cells. *Journal of Biological Chemistry*, **285**, 7028–7034.

271 Williams, I.R. (2004) Chemokine receptors and leukocyte trafficking in the mucosal immune system. *Immunologic Research*, **29**, 283–291.

272 Ito, T., Carson, W.F., IV, Cavassani, K.A., Connett, J.M., and Kunkel, S.L. (2011) CCR6 as a mediator of immunity in the lung and gut. *Experimental Cell Research*, **317**, 613–619.

273 Wu, Z.B., Hoover, D.M., Yang, D., Boulegue, C., Santamaria, F., Oppenheim, J.J., Lubkowski, J., and Lu, W.Y. (2003) Engineering disulfide bridges to dissect antimicrobial and chemotactic activities of human beta-defensin 3. *Proceedings of the National Academy of Sciences of the United States of America*, **100**, 8880–8885.

274 Tyrrell, C., De Cecco, M., Reynolds, N.L., Kilanowski, F., Campopiano, D., Barran, P., Macmillan, D., and Dorin, J.R. (2010) Isoleucine/leucine(2) is essential for chemoattractant activity of beta-defensin Defb14 through chemokine receptor 6. *Molecular Immunology*, **47**, 1378–1382.

275 Taylor, K., Clarke, D.J., McCullough, B., Chin, W., Seo, E., Yang, D., Oppenheim, J., Uhrin, D., Govan, J.R.W., Campopiano, D.J., MacMillan, D., Barran, P., and Dorin, J.R. (2008) Analysis and separation of residues important for the chemoattractant and antimicrobial activities of beta-defensin 3. *Journal of Biological Chemistry*, **283**, 6631–6639.

276 Rohrl, J., Yang, D., Oppenheim, J.J., and Hehlgans, T. (2008) Identification and biological characterization of mouse beta-defensin 14, the orthologue of human beta-defensin 3. *Journal of Biological Chemistry*, **283**, 5414–5419.

277 Taylor, K., Rolfe, M., Reynolds, N., Kilanowski, F., Pathania, U., Clarke, D., Yang, D., Oppenheim, J., Samuel, K., Howie, S., Barran, P., Macmillan, D., Campopian, D., and Dorin, J. (2009) Defensin-related peptide 1 (Defr1) is allelic to Defb8 and chemoattracts immature DC and CD4+ T cells independently of CCR6. *European Journal of Immunology*, **39**, 1353–1360.

278 Rohrl, J., Yang, D., Oppenheim, J.J., and Hehlgans, T. (2010) Human beta-defensin 2 and 3 and their mouse orthologs induce chemotaxis through interaction with CCR2. *Journal of Immunology*, **184**, 6688–6694.

279 Jin, G., Kawsar, H.I., Hirsch, S.A., Zeng, C., Jia, X., Feng, Z.M., Ghosh, S.K., Zheng, Q.Y., Zhou, A.M., McIntyre, T.M., and Weinberg, A. (2010) An antimicrobial peptide regulates tumor-associated macrophage trafficking via the chemokine receptor CCR2, a model for tumorigenesis. *PLoS ONE*, **5**, e10993.

280 Bennett, L.D., Fox, J.M., and Signoret, N. (2011) Mechanisms regulating chemokine receptor activity. *Immunology*, **134**, 246–256.

281 Struthers, M. and Pasternak, A. (2010) CCR2 antagonists. *Current Topics in Medicinal Chemistry*, **10**, 1278–1298.

282 Froy, O. (2005) Regulation of mammalian defensin expression by Toll-like receptor-dependent and independent signalling pathways. *Cellular Microbiology*, **7**, 1387–1397.

283 Akira, S. (2011) Innate immunity and adjuvants. *Philosophical Transactions of the Royal Society B*, **366**, 2748–2755.

284 Funderburg, N., Lederman, M.M., Feng, Z., Drage, M.G., Jacllowsky, J., Harding, C.V., Weinberg, A., and Sieg, S.F. (2007) Human beta-defensin-3 activates professional antigen-presenting cells via Toll-like receptors 1 and 2. *Proceedings of the National Academy of Sciences of the United States of America*, **104**, 18631–18635.

285 Biragyn, A., Ruffini, P.A., Leifer, C.A., Klyushnenkova, E., Shakhov, A., Chertov, O., Shirakawa, A.K., Farber, J.M., Segal, D.M., Oppenheim, J.J., and Kwak, L.W. (2002) Toll-like receptor 4-dependent activation of dendritic cells by beta-defensin 2. *Science*, **298**, 1025–1029.

286 Reis e Sousa, C. (2006) Dendritic cells in a mature age. *Nature Reviews Immunology*, **6**, 476–483.

287 Biragyn, A., Coscia, M., Nagashima, K., Sanford, M., Young, H.A., and Olkhanud, P. (2008) Murine beta-defensin 2 promotes TLR-4/MyD88-mediated and NF-kappa B-dependent atypical death of APCs via activation of TNFR2. *Journal of Leukocyte Biology*, **83**, 998–1008.

288 Funderburg, N.T., Jadlowsky, J.K., Lederman, M.M., Feng, Z.M., Weinberg, A., and Sieg, S.F. (2011) The Toll-like receptor 1/2 agonists Pam(3)CSK(4) and human beta-defensin-3 differentially induce interleukin-10 and nuclear factor-kappa B signalling patterns in human monocytes. *Immunology*, **134**, 151–160.

289 Niyonsaba, F., Ushio, H., Nakano, N., Ng, W., Sayama, K., Hashimoto, K., Nagaoka, I., Okumura, K., and Ogawa, H. (2007) Antimicrobial peptides human beta-defensins stimulate epidermal keratinocyte migration, proliferation and production of proinflammatory cytokines and chemokines. *Journal of Investigative Dermatology*, **127**, 594–604.

290 Niyonsaba, F., Ushio, H., Nagaoka, I., Okumura, K., and Ogawa, H. (2005) The human beta-defensins (-1,-2,-3,-4) and

cathelicidin LL-37 induce IL-18 secretion through p38 and ERK MAPK activation in primary human keratinocytes. *Journal of Immunology*, **175**, 1776–1784.

291 Donald, C.D., Sun, C.Q., Lim, S.D., Macoska, J., Cohen, C., Amin, M.B., Young, A.N., Ganz, T.A., Marshall, F.F., and Petros, J.A. (2003) Cancer-specific loss of beta-defensin 1 in renal and prostatic carcinomas. *Laboratory Investigation*, **83**, 501–505.

292 Niyonsaba, F., Ushio, H., Hara, M., Yokoi, H., Tominaga, M., Takamori, K., Kajiwara, N., Saito, H., Nagaoka, I., Ogawa, H., and Okumura, K. (2010) Antimicrobial peptides human beta-defensins and cathelicidin LL-37 induce the secretion of a pruritogenic cytokine IL-31 by human mast cells. *Journal of Immunology*, **184**, 3526–3534.

293 Semple, F., Webb, S., Li, H.N., Patel, H.B., Perretti, M., Jackson, I.J., Gray, M., Davidson, D.J., and Dorin, J.R. (2010) Human beta-defensin 3 has immunosuppressive activity *in vitro* and *in vivo*. *European Journal of Immunology*, **40**, 1073–1078.

294 Semple, F., MacPherson, H., Webb, S., Cox, S.L., Mallin, L.J., Tyrrell, C., Grimes, G.R., Semple, C.A., Nix, M.A., Millhauser, G.L., and Dorin, J.R. (2011) Human beta-defensin 3 affects the activity of pro-inflammatory pathways associated with MyD88 and TRIF. *European Journal of Immunology*, **41**, 3291–3300.

295 Motzkus, D., Schulz-Maronde, S., Heitland, A., Schulz, A., Forssmann, W.-G., Juebner, M., and Maronde, E. (2006) The novel beta-defensin DEFB123 prevents lipopolysaccharide-mediated effects *in vitro* and *in vivo*. *FASEB Journal*, **20**, 1701-+.

296 Pingel, L.C., Kohlgraf, K.G., Hansen, C.J., Eastman, C.G., Dietrich, D.E., Burnell, K.K., Srikantha, R.N., Xiao, X., Belanger, M., Progulske-Fox, A., Cavanaugh, J.E., Guthmiller, J.M., Johnson, G.K., Joly, S., Kurago, Z.B., Dawson, D.V., and Brogden, K.A. (2008) Human beta-defensin 3 binds to hemagglutinin B (rHagB), a non-fimbrial adhesin from *Porphyromonas gingivalis*, and attenuates a pro-inflammatory cytokine response. *Immunology and Cell Biology*, **86**, 643–649.

297 Hazrati, E., Galen, B., Lu, W., Wang, W., Yan, O., Keller, M.J., Lehrer, R.I., and Herold, B.C. (2006) Human alpha- and beta-defensins block multiple steps in herpes simplex virus infection. *Journal of Immunology*, **177**, 8658–8666.

298 Navid, F., Boniotto, M., Walker, C., Ahrens, K., Proksch, E., Sparwasser, T., Mueller, W., Schwarz, T., and Schwarz, A. (2012) Induction of regulatory T cells by a murine beta-defensin. *Journal of Immunology*, **188**, 735–743.

299 Ryan, L.K., Dai, J., Yin, Z., Megjugorac, N., Uhlhorn, V., Yim, S., Schwartz, K.D., Abrahams, J.M., Diamond, G., and Fitzgerald-Bocarsly, P. (2011) Modulation of human beta-defensin-1 (hBD-1) in plasmacytoid dendritic cells (PDC), monocytes, and epithelial cells by influenza virus, Herpes simplex virus, and Sendai virus and its possible role in innate immunity. *Journal of Leukocyte Biology*, **90**, 343–356.

300 McFarland, K.L., Klingenberg, J.M., Boyce, S.T., and Supp, D.M. (2008) Expression of genes encoding antimicrobial proteins and members of the toll-like receptor/nuclear factor-kappa B pathways in engineered human skin. *Wound Repair and Regeneration*, **16**, 534–541.

301 Supp, D.M., Karpinski, A.C., and Boyce, S.T. (2004) Expression of human beta-defensins HBD-1, HBD-2, and HBD-3, in cultured keratinocytes and skin substitutes. *Burns*, **30**, 643–648.

302 Smiley, A.K., Klingenberg, J.M., Aronow, B.J., Boyce, S.T., Kitzmiller, W.J., and Supp, D.M. (2005) Microarray analysis of gene expression in cultured skin substitutes compared with native human skin. *Journal of Investigative Dermatology*, **125**, 1286–1301.

303 Butmarc, J., Yufit, T., Carson, P., and Falanga, V. (2004) Human beta-defensin-2 expression is increased in chronic wounds. *Wound Repair and Regeneration*, **12**, 439–443.

304 Roupe, K.M., Nybo, M., Sjobring, U., Alberius, P., Schmidtchen, A., and Sorensen, O.E. (2010) Injury is a major inducer of epidermal innate immune responses during wound healing. *Journal of Investigative Dermatology*, **130**, 1167–1177.

305 Johnston, A., Gudjonsson, J.E., Aphale, A., Guzman, A.M., Stoll, S.W., and Elder, J.T. (2011) EGFR and IL-1 signaling synergistically promote keratinocyte antimicrobial defenses in a differentiation-dependent manner. *Journal of Investigative Dermatology*, **131**, 329–337.

306 McDermott, A.M., Redfern, R.L., Zhang, B., Pei, Y., Huang, L., and Proske, R.J. (2003) Defensin expression by the cornea: multiple signalling pathways mediate IL-1 beta stimulation of hBD-2 expression by human corneal epithelial cells. *Investigative Ophthalmology & Visual Science*, **44**, 1859–1865.

307 Vongsa, R.A., Zimmerman, N.P., and Dwinell, M.B. (2009) CCR6 regulation of the actin cytoskeleton orchestrates human beta defensin-2-and CCL20-mediated restitution of colonic epithelial cells. *Journal of Biological Chemistry*, **284**, 10034–10045.

308 Baroni, A., Donnarumma, G., Paoletti, I., Longanesi-Cattani, I., Bifulco, K., Tufano, M.A., and Carriero, M.V. (2009) Antimicrobial human beta-defensin-2 stimulates migration, proliferation and tube formation of human umbilical vein endothelial cells. *Peptides*, **30**, 267–272.

309 Lan, C.C.E., Wu, C.S., Huang, S.M., Kuo, H.Y., Wu, I.H., Liang, C.W., and Chen, G.S. (2012) High-glucose environment reduces human beta-defensin-2 expression in human keratinocytes: implications for poor diabetic wound healing. *British Journal of Dermatology*, **166**, 1221–1229.

310 Sorensen, O.E., Thapa, D.R., Rosenthal, A., Liu, L.D., Roberts, A.A., and Ganz, T. (2005) Differential regulation of beta-defensin expression in human skin by microbial stimuli. *Journal of Immunology*, **174**, 4870–4879.

311 Menzies, B.E. and Kenoyer, A. (2006) Signal transduction and nuclear responses in *Staphylococcus aureus*-induced expression of human beta-defensin 3 in skin keratinocytes. *Infection and Immunity*, **74**, 6847–6854.

312 Sorensen, O.E., Cowland, J.B., Theilgaard-Monch, K., Liu, L.D., Ganz, T., and Borregaard, N. (2003) Wound healing and expression of antimicrobial peptides/polypeptides in human keratinocytes, a consequence of common growth factors. *Journal of Immunology*, **170**, 5583–5589.

313 Lan, C.C.E., Wu, C.S., Huang, S.M., Kuo, H.Y., Wu, I.H., Wen, C.H., Chai, C.Y., Fang, A.H., and Chen, G.S. (2011) High-glucose environment inhibits p38MAPK signaling and reduces human beta-3 expression in keratinocytes. *Molecular Medicine*, **17**, 771–779.

314 Hackl, F., Kiwanuka, E., Nowinski, D., and Eriksson, E. (2011) The diabetic wound: a new experimental wound healing model in large animals. *Advances in Wound Care*, **2**, 166–170.

315 Hirsch, T., Spielmann, M., Zuhaili, B., Fossum, M., Metzig, M., Koehler, T., Steinau, H.-U., Yao, F., Onderdonk, A.B., Steinstraesser, L., and Eriksson, E. (2009) Human beta-defensin-3 promotes wound heating in infected diabetic wounds. *Journal of Gene Medicine*, **11**, 220–228.

316 Wingate, K.V., Torres, S.M., Silverstein, K.A.T., Hendrickson, J.A., and Rutherford, M.S. (2009) Expression of endogenous antimicrobial peptides in normal canine skin. *Veterinary Dermatology*, **20**, 19–26.

317 Erles, K. and Brownlie, J. (2010) Expression of β-defensins in the canine respiratory tract and antimicrobial activity against *Bordetella bronchiseptica*. *Veterinary Immunology and Immunopathology*, **135**, 12–19.

318 van Damme, C.M.M., Willemse, T., van Dijk, A., Haagsman, H.P., and Veldhuizen, E.J.A. (2009) Altered cutaneous expression of beta-defensins in dogs with atopic dermatitis. *Molecular Immunology*, **46**, 2449–2455.

319 Leonard, B.C., Affolter, V.K., and Bevins, C.L. (2012) Antimicrobial peptides: agents of border protection for companion animals. *Veterinary Dermatology*, **23**, 177-e3.

320 Schmutz, S.M. and Berryere, T.G. (2007) Genes affecting coat colour and pattern in domestic dogs: a review. *Animal Genetics*, **38**, 539–549.

321 Singh, A., Haslach, E., and Haskell-Luevano, C. (2010) Structure–activity relationships (SAR) of melanocortin and agouti-related (AGRP) peptides, in *Melanocortins: Multiple Actions and Therapeutic Potential* (ed. A. Catania), Springer, New York, pp. 1–18.

322 Little, C.C. (1957) *The Inheritance of Coat Color in Dogs*, Howell Book House, New York.

323 Kerns, J.A., Olivier, M., Lust, G., and Barsh, G.S. (2003) Exclusion of melanocortin-1 receptor (Mc1r) and Agouti as candidates for dominant black in dogs. *Journal of Heredity*, **94**, 75–79.

324 Kerns, J.A., Cargill, E.J., Clark, L.A., Candille, S.I., Berryere, T.G., Olivier, M., Lust, G., Todhunter, R.J., Schmutz, S.M., Murphy, K.E., and Barsh, G.S. (2007) Linkage and segregation analysis of black and brindle coat color in domestic dogs. *Genetics*, **176**, 1679–1689.

325 Kaelin, C.B., Candille, S.I., Yu, B., Jackson, P., Thompson, D.A., Nix, M.A., Binkley, J., Millhauser, G.L., and Barsh, G.S. (2008) New ligands for melanocortin receptors. *International Journal of Obesity*, **32**, S19–S27.

326 Candille, S.I., Kaelin, C.B., Cattanach, B.M., Yu, B., Thompson, D.A., Nix, M.A., Kerns, J.A., Schmutz, S.M., Millhauser, G.L., and Barsh, G.S. (2007) A beta-defensin mutation causes black coat color in domestic dogs. *Science*, **318**, 1418–1423.

327 Anderson, T.M., vonHoldt, B.M., Candille, S.I., Musiani, M., Greco, C., Stahler, D.R., Smith, D.W., Padhukasahasram, B., Randi, E., Leonard, J.A., Bustamante, C.D., Ostrander, E.A., Tang, H., Wayne, R.K., and Barsh, G.S. (2009) Molecular and evolutionary history of melanism in North American gray wolves. *Science*, **323**, 1339–1343.

328 Dreger, D.L. and Schmutz, S.M. (2010) The variant red coat colour phenotype of Holstein cattle maps to BTA27. *Animal Genetics*, **41**, 109–112.

329 Beaumont, K.A., Smit, D.J., Liu, Y.Y., Chai, E., Patel, M.P., Millhauser, G.L., Smith, J.J., Alewood, P.F., and Sturm, R.A. (2012) Melanocortin-1 receptor-mediated signalling pathways activated by NDP-MSH and HBD3 ligands. *Pigment Cell & Melanoma Research*, **25**, 370–374.

330 de Leeuw, E. and Lu, W. (2007) Human defensins: turning defense into offense? *Infectious Disorders Drug Targets*, **7**, 67–70.

2
Cationic Antimicrobial Peptides

Summary

Cationic antimicrobial peptides (CAMPs) are important components of the innate immune system of vertebrates, invertebrates, and plants. These peptides are active against bacteria, fungi, viruses, and pests such as insects. CAMPs use a variety of mechanisms to kill microbes but it is generally accepted that the interaction of these peptides with the membranes of target organisms is dependent on the membranes characteristics and hence is a "key" step in their mechanism of interaction. In response, we review the recent research into these peptides and identify the peptide characteristics required for antimicrobial action. These CAMPs show promise as candidates for development as novel antimicrobials.

2.1
Introduction

Primarily due to the emergence of microbial pathogens with multidrug resistance, infectious diseases remain the leading cause of mortality on a global scale [1, 2] and are increasingly being recognized as a public health issue in countries across the world [3, 4]. It is now generally accepted that one major factor contributing to the rise of multidrug-resistant (MDR) pathogens is the overprescription and misuse of antibiotics [5], and exacerbating this situation is an ever-decreasing supply of new agents due to a serious decline in the level and quality of industrial research aimed at developing novel antimicrobial drugs [3, 6–8]. Since the late 1970s, only two new classes of antibiotics have been developed: cyclic lipopeptides and oxazolidinones, of which neither was effective against Gram-negative bacteria. Moreover, a survey of the top 15 pharmaceutical companies has revealed that only a small number of drugs currently under research and development are antibacterial agents [9]. Taken with the fact that a number of social, political, economic, and environmental factors have negatively impacted on the spread of infection, the treatment of infectious diseases has become a heavy burden on healthcare services [8, 10–12]. In response, resolutions were adopted by the World Health Organization (WHO) in 2001, which asserted

Antimicrobial Peptides, First Edition. David A. Phoenix, Sarah R. Dennison, and Frederick Harris.
© 2013 Wiley-VCH Verlag GmbH & Co. KGaA. Published 2013 by Wiley-VCH Verlag GmbH & Co. KGaA.

that microbial resistance to drugs is a major threat to human health and a concerted global response is needed to counter this threat [11]. A decade later, the problem of antimicrobial drug resistance was chosen by the WHO as its theme for the World Health Day of 2011 and it was strongly recommended that governments implement antibiotic stewardship programs for the containment of this problem [13]. Based on these resolutions and recommendations, stewardship programs and infection control regimes have been adopted to minimize the use of antimicrobial drugs while simultaneously optimizing treatment, outcome, and cost [14, 15]. For example, antimicrobial stewardship programs developed in the United States have introduced methodologies to measure and monitor antibiotic use in health institutions in order to address hospital acquired infections and antimicrobial drug resistance [16]. In Taiwan, antimicrobial stewardship programs have been developed, which includes the establishment of a national task force with a remit to implement a variety of antimicrobial-resistance management strategies such as the enforcement of appropriate regulations and the audit of antimicrobial use through hospital accreditation and inspection [17]. On the research front, the recent availability of whole-genome sequences for numerous pathogenic bacteria has enabled the identification of a large number of essential genes that are potential targets of antibacterial agents [18]. Taken with the availability of the human genome sequence, this has raised the possibility of targeting unique bacterial genes either not found in humans or without significant homology to human genes. This approach of comparative genomics has been extensively pursued and has revealed a number of targets for new antimicrobial agents, including enzymes involved in biosynthetic pathways, protein synthesis, and DNA replication [19, 20]. As a recent example, the comparative genomics approach was used to identify novel drug and vaccine targets that were absent in humans but are essential for *Mycoplasma genitalium*, a human parasitic pathogen that is associated with several sexually transmitted diseases. These studies indicated that enzymes from thiamine biosynthesis, protein biosynthesis, and foliate biosynthesis may act as potential targets for the development of drugs with activity against these bacterial pathogens [21]. The availability of bacterial genomes has also supported a host of other novel approaches to the antibacterial discovery process [22–28] and underpinned structural genomics, thereby facilitating structure guided drug discovery for the development of new antimicrobial agents [29–31]. Other novel approaches to the search for antimicrobial molecules have been adopted [32–36] with current examples including identifying agents that target bacterial virulence factors [37, 38] and synergizing the action of conventional antimicrobial agents with inhibitors of efflux pumps responsible for MDR [39]. There has also been renewed interest in antimicrobial strategies that had fallen into relative disuse, such as photodynamic antimicrobial chemotherapy (PACT) and phage therapy [40–42]. Phage therapy is the therapeutic use of lytic bacteriophages to treat bacterial infections and had been utilized in parts of Eastern Europe for over 60 years before the discovery of antibiotics in the early 1940s [43, 44]. PACT has its origins in the 1930s and involves the utilization of photosensitizers that are activated by exposure to visible light in order to eradicate microbes [45, 46]. As with phage therapy, PACT was relinquished in favor of antibiotic therapy [47], but has been brought back into public prominence over the last two decades by work showing that it was able to eradicate methicillin-resistant *Staphylococcus aureus* (MRSA) [48–50], currently the highest profile MDR bacterial pathogen [51]. PACT is now one of the major areas of research for the development of novel antimicrobial compounds and is increasingly attracting interest as a potential alternative or adjunct therapy for the treatment of infections that are difficult to treat with conventional drugs and regimes [45, 46, 52]. For example, PACT based on phenothiaziniums, which are cationic heteroaromatic dyes and among the most commonly used photosensitizers [53, 54], has recently been shown to have efficacy against intractable fungal infections, such as onychomycosis [50, 55], and parasitic infections responsible for malaria, leishmaniasis and trypanosomiasis [56]. PACT also shows potential for the treatment of sessile bacteria whose community association in biofilms provides them with a dense protective environment that offers resistance to antibiotics that is up to 1000-fold higher than that of their planktonic counterparts [57]. In addition to these mainstream PACT regimes, a unique form of this therapy has also been developed based on the use of 5-aminolevulinic acid (ALA) [58]. ALA itself is not photodynamically active; however, when taken up by target cells, it can act as a prodrug, inducing the accumulation of endogenous photosensitizers whose light activation can lead to a range of antimicrobial effects with wide potential therapeutic application [59]. For example, ALA-based PACT has shown efficacy in treating a variety of recalcitrant fungal infections such as *Malassezia* folliculitis [52] and bacterial infections such as acne vulgaris [60, 61].

Along with PACT, one of the most promising approaches towards developing novel anti-infective compounds lies with antimicrobial peptides (AMPs) of the innate immune system [62–64]. It has been established that these peptides are evolutionarily conserved, multifunctional components of innate immune systems across the eukaryotic kingdom [65–72], with around 300 [26] identified in plants [73–75], and approximately 1700 [76, 77] reported in vertebrates [78–82] and invertebrates [83–87] (Chapter 1). The expression of AMPs may be constitutive or induced by an appropriate trigger such as bacterial challenge and, consistent with a defense role, these peptides are generally located at sites exposed to microbial invasion [88, 89]. For example, the surfaces of leaves [73–75], the epithelium of the mammalian gut [90–92], and the outer layers of amphibian skin [93–95]. Numerous studies have shown that AMPs possess the ability to discriminate healthy host cells over microbial and kill a remarkably wide spectrum of these latter cells, including parasites [96–98], fungi [99–101], and viruses [102–104], along with most Gram-negative and Gram-positive bacteria [85, 105, 106]. These studies have also shown that AMPs are generally able to lyse cancer cells (Chapter 6) [107, 108] and microbial cells that are recalcitrant to conventional antibiotics, such as those in a sessile form and those with MDR mechanisms [109]. Taken with the fact that microbial resistance to the action of these peptides is rare [110,

111], these observations clearly suggested that AMPs have the potential to be developed for therapeutic application [112–115]. Efforts to fulfill this potential have shown that approximately 10% of AMPs are anionic (discussed in Chapter 3) while the vast majority of those remaining are cationic and have been the predominant focus of research attention. A key finding of this research was that to fully characterize the antimicrobial activities of cationic antimicrobial peptides (CAMPs) consideration must be given to the characteristics of both the target membrane and the peptides involved. A full discussion of the membrane characteristics involved in these antimicrobial activities is given in Chapter 5; those that contribute to the anticancer action of CAMPs are described in Chapter 6. In this chapter, we primarily consider the characteristics of these peptides that contribute to their antimicrobial action and review the current research position in this area.

2.2
CAMPs and Their Antimicrobial Action

In general, CAMPs are formed from amino acid residues found in their L-configuration, although exceptions are known [116]. For example, CAMPs that include D-amino acids are found in bombinins from amphibian skin secretions [117] and defensin-like peptides from platypus venom [118]. The enzymes responsible for generating these residues these have been isolated [119–121], and based on their occurrence, it has been predicted that other amphibians and mammals may produce CAMPs, which include D-amino acids [122–124]. A number of CAMPs have also been identified that include non-standard amino acid residues within their sequences [125, 126], such as styelin D from tunicates, which remarkably includes 12 post-translationally modified residues such as bromotryptophan within its sequence [127]. These findings have inspired recent studies on the design of novel CAMPs that incorporate unnatural amino acid residues into their primary structure to improve microbial selectivity and toxicity as well as metabolic stability [128–131].

CAMPs share the property of richness in basic amino acids such as arginine and/or lysine [76, 77], which clearly gives these peptides a net positive charge, although many of these peptides possess other positively charged moieties such as a C-terminal amide group [132, 133], typically found in many amphibian CAMPs [134–137]. The presence of these basic residues and other charged moieties is generally accepted to be the primary driver underpinning the ability of CAMPs to target the membranes of bacteria, fungi, and other microbes. The vast majority of CAMPs also possess amphiphilic structures, manifested as a spatial segregation of hydrophobic and hydrophilic residues on their molecular surface, which allows them to partition into microbial membranes. On partitioning into membranes, CAMPs orientate such that their apolar regions interact with the membrane lipid core while their polar regions concomitantly engage in electrostatic interactions with the membrane lipid head-group region (Chapter 5) [138–140]. However, in relation to other structural properties, CAMPs show great diversity and can be classified into many subgroups depending upon the criteria chosen, including origin, size, structure, amino acid sequence, biological action, and antimicrobial mechanism [77, 141]. Most classifications are based on the secondary structure of these peptides and the most commonly used recognize three broad classes of CAMPs: those that adopt an α-helical structure, those with a cysteine-stabilized architecture, and extended peptides that are rich in specific residues [142]. Nonetheless, as shown by examination of the APD2 database [76, 77], of the around 2000 CAMPs listed, approximately 3% possess more than one domain such as an α-helical and β-sheet structure while around 70% of these peptides cannot be classified or have not been structurally characterized. Here, we describe several groups of CAMPs that are less well discussed, namely those that adopt extended structures such as histsatins and those that possess a cystine knot architecture such as cyclotides. In addition, we describe aspects of α-helical CAMPs that complement the fuller discussion of these peptides given in Chapter 5.

2.3
CAMPs That Adopt an α-Helical Structure

A typical example of CAMPs that adopt an α-helical structure (α-CAMPs) is depicted in Figure 2.1, which shows magainin from the frog, *Xenopus laevis*, generally regarded as the prototypic example of these peptides (Chapter 1) [143]. Studies on magainin and many other α-CAMPs, such as aureins and citropins from frogs of the *Litoria* species [134, 135], have shown that these peptides exhibit little secondary structure in aqueous solution and generally require the amphiphilic environment of the interface to adopt α-helical architecture [146]. Indeed, it is generally accepted that the insertion of α-CAMPs into membranes is an enthalpy-driven process that derives from the random coil to α-helix transition, the reorganization of membrane lipid, and interaction of the apolar residues present in AMPs with the hydrophobic bilayer core [146, 147]. As described in Chapter 4, the spatial regularity of the residues forming the secondary structure of α-CAMPs supports the quantification of a number of structurally dependent parameters [133, 148–153]. Based on this facility, numerous studies have established that residue composition, sequence length, molecular weight, isoelectric point, net charge, hydrophobicity, hydrophobic arc size, and amphiphilicity are all able to influence the ability of α-CAMPs to insert into microbial membranes and kill target microbes [138, 154–158]. These studies have led to a number of models to describe the antimicrobial mechanisms of α-CAMPs, which are described in Chapter 5 and have been extensively reviewed elsewhere [116, 132, 133, 139, 140]. However, it is well established that many α-CAMPs exhibit specificity for one or more groups of microbes and relatively few studies have investigated the influence of the structural properties of α-CAMPs on the ability of these peptides to discriminate between microbial groups. In response, we undertook several studies in pursuit of this aim and here present key results from this work.

(a) A typical α-helical CAMP Magainin

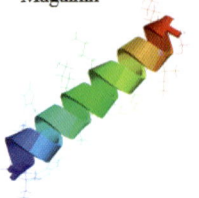

(b) A typical cysteine stabilised β-sheet protegrin (PG-1)

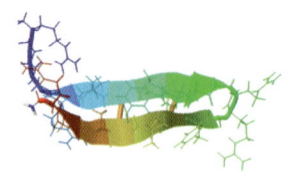

(c) A typical CAMP rich in specific residues Indolicidin

Figure 2.1 Structures of CAMPs representative of those described within this chapter. (a) Magainin from frogs [143], which is generally regarded as the prototypic example of α-CAMPs (Chapter 1). (b) Protegrin-1 [144], which is representative of β-CAMPs. (c) Indolicidin, which is one of the best characterized extended CAMPs [145].

In the first of these studies, a database of α-CAMPs was constructed and these peptides divided into four datasets, which showed activity against both Gram-positive and Gram-negative bacteria [G+, G−], against Gram-positive and Gram-negative bacteria, and fungi [G+, G−, F], solely against Gram-positive bacteria [G+], and solely against Gram-negative bacteria [G−] [159]. A compositional analysis of the entire database of α-CAMPs [159] showed that these peptides contained close to 50% apolar and 50% polar residues consistent with their possession of an amphiphilic structure and as reported by previous studies [160]. To ascertain if the residue composition of α-CAMPs varied with microbial specificity, the relative residue frequencies of the four datasets were determined and comparisons undertaken (Figure 2.2) [159]. These comparisons suggested that the overall residue compositions of α-CAMPs across these datasets were generally similar, although

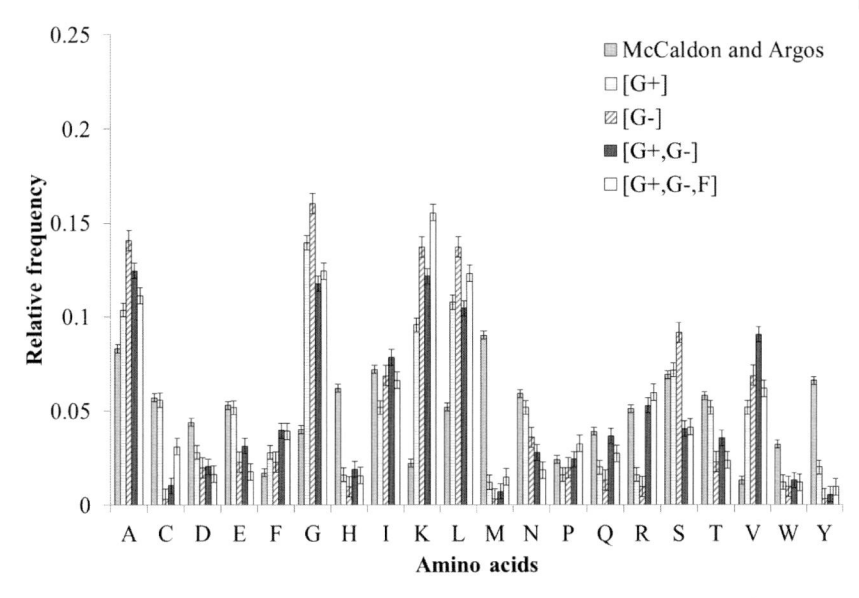

Figure 2.2 Relative frequency of occurrence of amino acid residues in the primary structures of α-CAMPs in four datasets [133]. These datasets include peptides with unique activity against Gram-positive bacteria [G+], Gram-negative bacteria [G−], Gram-positive and Gram-negative bacteria [G+, G−], and Gram-positive and Gram-negative bacteria, and fungi [G+, G−, F]. Also shown for comparative purposes are the frequencies of residues in a random sample of oligopeptides has been reproduced from McCaldon and Argos [161].

when the relative frequencies of amino acid residues in these α-CAMPs were compared to those of the dataset of McCaldon and Argos, which is a random sample of oligopeptides [161], clear differences in residue relative frequency of occurrence were shown by certain residues, suggesting functional relevance (Figure 2.2 and Chapter 6) [159]. However, a key result was that, compared to the dataset of McCaldon and Argos, α-CAMPs were generally more abundant in arginine, lysine, alanine, phenylalanine, leucine, and isoleucine [159], which is further discussed in Chapter 6 where amino acid function is considered. One aspect of this preference over other residues would seem to be that they possess a strong α-helix-forming propensity [108], although other functions would appear to be associated with these residues, as described below.

For each of these datasets of α-CAMPs studied, ranges of minimum inhibitory concentrations (MICs) when directed against target organisms were obtained and statistically compared both within and between datasets (Table 2.1) [159]. The MICs of all these peptides lay in the micromolar range, reinforcing the view that the interaction of CAMPs with the microbial membrane is not generally a receptor-mediated process [160, 162]. In the case of bacteria, these analyses showed that there were no significant differences between the MICs of α-CAMPs across

Table 2.1 Toxicity and structural properties of α-CAMPs (derived from [159]).

Dataset	MIC (µM)	Length (residues)	Net charge	Arc size (°)
[G+]	7.8–40 (15.6)	21–43 (31.38)	+1 to +4	100–180 (135)
[G−]	0.4–140 (17.7)	19–27 (23.88)	+1 to +3	220–260 (232)
[G+, G−]	0.2–40 (7.9)	10–68 (32.72)	+1 to +16	40–300 (161)
[G+, G−, F]	0.4–140 (14.5)	13–46 (27.31)	0 to +10	40–240 (161)

The ranges of toxicities and structural properties of α-CAMPs in four datasets: peptides with unique activity against Gram-positive bacteria [G+]; Gram-negative bacteria [G−]; Gram-positive and Gram-negative bacteria [G+, G−]; and Gram-positive and Gram-negative, bacteria and fungi [G+, G−, F]. Figures in brackets represent the mean value for the accompanying quoted range.

datasets, indicating that irrespective of the specificity shown by these peptides, Gram-positive and Gram-negative bacteria show similar sensitivity to their antimicrobial action. However, these studies also showed that fungi require higher concentrations of α-CAMPs to achieve levels of growth inhibition similar to those observed with bacteria and it was suggested that this may reflect the fact that the fungal cells wall provide a highly effective barrier to the action of these peptides [159, 163]. Indeed, in addition to the fungal envelope described above, α-CAMPs may encounter the glycocalyx of fungi, which is a macromolecular coating of polysaccharide that encloses the plasma membrane and can provide an additional barrier to activity [164].

Length or the number of residues in the primary structure of a peptide is one of its most fundamental properties [140]. Currently, examples of the shortest and longest AMPs reported are found in earthworms, including F1, F2, and OEP3121, which are five-residue uncharged peptides, and lysenins, which are 300-residue anionic proteins [165]. Examination of the APD2 database [76, 77] shows that examples of such peptides in the case of α-CAMPs include JCpep7, which is a seven-residue peptide from plants [166], and antiviral protein Y3, which is a 97-residue peptide from mushrooms [167]. The relationship between sequence length and the bacterial specificity of α-CAMPs showed that peptides in the [G−] dataset exhibited the narrowest range of sequence lengths, 19–27 amino acids, while the [G+, G−] peptides exhibited the widest variation in this parameter with sequence lengths between 10 and 68 residues. Studies on designed α-CAMPs have indicated that increasing the length of these peptides up to an optimum of 24 residues could enhance their antimicrobial activity by optimizing the characteristics of their hydrophilic and hydrophobic faces to interact with membranes [168]. It has also been observed that the length of α-CAMPs is likely to influence their mode of membrane interaction in that those of over 20 residues are able to span the bilayer and form transmembrane (TM) pores, whereas shorter peptides are more likely to use carpet-type membranolytic mechanisms (Chapter 5) [140, 169]. In addition, it can be seen from the APD2 database that across the eukaryotic kingdom, the vast majority of α-CAMPs are less than 50 residues in length, which implies biological relevance to their role as defense agents [76, 77]. It has been suggested that this importance may derive from the fact that short peptides are metabolically economical to the host and more easily stored in quantity, thereby enhancing the efficiency of the host immune response [170]. This principle appears to have been incorporated into the design of synthetic CAMPs by the tendency to produce peptides of four or five residues that fulfill the minimum range of properties necessary for efficient antimicrobial activity. The use of these peptides is advantageous in terms of manufacture, with reduced production costs and synthesis times as compared to naturally occurring CAMPs, which are larger and more expensive [105, 171].

Clearly, net positive charge is a fundamental property of α-CAMPs and given the preponderance of these peptides over other classes of AMPs [76, 77, 165] there has been much research aimed at a detailed understanding of the role of this property in their antimicrobial mechanisms [172–176]. For example, increasing net positive charge has been shown to maintain or increase the antimicrobial activity without increasing the hemolytic activity of a number of α-CAMPs [177, 178] such as FALL-39 from humans [179]. Another study found that decreasing the net charge of designed α-CAMPs to less than +4 rendered these peptides inactive against bacteria and erythrocytes. However, systematically increasing the net charge from +4 to +8 promoted increased antimicrobial activity while maintaining low-level hemolytic activity. Increasing net charge to +9 and +10 further improved antimicrobial activity, but induced high hemolytic activity, suggesting that net positive charge was critical to optimizing the antimicrobial selectivity and efficacy of α-CAMPs [176]. Based on such research it is generally accepted that varying the net charge and positively charged residues on the polar face of individual α-CAMPs can have a significant effect on their antimicrobial and hemolytic activity [155, 160], although studies have shown no general correlation between net charge and MICs. Studies on peptides derived from human β-defensin 1 indicated that although net charge is a major factor in the antimicrobial activity of CAMPs, the relationship between net positive charge and antimicrobial efficacy is likely to be complex, involving interplay with other factors [174]. Consistent with this view, more recent investigations on designed α-CAMPs showed that an appropriate balance between the hydrophobicity and charge distribution of these peptides was necessary for efficient disruption of the bacterial membrane without inducing damage to mammalian membranes [173]. Based on these latter studies, a number of α-CAMPs with high specificity and toxicity towards Gram-negative pathogens have been produced as viable antimicrobials with therapeutic potential [180].

To investigate the effect of net charge on the bacterial specificity of α-CAMPs at the level of amino acid residues, a compositional analysis has been undertaken to determine the frequency of arginine and lysine residues in a range of AMPs (Figure 2.2). As may be expected, anionic residues such as aspartic acid and glutamic acid were relatively scarce in the α-CAMPs and are more usually found in anionic AMPs, as discussed in Chapter 3. Arginine was found to be present in these datasets and that of McCaldon and Argos [161] at comparable levels; however, compared to arginine, lysine was present at far higher levels in α-CAMPs across the [G−], [G+], [G+, G−], and [G+, G−, F] datasets (Figure 2.2). These results clearly suggested that α-CAMPs have a preference for lysine, and recent studies on a variety of AMPs have shown that the contributions of the latter residue and arginine to their antimicrobial activity can differ greatly. For example, arginine to lysine substitutions in a cyclic β-stranded peptide [181] along with lactoferrin B and bactenecin 5 [182], which are both from bovine sources [183, 184], led to decreased activity against Gram-negative bacteria in each case. However, similar substitutions in protegrin-1, which is derived from porcine sources [144], led to a 4-fold increase in activity against Gram-negative bacteria [185]. More recently, arginine to lysine substitutions in tritrpticin, which is also derived from porcine sources [186], were found to induce 2-fold increases in antimicrobial efficacy against both Gram-positive and Gram-negative bacteria [187]. Most recently, it was found that replacing the N-terminal lysine of parasin 1, an α-CAMP from catfish, with a polar, neutral, or acidic residue resulted in a loss of antimicrobial activity, but replacement with arginine fully restored this activity [188]. In combination, these studies clearly showed that arginine and lysine may not always be taken as functionally equivalent, and properties in addition to their positive charge play roles in their contributions to the antimicrobial activity of CAMPs. Consistent with this view, recent studies have shown that differences in the length and flexibility of basic residue side-chains can have a significant effect on the potency and selectivity of AMPs [130]. There is also evidence to suggest that differences between the localization of positive charge on arginine and lysine side-chains [186, 189] may be linked to membrane lipid selectivity, and thereby contribute to differences in the antimicrobial action of AMPs [187], which is further discussed in Chapter 5.

Hydrophobicity is another fundamental property of AMPs and may be viewed as a measure of their affinity for the membrane core [138, 140, 151]. In the case of α-CAMPs, this structural property is often quantified as the angle subtended by their hydrophobic arc (θ) when it is represented as a two-dimensional axial projection [155, 156] (Chapter 4). The θ for α-CAMPs varies widely across these peptides between 40° and 300° (Table 2.1) while no statistically significant association has been identified between θ and the MICs, low values of θ tend to correspond to high MICs and vice versa [159]. Studies on designed α-CAMPs showed that increasing hydrophobicity up to an optimum window length of 11 endowed these peptides with high activity against Gram-negative bacteria and low levels of hemolysis. However, increasing hydrophobicity above this window had the effect of significantly reducing the antibacterial activity of these α-CAMPs while inducing their hemolytic ability, thereby reducing their specificity for bacterial cells over eukaryotic cells [190]. A key result from our second study was that although increasing hydrophobicity was likely to influence the antimicrobial activity of the more virulent α-CAMPs found in the [G+, G−] and [G+, G−, F] datasets, the spatial distribution of their residues was also likely to be influential [191]. Supporting these results, studies on designed α-CAMPs [190] have shown that the hydrophobicity of these peptides could be optimized for activity against not only Gram-negative bacteria, but also Gram-positive bacteria and a variety of Zygomycota fungi while maintaining low hemolytic ability. However, as before, increasing the hydrophobicity of these α-CAMPs above optimal values led to reduced antimicrobial activity and the elevation of their hemolytic ability [192]. These peptides were shown able to self-associate as their hydrophobicity was increased, which was proposed to explain the observed variations in their antimicrobial activity with changes in their hydrophobicity. It was suggested that in their oligomeric forms, these peptides may be unable to pass through the cell walls of bacteria and fungi, but still have the ability to access eukaryotic membranes [190, 192]. Interestingly, it was also found that in the case of Ascomycota fungi, increasing the hydrophobicity of these α-CAMPs above optimal values led to improved antifungal activity, which suggested that the barrier function of the fungal cell wall [163] to these self-associated α-CAMPs varied with fungal type [192]. It has previously been observed that in the design of AMPs, consideration should be given not only to the characteristics of these peptides but also those of the membranes and outer structures of their target organisms (Chapter 5) [193].

Numerous studies have established that amphiphilicity plays a major role in the antimicrobial action of α-CAMPs [154–157, 194], which in the case of α-CAMPs and β-sheet AMPs [154] is generally quantified as $<\mu_H>$ according to hydrophobic moment methodology (Chapter 4) [195]. Indeed, the value of the hydrophobic moment as a measure of this parameter is recognized by the fact that it has been included in a recently published database for the structural properties of AMPs [196]. In a recent study, we computed $<\mu_H>$ for α-CAMPs in the [G+], [G−], [G+, G−], and [G+, G−, F] datasets to determine if any association between this parameter and the MICs of these peptides existed [133]. It was found that although there was no statistically significant association between these parameters ($p > 0.05$), high values of $<\mu_H>$ tended to correspond to high MICs and vice versa (Table 2.1). This trend was the inverse of that observed by our study for the MICs of these peptides and $<H>$ [133], which is a common measure of hydrophobicity and represents the mean hydrophobicity of a protein segment (Chapter 6) [151]. These observations implied that there may be a negative linear correlation between $<\mu_H>$ and $<H>$ for α-CAMPs, which led us to investigate if such relationships existed for peptides of the [G+, G−] and [G+, G−, F] datasets, along with a dataset of amphibian α-CAMPs, designated [G+, G−, F, P] to indicate their additional antiparasitic activity. Regression analysis confirmed that a negative linear relationship between $<\mu_H>$ and $<H>$ existed for these three datasets (Figure 2.3), which was shown to be highly statistically significant in each case ($p < 0.05$). These linear relationships implied that a characteristic balance between amphiphilicity and hydrophobicity may be required by these datasets of α-CAMPs to invade the membranes of their respective target organisms [133]. These observations are supported by studies on tilted peptides [148, 153], protein anchor regions [197], a variety of AMPs [198], and in particular defensins, where it was found that the ability to invade different microbial membranes depended on appropriate balances between amphiphilicity and hydrophobicity [199].

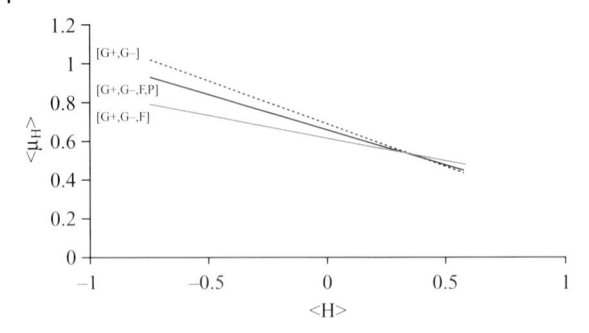

Figure 2.3 Hydrophobic moment plots of $\langle\mu_H\rangle$ versus $\langle H\rangle$ for α-CAMPs in datasets [133]. These datasets include peptides with unique activity against Gram-positive and Gram-negative bacteria [G+, G−], Gram-positive and Gram-negative bacteria, and fungi and parasites [G+, G−, F, P], and Gram-positive and Gram-negative bacteria, and fungi [G+, G−, F]. For each of these groups, a negative linear relationship can be seen to exist between $\langle\mu_H\rangle$ and $\langle H_0\rangle$. The decreasing gradients of these relationships indicates decreasing sensitivity of $\langle\mu_H\rangle$ to changes in $\langle H_0\rangle$, thereby implying that hydrophobicity assumes increasing importance to the antimicrobial action of these groups of α-CAMPs in the order [G+, G−] → [G+, G−, F, P] → [G+, G−, F].

Interestingly, these regression lines have a common point at around $\langle\mu_H\rangle = 0.5$ and $\langle H\rangle = 0.3$, suggesting that α-CAMPs with moderate amphiphilicity and hydrophobicity may be optimal for broad range antimicrobial activity and specificity [133]. Consistent with this suggestion, previous studies have shown that the hemolytic activity of α-CAMPs increases with increasing amphiphilicity as measured by the $\langle\mu_H\rangle$ [200]. More recent studies on designed α-CAMPs strongly supported our suggestion, when it was shown that moderate amphiphilicity and hydrophobicity optimized the activity of designed α-CAMPs against a variety of pathogenic fungi, Gram-negative bacteria and Gram-positive bacteria [192]. Our studies also suggested that sensitivity of $\langle\mu_H\rangle$ to changes in $\langle H\rangle$ decreased significantly in the order [G+, G−] → [G+, G−, F, P] → [G+, G−, F], which would seem to imply that hydrophobicity is of increasing importance in the order of this progression and therefore, to α-CAMPs with broader-range antimicrobial activity that includes the invasion of eukaryotic membranes [133]. These observations receive general support from a number of previous studies, which have shown that increasing hydrophobicity promotes the selectivity of α-CAMPs for eukaryotic membranes over prokaryotic membranes, partly due to an enhanced affinity for zwitterionic lipid [160], which is generally far more abundant in these former membranes than the latter [138, 201]. Nonetheless, as described above, in general, increasing the hydrophobicity of α-CAMPs can also lead to elevated levels of hemolysis and hence decreased microbial specificity, suggesting that for these peptides there is an optimum interplay between their structural characteristics for efficient antimicrobial activity.

2.4
CAMPs That Adopt a β-Sheet Structure

CAMPs that adopt a β-sheet structures (β-CAMPs) encompass a range of structural types, and include a number of linear peptides that are conformationally unconstrained and that similarly to α-CAMPs require the membrane interface to adopt a β-sheet structure [133]. A typical example of such peptides is cateslytin, which is a peptide that is enzymatically cleaved from the hormone chromogranin A, and exhibits low hemolytic ability accompanied by potent antibacterial, antiparasitic, and antifungal activity [202]. Studies on the antifungal activity of the peptide have shown that it adopts an aggregated antiparallel β-sheeted structure in the presence of fungal membranes and appears to interact with ergosterol, thereby inducing membrane permeabilization via boundary effects resulting from lateral lipid phase separation and domain formation of lipid domains [203]. Three sequences derived from the C-terminal region of arthropod defensins have been reported to adopt β-sheet conformations in the presence of bacterial membranes and to exert activity against Gram-negative bacteria by permeabilizing outer membranes of these bacteria [204]. A number of synthetically designed linear β-CAMPs have also been reported [205–207], including a series of hexameric repeat peptides $(KIGAKI)_3$, which were based on the amphibian peptide PGLa [208] and developed to exhibit enhanced selectivity for bacterial lipids [154, 209]. Several of these peptides exhibit low levels of hemolysis accompanied by broad range membranolytic antibacterial activity, which appears to be due to the peptide exhibiting preferential binding to membrane phosphatidylethanol (PE). It was suggested that these peptides may form candidates for development as antimicrobial agents for topical applications [154, 209].

Most β-CAMPs are rich in cysteine residues that stabilize the structure of the parent molecule through the formation of disulfide bonds [142]. Among the best characterized of these β-CAMPs are the protegrins, which form an amphiphilic antiparallel β-hairpin structure exhibiting two disulfide bonds between the two strands of the hairpin. These peptides have potent activity against Gram-positive and Gram-negative bacteria along with fungi and viruses [144, 210–212]. It is believed that protegrins oligomerize on association with the microbial membrane and induce cell death via pore formation [213–216], which recent studies have suggested may involve the use of amyloid type structures (Chapter 5) [217, 218] as predicted for a number of other CAMPs [219]. Probably the biggest known group of β-CAMPs with cysteine-stabilized structures are defensins, which have been identified in reptiles, fish, mollusks, arthropods [220, 221], plants [222, 223], fungi [101, 224], insects [225], birds [226], and mammals [227–231], including humans [232–234]. The 10 known human defensins are among the best characterized of this group of peptides, which are typically 29–45 residues in length, and based on the differing structural arrangement of their disulfide bonds (Chapter 1) can be classed as either α-defensins or β-defensins [232]. The overall architecture of these peptides is an amphiphilic, arginine-rich molecule comprising a triple-stranded β-sheet arrangement stabilized by intramolecular disulfide bonds [235, 236], as depicted in Figure 2.1b. These defensins have potent activity against bacteria, fungi, parasites, and viruses [232, 237], which has been demonstrated in an animal model for human α-defensin 5 where transgenic mice expressing the peptide were found to be less susceptible to orally administered challenges with *Salmonella typhimurium* [238]. The antimicrobial efficacy of other mammalian defensins have also been demonstrated in mice models [235, 239] as shown when deletion of the gene encoding mouse β-defensin 1 led to increased *Staphylococcus* species in the normally sterile urine [240] and an inability to clear *Haemophilus influenza* from the airway after challenge [241]. As described above for other defensins [224], there is evidence to suggest that some human defensins may utilize lipid receptors in their antibacterial action as in the case of HNP1 from the α-defensins and human β-defensin 3, which appear to target lipid II in the bacterial cell wall, thereby inhibiting cell wall synthesis [242]. In general, the mechanisms underpinning the antimicrobial activity of human defensins are not well understood, although cationicity and hydrophobicity have been shown to be important determinants for the antibacterial activity of both human α-defensins and β-defensins. It is generally believed that the antimicrobial activity of human defensins involves the compromise of microbial membrane integrity and a number of the models discussed in Chapter 5 have been proposed to describe this activity, including the carpet mechanism and pore formation via the self-association of these peptides. Compromised membrane integrity then leads to collapse of the membrane's electrochemical potential and proton motive force, allowing the leakage of cellular contents and the entry of defensins into the cell to interact with cytoplasmic targets, including DNA and RNA [132, 232, 236, 237, 239].

Among the most researched disulfide bridge stabilized β-CAMPs are the cyclotides, which were first reported in the early 1970s when they were identified in *Oldenlandia affinis* of the Rubiaceae [243–245]. Kalata B1 is the prototype of these peptides [243–245], which derive their name from the fact that they are components of the uterotonic agent in a medicinal preparation, *kalata-kalata* [245, 246], obtained from *O. affinis* and used in parts of Africa to accelerate child birth [247, 248]. Many more cyclotides have since been identified in the Rubiaceae and Violaceae [249, 250], with others recently reported in the Cucurbitaceae [250, 251], Fabaceae [252–255], and Apocynaceae plant families [249]. Currently, there are approaching 400 cyclotides listed in various databases [26, 76, 256, 257] although it has been predicted that as many as 50 000 of the peptides may exist [249]. Cyclotides expressed by individual plants can include up to 100 isoforms and thus these peptides make a major contribution to the plant's arsenal of defense toxins [258]. It is generally accepted that plants manufacture large arsenals of toxins to offset their lack of some of the defense mechanisms possessed by mammals, such as mobility and a somatic adaptive immune system [259]. Consistent with cyclotides possessing a role in plant defense [260], these peptides have been shown to possess activity against bacteria [248, 261], fungi [262], insects [263, 264], worms [265], and snails [266]. However, numerous studies have shown that cyclotides possess other biological properties [267, 268] whose relevance to plant defense is not always clear

[269], including activity against the HIV [258, 270, 271], hemolytic ability [268, 269], and toxicity to cancer cells [272].

Numerous investigations have shown that cyclotides share a general molecular architecture, which comprises around 30 amino acid residues that are arranged as a continuous peptide backbone. In addition to their continuous backbone, these peptides also possess a cystine knot motif, which consists of three conserved disulfide bonds, two of which, along with their connecting backbone segments, form an embedded ring in the structure that is penetrated by the third disulfide bond [250]. This molecular framework gives cyclotides a rigid topology that is essential to the maintenance of their biological activity while residues that form the six backbone loops of this framework define surface regions that act as major determinants of their biological functions. However, based on structural differences in their molecular framework, the majority of cyclotides can be divided into two subfamilies: Möbius peptides, which possess a *cis*-proline peptide bond in loop 5, and bracelet peptides, which do not possess this proline bond [250] (Figure 2.4). Möbius and bracelet subfamilies of cyclotides can also be differentiated by their proportion of hydrophobic residues and location of these residues on the molecular surface. For example, around 60% of the surface residues in cycloviolacin O2, which is generally taken as the prototypic bracelet cyclotide and was identified in *Viola odorata* [258], are hydrophobic, and are located primarily in loops 2 and 3. In contrast, only 40% of the surface residues in kalata B1, which is a Möbius cyclotide, are hydrophobic and are mainly localized in loops 2, 5, and 6 [273]. In both cases, these peptides have potent broad range toxicity to microbes and other organisms, and are the most extensively studied of the cyclotides to date [269]. Recently, a number of cyclotides have been reported that share properties of both Möbius and bracelet subgroups, which are generally referred to as hybrid cyclotides [274] such as kalata B8, which is found in *O. affinis* [275].

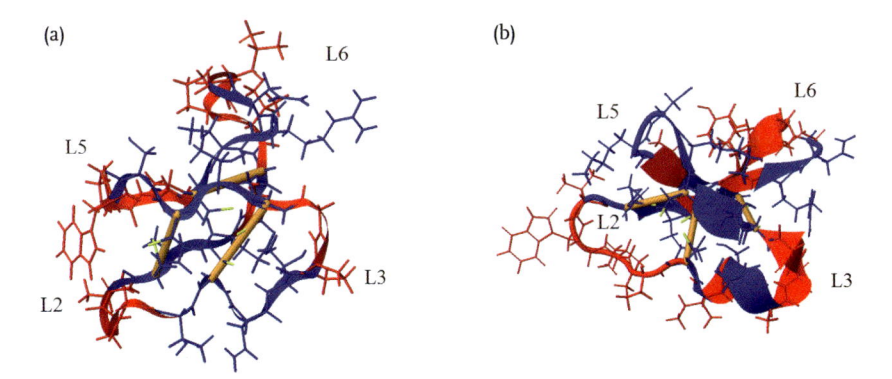

Figure 2.4 Structures of cyclotides: prototypical Möbius kalata B1 (a) and bracelet cycloviolacin O2 (b) [250]. The loops 2, 3, 5, and 6 are indicated as L2, L3, L5, and L6. For both structures, hydrophilic and charged regions are indicated in red.

There is strong evidence to suggest that the insecticidal, antiviral, hemolytic, and other biological activities of cyclotides is underpinned by the ability to disrupt membranes [276], via mechanisms that do not appear to involve the use of membrane based receptors [277, 278]. Studies on the membrane disruptive mechanisms of cyclotides have shown that the vast majority of these peptides possess a conserved glutamate residue in loop 1 of their molecular framework that is essential for activity and is thought to participate in a structurally important network of hydrogen bonds with the adjacent residues in loop 3. The sole exception appears to be kalata B12, which is an anionic cyclotide that has an aspartate residue substituted in the otherwise conserved glutamate position [279]. Several studies have suggested that these residues may contribute to sites that bind cationic metal ions, thereby forming a bridge to interact with anionic microbial membranes [280–282]. Cyclotides also contain a highly conserved asparagine or aspartic acid in loop 6 (Figure 2.5), which is the natural site of cyclization for these peptides [274], although linear cyclotides (uncyclotides) are increasingly being reported [283, 284], with examples including violacin A from *V. odorata* [285] and hedyotide B2 from *Hedyotis biflora* [286]. Structure–function studies on a number of cyclotides have shown that their net positive charge, the size and distribution of their hydrophobic surface, and their intramolecular hydrogen-bonding potential are all factors that influence their ability to disrupt membranes. This ability was also shown to be dependent upon the lipid composition of the target membrane, namely the presence of cholesterol, anionic lipids, and in particular PE, whose affinity for these peptides appears to modulate their antibacterial, antiviral, and

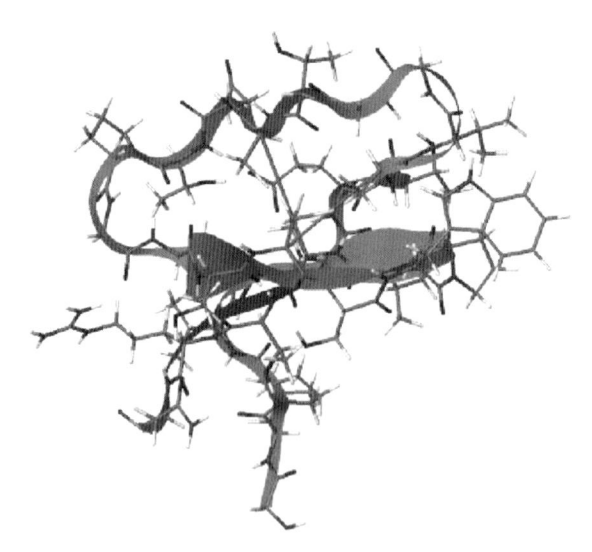

Figure 2.5 The structure of kalata B1 loop 6, which is the natural site of cyclization for cyclotides [250].

hemolytic activities and specificities [272, 277, 287]. A general model for membrane interaction by cyclotides has been presented where the apolar residues clustered primarily on loops 5 and 6 of Möbius peptides played a major role in the membrane partitioning of these peptides by interacting with the hydrophobic bilayer core [273]. In the case of bracelet cyclotides, these interactions with the hydrophobic bilayer core involve the apolar residues clustered of on loops 2 and 3 of their structures. For both Möbius and bracelet peptides these hydrophobic interactions are stabilized by electrostatic associations between positively charged residues in loops 5 and 6 of these cyclotides and the membrane head-group region [272, 273]. Upon membrane partitioning, these peptides diffuse laterally through the bilayer to form oligomeric species, which leads to pore formation and disruption of the target cell membrane, thereby initiating the toxic activity of these peptides [258].

2.5
CAMPs That Adopt Extended Structures Rich in Specific Residues

There are a number of CAMPs that do not possess a specific archetype or motif, but rather are defined by a high content of one or more residues that impose particular constraints to their structure [133]. Glycine-rich CAMPs are increasingly being reported, as witnessed by recent studies that have identified these peptides in a broad range of species, including fish [288, 289], plants [290, 291], arthropods [292–297], and amphibians [288, 298–300]. In general, it would appear that the conformational flexibility given to these peptides by the presence of multiple glycine residues is important to their antimicrobial activity, which most often appears to be primarily the inhibition of fungal growth [288–300]. Other than the latter peptides, most CAMPs that are rich in specific residues fall into three major groups and are of particular interest in that they provide insight into the roles of these residues in the antimicrobial action of AMPs, indicating those that are membranolytic and those with intracellular targets [145, 301, 302]. The first of these groups are rich in histidine residues and include CAMPs from a number of species, such as clavanins from tuna [126], alloferons from flies [303], and histatins from humans and higher primates [304]. These latter peptides are the most studied, histidine-rich CAMPs and form a family of peptides that are secreted by parotid and submandibular glands, and are effective against parasites, bacteria, fungal, and yeast species [301, 305]. Histatin 5 (His 5; Figure 2.6) is the by-far most studied and shows potent activity against a range of fungi [306–308], including *Candida albicans*, which is a significant human pathogen [309, 310]. It has been shown that the net positive charge carried by the peptide is necessary for efficient antifungal activity [305] while a structural requirement for this activity seems to be that the peptide adopts an extended conformation with low amphiphilicity involving contributions from β-turn/polyproline II architecture [311]. The peptide does not appear to induce fungal death through membrane pore formation [312],

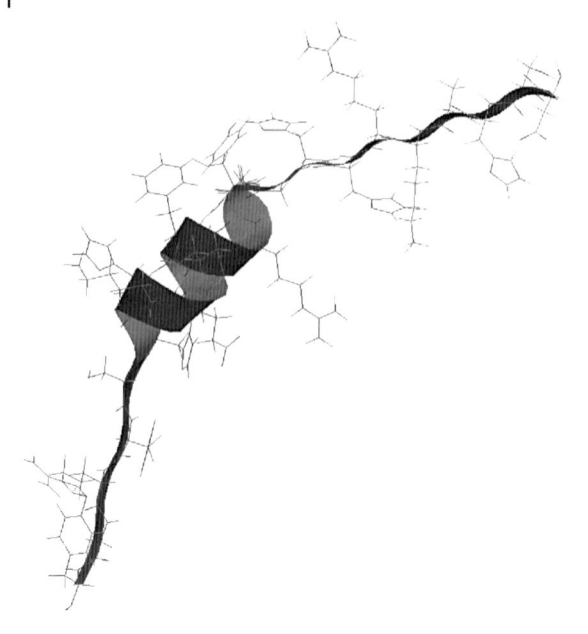

Figure 2.6 Molecular architecture of the His 5 peptide. Residues 1–10 are random coil, 11–16 α-helical, 17–19 α-helical, and 20–24 random coil [306].

but appears to translocate across the membrane via a number of possible pathways, including receptor-mediated endocytosis [313], mechanisms driven by the TM potential [313, 314], and internalization via the use of specific cell surface receptor(s) and fungal transporter systems [315–318]. Upon internalization, subsequent steps in the antifungal mechanism of His 5 are poorly understood, but it has been proposed that the peptide may act via a number of possible mechanisms involving altered mitochondrial function. The peptide appears to specifically target the energized mitochondrion via interaction with anionic lipids such as cardiolipin, which leads to collapse of the mitochondrial TM potential and the His 5-mediated release of mitochondrial ATP into the cytoplasm [319]. The non-lytic release of this ATP from the cell, possibly through the activation of various transporter proteins, is accompanied by the loss of other intracellular nucleotides and essential energy storage molecules, which leads to fungal cell death [320]. It has also been suggested that the released ATP may activate purigenic receptors, which comprise a family of TM receptors that are activated by extracellular nucleosides and nucleotides, thereby inducing fungal cell death via apoptosis. In addition, it has been proposed that the His 5-mediated loss of mitochondrial TM potential may lead to the generation of reactive oxygen species (ROS) by inhibition of the respiratory chain, which could potentially induce fungal cell death through damage to membranes, DNA or organelles [301, 319, 321]. The mechanisms responsible for the production of ROS by His 5 are unclear, but it has been suggested that this ability may relate to the multiple histidine residues present in histatins, which all possess a segment (KFHEKHHSHRGY) that appears to be essential for their antimicrobial activity [301]. The side-chains of histidine are known to participate in metal coordination, and the sequence of His 5 includes structural motifs that facilitate the binding of copper, zinc, and other metal ions [322–324], which appear to facilitate oxidative mechanisms that contribute to the antifungal activity of these peptides [325–328].

The second major group of CAMPs rich in specific residues contain peptides that possess multiple tryptophan and arginine residues (TA-CAMPs), and research has shown that the structural and electrochemical properties of these residues are of great importance to the membrane interactions and antimicrobial mechanisms of these peptides [142, 186]. Tryptophan possesses an indole side-chain with a π-electron system that gives rise to negatively charged electron clouds above and below the aromatic ring of the side-chain that are able to interact with cationic species [329]. Indeed, it is well established that based on energetic and entropic considerations, tryptophan residues have a preference for the interfacial region of lipid bilayers, interacting with positively charged moieties such as choline headgroups [330]. However, these negatively charged electron clouds are also able to interact with the side-chain of arginine, making entry of this residue into the lipidic region of the bilayer energetically more favorable as it is effectively shielded from the highly hydrophobic nature of this region by its associations with tryptophan [189]. These π–cation interactions feature in the membrane interactions of a number of TA-CAMPs such as puroindoline A [186], which is found in wheat seeds, and possesses potent antibacterial and antifungal activity [331–333]. This peptide forms a coil structure wherein the side-chains of some arginine residues are sandwiched between the indole rings of flanking tryptophan residues, thereby allowing puroindoline A to achieve deeper levels of membrane penetration [334]. However, other TA-CAMPs appear to utilize different properties of tryptophan in their membrane interactions as in the case of lactoferricin B, which is a TA-CAMP from bovine sources with antibacterial and antifungal activity [184]. This peptide appears to use the tryptophan residue solely as a stabilizing hydrophobic anchor to facilitate its passage through the bilayer to attack intracellular targets [335, 336]. In general, the membrane interactive mechanisms used by TA-CAMPs are poorly understood although it appears that they vary from peptide to peptide [186]. The two most studied TA-CAMPs are tritrpticin and indolicidin, which are from porcine and bovine leukocytes, respectively, and exhibit broad range antimicrobial activity [105, 138, 186, 337, 338]. The extended structure of indolicidin is depicted in Figure 2.1c, and both this peptide and tritrpticin appear to form a wedge-type shape with tryptophan residues in a trough that are flanked by positively charged regions [142, 339], thereby endowing their molecular architecture with a distribution of hydrophobicity that appears to be important for their membrane orientation and antimicrobial action [133]. Currently, it is believed that both tritrpticin and indolicidin insert into the membrane with their central tryptophan-rich core making contact with the interfacial region of the membrane while their flanking cationic parts interact with negatively charged moieties in the lipid head-group

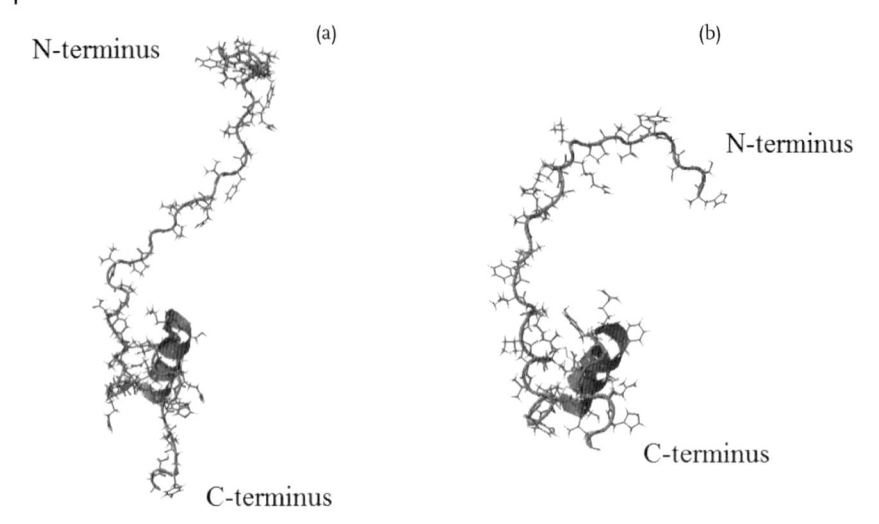

Figure 2.7 Structures of typical PR-CAMPs penaeidin 3 (a) and penaeidin 4 (b) rich in proline residues at the N-terminus and cysteine-rich at the C-terminus [347].

region [186]. However, in the case of both peptides, their modes of antimicrobial action are unclear and again multiple mechanisms appear to be involved in activity, including interactions with membrane lipids and attack on intracellular targets such as DNA [142, 337].

The most studied group of extended CAMPs are those characterized by a high proline content and the presence of multiple arginine residues (PR-CAMPs) [82, 145, 340–342]. These peptides have been identified in insect and mammalian species [82, 341, 343], along with amphibians [344], crustaceans [340, 342, 345], and mollusks [346]. Evidence suggests that similarities between these PR-CAMPs most likely arose through convergent evolution [145]. Most PR-CAMPs have a propensity to adopt a poly-L-proline type II conformation, which is an extended structure with three residues per turn [145, 341] (Figure 2.7). This conformational preference has been established for a number of peptides, which have provided the bulk of current understanding on the antimicrobial action of PR-CAMPs, including PR-39, which is derived from porcine sources [82], the bactenecins, Bac5 and Bac7, which are of bovine origin [183], pyrrhocoricins, apidaecins, and drosocin, which are found in various insects [340, 341], along with pen-3 and pen-4, which are from shrimps [342, 347]. These latter studies along with those on the sheep bactenecins, OaBac5 or OaBac7.5 [348], have suggested that the N-terminal region of PR-CAMPs is important to their antimicrobial activity and, in particular, in addition to charge, specific properties of the arginine residues within this region such as side-chain stereochemistry and hydrogen-bonding potential [145]. Nonetheless, it has also been shown that specific segments and

2.5 CAMPs That Adopt Extended Structures Rich in Specific Residues | 59

residues within the sequences of these peptides can be important determinants of their antimicrobial activity for both mammals [349] and insects, as in the case of drosocin where the presence of post-translational C-terminal modifications such as glycosylation were found to be required for efficient antibacterial activity [340]. The antimicrobial mechanisms employed by PR-AMPs have been the subject of intensive investigation, which has suggested that these peptides inactivate bacteria via two different mechanisms, depending on the concentration of the peptide (Figure 2.8) [145]. Essentially, the first of these mechanisms is stereo-specific and involves uptake of the peptide at concentrations near its MIC value via a transport system involving the inner membrane protein, SbmA, which is predicted to be part of an ABC transporter. Having gained entry to the bacterial cell, PR-AMPs are able to interact with the bacterial chaperone, DnaK, thereby affecting ATPase

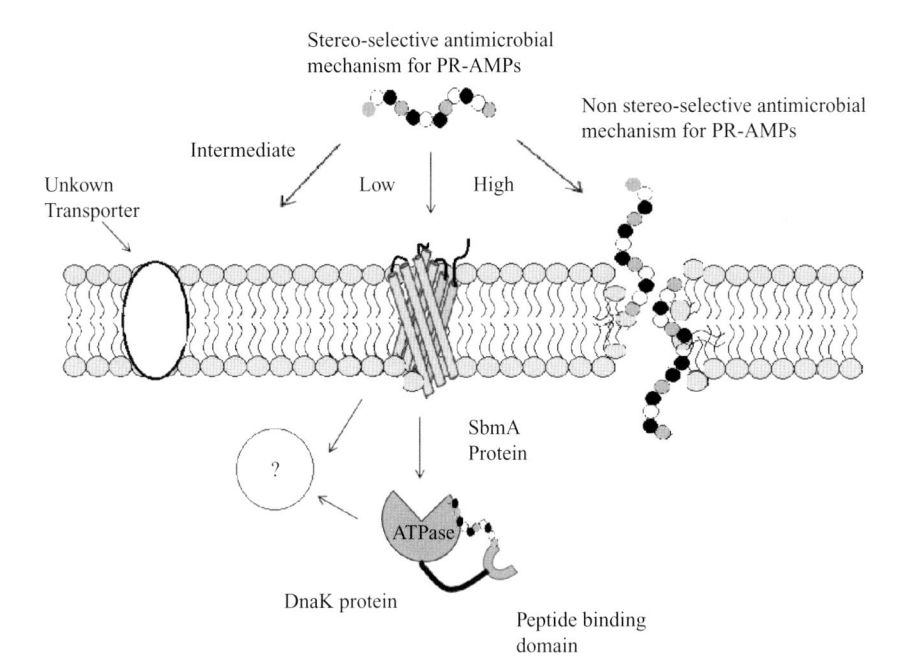

Figure 2.8 Putative model for the antibacterial mechanism of PR-AMPs [145]. According to this model, these peptides enter bacterial cells in a stereo-selective manner using a transport system involving the inner membrane protein SbmA, which is predicted to be part of an ABC transport system. At significantly higher concentrations, PR-AMPs are able to kill bacteria via a non-stereo-selective membranolytic mechanism. At intermediate concentrations, these peptides may be internalized by other, as yet unknown, transporters. Upon bacterial cell entry, PR-AMPs are able to interact with the bacterial chaperone DnaK, thereby affecting ATPase activity, the peptide-binding domain of DnaK, or both. However, there is evidence to suggest that other intracellular targets may also be involved in the antimicrobial action of PR-AMPs, which are currently unknown.

activity, the DnaK peptide-binding domain, or both. However, there is evidence to suggest that other intracellular targets may also be involved in the antimicrobial action of PR-AMPs, which are currently unknown. In the second of these mechanisms, bacterial inactivation is non-stereo-selective and proceeds via membranolytic mechanism at concentrations significantly higher than the MIC value of the peptide (Figure 2.8) [145]. Although not fully elucidated, a deeper understanding of the antimicrobial mechanisms utilized by PR-CAMPs has led to the development of designer peptides that may form viable preclinical candidates for the treatment of microbial infections [350, 351] such as A3-APO [352] and oncocin [353–356].

2.6
Discussion

In this chapter, we have reviewed recent work aimed at understanding the antimicrobial mechanisms of various groups of CAMPs. Some of these peptides are believed to interact with the cytoplasmic membrane (CM) of microbes via multiple mechanisms as in the case of those rich in specific residues such as histatins and indolicidin. These CAMPs appear able to utilize membranolytic mechanisms in some cases and host transport uptake systems in other cases to traverse the microbial membrane, thereby facilitating attack on intracellular targets such as fungal mitochondria or bacterial chaperone proteins. However, in most cases CAMPs appear to interact with microbial membranes in a relatively non-specific manner, which depends upon the lipid composition and other characteristics of the membrane, including the TM potential, the presence of anionic lipids, and in the case of eukaryotes sterols, which mediate membrane fluidity. All of these factors appear to be major determinants of the ability of CAMPs to partition into membranes, and thereby influence their selectivity and toxicity to microbes.

In order to gain insight into the characteristics of CAMPs that contribute to their antimicrobial mechanisms, we undertook a series of investigations into the relationship between these properties and four datasets of these α-CAMPs, each with differing microbial specificity. The toxicity of these peptides as measured by their MICs all lay in the micromolar range, consistent with the view that the interaction of CAMPs with the microbial membrane is not generally a receptor-mediated process (Table 2.1). Irrespective of the specificity shown by these peptides, these studies revealed that Gram-positive and Gram-negative bacteria show similar sensitivity to the antimicrobial action of α-CAMPs. However, these studies also showed that fungi were more resistant than bacteria to the action of these peptides, which appeared to be related, in part, to specific properties of these eukaryotes. For example, the differing characteristics of fungal cell walls as compared to bacteria and the presence of ergosterol in the fungal CM, which is absent from those of bacteria.

The residue composition of α-CAMPs across these datasets were determined and were generally similar, but showed clear differences to a dataset comprising a random sample of oligopeptides, suggesting functional relevance. In particular, compared to the dataset of oligopeptides, each of the datasets was found to be rich in cationic residues, indicating the importance of net positive charge to their antimicrobial actions. In further contrast to the dataset of oligopeptides, α-CAMPs across datasets showed comparable levels of polar and apolar residues, indicating the importance of hydrophobicity and amphiphilicity to their antimicrobial actions. To investigate the importance of these peptide properties further, statistical analyses of the α-CAMPs in our datasets were undertaken, which showed that, in general, within and across datasets, there was no significant association between the MICs of these peptides and their length, net charge, hydrophobicity, and amphiphilicity. However, although not statistically significant, some trends did appear to be present in our data, such as the observation that increasing hydrophobicity may enhance the antimicrobial efficacy of α-CAMPs with narrow bacterial specificity. Nonetheless, taken together, our results indicated that length, net charge, hydrophobicity, and amphiphilicity do not alone determine the microbial specificity of α-CAMPs, but clearly suggested that these structural properties make major contributions to the ability of these peptides to partition into membranes and hence the efficacy of these peptides. Indeed, a number of experimental studies reported in this chapter confirm this suggestion, and indicate that the interplay between these and other peptide properties contribute to the antimicrobial mechanisms of CAMPs. However, these studies also indicated that optimal balances of these contributions, which vary according to the peptide, are required for efficient action against microbes. Consistent with these results our studies suggested that an appropriate balance between hydrophobicity and amphiphilicity may be required for broad-range antimicrobial activity of α-CAMPs, although the distribution of their residues and final molecular architecture was also likely to be influential.

A more detailed compositional analysis of our datasets of α-CAMPs showed that these peptides had a strong preference for lysine over arginine, which appeared to be related to the differing structural characteristics of these residues such as chain length and charge localization. Presumably, this preference enhances the ability of the α-helical structure possessed by these peptides to partition into the membrane. Interestingly, many of the other groups of CAMPs reviewed here such as defensins were rich in arginine residues, although the reasons for this preference are unclear. The α-CAMPs of our datasets were also found to be rich in strongly hydrophobic residues such as phenylalanine, alanine, leucine, and isoleucine, which suggested that these residues were the main drivers of the propensity of α-CAMPs to penetrate the hydrophobic core of membranes. The roles of other individual residues were revealed by our review of extended CAMPs rich in specific residues. Studies on TA-CAMPs such as indolicidin have shown that tryptophan appears to be instrumental in anchoring some of these peptides to the membrane but also associates with arginine residues via π–cation interactions, effectively shielding these residues from the highly hydrophobic nature of the bilayer core and facilitating their membrane insertion. In the case of PR-AMPs such as PR-39 and bactenecins, it appears that the conformational flexibility of their extended structures due to the presence of multiple proline residues is essential for their antimicrobial mechanisms. There is also some evidence to suggest that histidine participates in metal coordination to facilitate the generation of ROS, thereby facilitating the antifungal action of the histidine-rich CAMPs, histatins.

Taken as a whole, the studies reviewed in this chapter would appear to indicate that a range of membrane-based factors and peptide-based factors contribute to the antimicrobial action of CAMPs, but no single factor(s) makes an over-riding contribution to their overall selectivity and toxicity. Rather, it would seem that differences between the antimicrobial actions of CAMPs are related to different levels of interplay between these peptide and membrane-based factors. Based on these results, it is clear that there has been some progress towards the goal of unraveling the mechanisms that underpin the antimicrobial mechanisms of CAMPs, although it is also clear that we are far from achieving that goal due to the number of variations, observations, and the range of CAMPs identified. Nonetheless, this chapter has also shown there has been some progress towards achieving the promise offered by CAMPs for development as novel anti-infectives.

References

1 Arias, C.A. and Murray, B.E. (2009) Antibiotic-resistant bugs in the 21st century – a clinical super-challenge. *The New England Journal of Medicine*, **360**, 439–443.

2 Livermore, D.M. (2009) Has the era of untreatable infections arrived? *The Journal of Antimicrobial Chemotherapy*, **64** (Suppl. 1), i29–i36.

3 Boucher, H.W., Talbot, G.H., Bradley, J.S., Edwards, J.E., Jr, Gilbert, D., Rice, L.B., Scheld, M., Spellberg, B., and Bartlett, J. (2009) Bad bugs, no drugs: no ESKAPE! An update from the infectious diseases society of America. *Clinical Infectious Diseases*, **48**, 1–12.

4 Berger, R.E. (2011) Emergence of a new antibiotic resistance mechanism in India, Pakistan, and the UK: a molecular, biological, and epidemiological study editorial comment. *Journal of Urology*, **185**, 154–154.

5 Costelloe, C., Metcalfe, C., Lovering, A., Mant, D., and Hay, A.D. (2010) Effect of antibiotic prescribing in primary care on antimicrobial resistance in individual patients: systematic review and meta-analysis. *British Medical Journal*, **340**, c2096.

6 Overbye, K.M. and Barrett, J.F. (2005) Antibiotics: where did we go wrong? *Drug Discovery Today*, **10**, 45–52.

7 Projan, S.J. and Shlaes, D.M. (2004) Antibacterial drug discovery: is it all downhill from here? *Clinical Microbiology and Infection*, **10**, 18–22.

8 Morel, C.M. and Mossialos, E. (2010) Stoking the antibiotic pipeline. *BMJ (Clinical Research Edition)*, **340**, c2115.

9 Alvan, G., Edlund, C., and Heddini, A. (2011) The global need for effective antibiotics – a summary of plenary presentations. *Drug Resistance Updates*, **14**, 70–76.

10 Sosa, A.J., Byarugaba, D.K., Amábile-Cuevas, C.F., Hsueh, P.-R., Kariuki, S., and Okeke, I.N. (eds) (2009) *Antimicrobial Resistance in Developing Countries*, Springer, New York.

11 Cars, O., Hogberg, L.D., Murray, M., Nordberg, O., Sivaraman, S., Lundborg, C.S., So, A.D., and Tomson, G. (2008) Meeting the challenge of antibiotic resistance. *British Medical Journal*, **337**, a1438.

12 Weiss, R.A. and McMichael, A.J. (2004) Social and environmental risk factors in the emergence of infectious diseases. *Nature Medicine*, **10**, S70–S76.

13 World Health Organization (2011) *WHO Global Strategy for Containment of Antimicrobial Resistance*. WHO, Geneva.

14 DiazGranados, C.A., Cardo, D.M., and McGowan, J.E., Jr (2008) Antimicrobial resistance: international control strategies, with a focus on limited-resource settings. *International Journal of Antimicrobial Agents*, **32**, 1–9.

15 Owens, R.C., Jr (2008) Antimicrobial stewardship: concepts and strategies in the 21st century. *Diagnostic Microbiology and Infectious Disease*, **61**, 110–128.

16 Ibrahim, O.M. and Polk, R.E. (2012) Benchmarking antimicrobial drug use in hospitals. *Expert Review of Anti-Infective Therapy*, **10**, 445–457.

17 Tseng, S.-H., Lee, C.-M., Lin, T.-Y., Chang, S.-C., Chuang, Y.-C., Yen, M.-Y., Hwang, K., Leu, H.-S., Yen, C.-C., and Chang, F.-Y. (2012) Combating antimicrobial resistance: antimicrobial stewardship program in Taiwan. *Journal of Microbiology, Immunology, and Infection [Wei mian yu gan ran za zhi]*, **45**, 79–89.

18 Guzman, E., Romeu, A., and Garcia-Vallve, S. (2008) Completely sequenced genomes of pathogenic bacteria: a review. *Enfermedades Infecciosas y Microbiologia Clinica*, **26**, 88–98.

19 Rossolini, G.M. and Thaller, M.C. (2010) Coping with antibiotic resistance: contributions from genomics. *Genome Medicine*, **2**, 15.

20 Amini, S. and Tavazoie, S. (2011) Antibiotics and the post-genome revolution. *Current Opinion in Microbiology*, **14**, 513–518.

21 Butt, A.M., Tahir, S., Nasrullah, I., Idrees, M., Lu, J., and Tong, Y. (2012) *Mycoplasma genitalium*: a comparative genomics study of metabolic pathways for the identification of drug and vaccine targets. *Infection Genetics and Evolution*, **12**, 53–62.

22 Alksne, L.E. and Dunman, P.A. (2008) Target-based antimicrobial drug discovery. *Methods in Molecular Biology*, **431**, 271–283.

23 Pucci, M.J. (2006) Use of genomics to select antibacterial targets. *Biochemical Pharmacology*, **71**, 1066–1072.

24 Pucci, M.J. (2007) Novel genetic techniques and approaches in the microbial genomics era: identification and/or validation of targets for the discovery of new antibacterial agents. *Drugs in R&D*, **8**, 201–212.

25 Mills, S.D. (2006) When will the genomics investment pay off for antibacterial discovery? *Biochemical Pharmacology*, **71**, 1096–1102.

26 Hammami, R. and Fliss, I. (2010) Current trends in antimicrobial agent research: chemo- and bioinformatics approaches. *Drug Discovery Today*, **15**, 540–546.

27 Wecke, T. and Mascher, T. (2011) Antibiotic research in the age of omics: from expression profiles to interspecies communication. *Journal of Antimicrobial Chemotherapy*, **66**, 2689–2704.

28 Relman, D.A. (2011) Microbial genomics and infectious diseases. *New England Journal of Medicine*, **365**, 347–357.

29 Charifson, P.S., Grossman, T.H., and Mueller, P. (2009) The use of structure-guided design to discover new antimicrobial agents: focus on antibacterial resistance. *Anti-Infective Agents in Medicinal Chemistry*, **8**, 73–86.

30 Barker, J.J. (2006) Antibacterial drug discovery and structure-based design. *Drug Discovery Today*, **11**, 391–404.

31 Schmid, M.B. (2006) Crystallizing new approaches for antimicrobial drug discovery. *Biochemical Pharmacology*, **71**, 1048–1056.

32 Jagusztyn-Krynicka, E.K. and Wyszynska, A. (2008) The decline of antibiotic era – new approaches for antibacterial drug discovery. *Polish Journal of Microbiology*, **57**, 91–98.

33 Wright, G.D. and Sutherland, A.D. (2007) New strategies for combating multidrug-resistant bacteria. *Trends in Molecular Medicine*, **13**, 260–267.

34 Coates, A.R.M. and Hu, Y.M. (2008) Targeting non-multiplying organisms as a way to develop novel antimicrobials. *Trends in Pharmacological Sciences*, **29**, 143–150.

35 Vicente, M., Hodgson, J., Massidda, O., Tonjum, T., Henriques-Normark, B., and Ron, E.Z. (2006) The fallacies of hope: will we discover new antibiotics to

combat pathogenic bacteria in time? *FEMS Microbiology Reviews*, **30**, 841–852.

36 Oancea, S. (2010) An overview of conventional and alternative strategies for developing new antibacterial agents. *Acta Chimica Slovenica*, **57**, 630–642.

37 Escaich, S. (2008) Antivirulence as a new antibacterial approach for chemotherapy. *Current Opinion in Chemical Biology*, **12**, 400–408.

38 Moellering, R.C., Jr (2011) Discovering new antimicrobial agents. *International Journal of Antimicrobial Agents*, **37**, 2–9.

39 Tegos, G.P., Haynes, M., Strouse, J., Khan, M.M.T., Bologa, C.G., Oprea, T.I. and Sklar, L.A. (2011) Microbial efflux pump inhibition: tactics and strategies. *Current Pharmaceutical Design*, **17**, 1291–1302.

40 Almeida, A., Cunha, A., Newton, M.C.M., Alves, E., Costa, L., and Faustino, M.A.F. (2009) Phage therapy and photodynamic therapy: low environmental impact approaches to inactivate microorganisms in fish farming plants. *Marine Drugs*, **7**, 268–313.

41 Maisch, T. (2007) Revitalized strategies against multi-resistant bacteria: antimicrobial photodynamic therapy and bacteriophage therapy. *Anti-Infective Agents in Medicinal Chemistry*, **6**, 145–150.

42 Ahmed, K., Kaderbhai, N.N., and Kaderbhai, M.A. (2012) Bacteriophage therapy revisited. *African Journal of Microbiology Research*, **6**, 3366–3379.

43 Gorski, A., Miedzybrodzki, R., Borysowski, J., Weber-Dabrowska, B., Lobocka, M., Fortuna, W., Letkiewicz, S., Zimecki, M., and Filby, G. (2009) Bacteriophage therapy for the treatment of infections. *Current Opinion in Investigational Drugs*, **10**, 766–774.

44 Brussow, H. (2007) Phage therapy: the western perspective, in *Bacteriophage: Genetics and Molecular Biology* (eds S. McGrath and D. van Sinderen), Caister Academic Press, Wymondham, pp. 159–192.

45 Ryskova, L., Buchta, V., and Slezak, R. (2010) Photodynamic antimicrobial therapy. *Central European Journal of Biology*, **5**, 400–406.

46 Kharkwal, G.B., Sharma, S.K., Huang, Y.Y., Dai, T.H., and Hamblin, M.R. (2011) Photodynamic therapy for infections: clinical applications. *Lasers in Surgery and Medicine*, **43**, 755–767.

47 Juzeniene, A., Peng, Q., and Moan, J. (2007) Milestones in the development of photodynamic therapy and fluorescence diagnosis. *Photochemical and Photobiological Sciences*, **6**, 1234–1245.

48 Zolfaghari, P.S., Packer, S., Singer, M., Nair, S.P., Bennett, J., Street, C., and Wilson, M. (2009) In vivo killing of *Staphylococcus aureus* using a light-activated antimicrobial agent. *BMC Microbiology*, **9**, 27.

49 Wainwright, M., Phoenix, D.A., Laycock, S.L., Wareing, D.R.A., and Wright, P.A. (1998) Photobactericidal activity of phenothiazinium dyes against methicillin-resistant strains of *Staphylococcus aureus*. *FEMS Microbiology Letters*, **160**, 177–181.

50 Kashef, N., Ravaei Sharif Abadi, G., and Djavid, G.E. (2012) Phototoxicity of phenothiazinium dyes against methicillin-resistant *Staphylococcus aureus* and multi-drug resistant *Escherichia coli*. *Photodiagnosis and Photodynamic Therapy*, **9**, 11–15.

51 Chambers, H.F. and Deleo, F.R. (2009) Waves of resistance: *Staphylococcus aureus* in the antibiotic era. *Nature Reviews Microbiology*, **7**, 629–641.

52 Calzavara-Pinton, P., Rossi, M.T., Sala, R., and Venturini, M. (2012) Photodynamic antifungal chemotherapy. *Photochemistry and Photobiology*, **88**, 512–522.

53 Garland, M.J., Cassidy, C.M., Woolfson, D., and Donnelly, R.F. (2009) Designing photosensitizers for photodynamic therapy: strategies, challenges and promising developments. *Future Medicinal Chemistry*, **1**, 667–691.

54 Harris, F., Chatfield, L.K., and Phoenix, D.A. (2005) Phenothiazinium based photosensitisers–photodynamic agents with a multiplicity of cellular targets and clinical applications. *Current Drug Targets*, **6**, 615–627.

55 Smijs, T.G.M. and Pavel, S. (2011) The susceptibility of dermatophytes to photodynamic treatment with special focus on *Trichophyton rubrum*. *Photochemistry and Photobiology*, **87**, 2–13.

56 Baptista, M.S. and Wainwright, M. (2011) Photodynamic antimicrobial chemotherapy (PACT) for the treatment of malaria, leishmaniasis and trypanosomiasis. *Brazilian Journal of Medical and Biological Research*, **44**, 1–10.

57 Vera, D.M.A., Haynes, M.H., Ball, A.R., Dai, T.H., Astrakas, C., Kelso, M.J., Hamblin, M.R., and Tegos, G.P. (2012) Strategies to potentiate antimicrobial photoinactivation by overcoming resistant phenotypes. *Photochemistry and Photobiology*, **88**, 499–511.

58 Phoenix, D.A. and Harris, F. (2006) Light activated compounds as antimicrobial agents – patently obvious? *Recent Patents on Anti-Infective Drug Discovery*, **1**, 181–199.

59 Harris, F. and Pierpoint, L. (2011) Photodynamic therapy based on 5-aminolevulinic acid and its use as an antimicrobial agent. *Medicinal Research Reviews*, doi: 10.1002/med.20251.

60 Elsaie, M.L. and Choudhary, S. (2010) Photodynamic therapy in the management of acne: an update. *Journal of Cosmetic Dermatology*, **9**, 211–217.

61 Sakamoto, F.H., Lopes, J.D., and Anderson, R.R. (2010) Photodynamic therapy for acne vulgaris: a critical review from basics to clinical practice. Part I. Acne vulgaris: when and why consider photodynamic therapy? *Journal of the American Academy of Dermatology*, **63**, 183–193.

62 Ajesh, K. and Sreejith, K. (2009) Peptide antibiotics: an alternative and effective antimicrobial strategy to circumvent fungal infections. *Peptides*, **30**, 999–1006.

63 Hirsch, T., Jacobsen, F., Steinau, H.U., and Steinstraesser, L. (2008) Host defense peptides and the new line of defence against multiresistant infections. *Protein Pept Lett*, **15**, 238–243.

64 Mookherjee, N. and Hancock, R.E.W. (2007) Cationic host defence peptides: innate immune regulatory peptides as a novel approach for treating infections. *Cellular and Molecular Life Sciences*, **64**, 922–933.

65 Bernard, J.J. and Gallo, R.L. (2011) Protecting the boundary: the sentinel role of host defense peptides in the skin. *Cellular and Molecular Life Sciences*, **68**, 2189–2199.

66 Steinstraesser, L., Kraneburg, U., Jacobsen, F., and Al-Benna, S. (2011) Host defense peptides and their antimicrobial-immunomodulatory duality. *Immunobiology*, **216**, 322–333.

67 Maróti, G., Kereszt, A., Kondorosi, É., and Mergaert, P. (2011) Natural roles of antimicrobial peptides in microbes, plants and animals. *Research in Microbiology*, **162**, 363–374.

68 Uvell, H. and Engstrom, Y. (2007) A multilayered defense against infection: combinatorial control of insect immune genes. *Trends in Genetics*, **23**, 342–349.

69 Cerenius, L., Jiravanichpaisal, P., Liu, H.P., and Soderhall, I. (2010) Crustacean immunity, in *Invertebrate Immunity* (ed. K. Soderhall), Springer, Berlin, pp. 239–259.

70 Octavio Luiz, F. (2011) Peptide promiscuity: an evolutionary concept for plant defense. *FEBS Letters*, **585**, 995–1000.

71 Dodds, P.N. and Rathjen, J.P. (2010) Plant immunity: towards an integrated view of plant–pathogen interactions. *Nature Reviews Genetics*, **11**, 539–548.

72 Benko-Iseppon, A.M., Galdino, S.L., Calsa, T., Jr, Kido, E.A., Tossi, A., Belarmino, L.C., and Crovella, S. (2010) Overview on plant antimicrobial peptides. *Current Protein & Peptide Science*, **11**, 181–188.

73 Barbosa Pelegrini, P., Del Sarto, R.P., Silva, O.N., Franco, O.L., and Grossi-de-Sa, M.F. (2011) Antibacterial peptides from plants: what they are and how they probably work. *Biochemistry Research International*, **2011**, 250349.

74 Ribeiro, S.M., Dias, S.C., and Franco, O.L. (2011) Plant antimicrobial peptides: from basic structures to applied research, in *Peptide Drug Discovery and*

75 Tavares, L.S., Santos Mde, O., Viccini, L.F., Moreira, J.S., Miller, R.N., and Franco, O.L. (2008) Biotechnological potential of antimicrobial peptides from flowers. *Peptides*, **29**, 1842–1851.

76 Wang, G., Li, X., and Wang, Z. (2009) APD2: the updated antimicrobial peptide database and its application in peptide design. *Nucleic Acids Research*, **37**, D933–D937.

77 Wang, G., Li, X., and Zasloff, M. (2010) A database view of naturally occurring antimicrobial peptides: nomenclature, classification and amino acid sequence analysis. *Advances in Molecular and Cellular Microbiology*, **18**, 1–21.

78 Calderon, L.D., Silva, A.D.E., Ciancaglini, P., and Stabeli, R.G. (2011) Antimicrobial peptides from Phyllomedusa frogs: from biomolecular diversity to potential nanotechnologic medical applications. *Amino Acids*, **40**, 29–49.

79 Bruhn, O., Grotzinger, J., Cascorbi, I., and Jung, S. (2011) Antimicrobial peptides and proteins of the horse – insights into a well-armed organism. *Veterinary Research*, **42**, 98.

80 van Dijk, A., Molhoek, E.M., Bikker, F.J., Yu, P.L., Veldhuizen, E.J.A., and Haagsman, H. (2011) Avian cathelicidins: paradigms for the development of anti-infectives. *Veterinary Microbiology*, **153**, 27–36.

81 Rajanbabu, V. and Chen, J.Y. (2011) Applications of antimicrobial peptides from fish and perspectives for the future. *Peptides*, **32**, 415–420.

82 Sang, Y. and Blecha, F. (2009) Porcine host defense peptides: expanding repertoire and functions. *Developmental and Comparative Immunology*, **33**, 334–343.

83 Salzet, M. (2005) Neuropeptide-derived antimicrobial peptides from invertebrates for biomedical applications. *Current Medicinal Chemistry*, **12**, 3055–3061.

84 Otero-Gonzalez, A.J., Magalhaes, B.S., Garcia-Villarino, M., Lopez-Abarratequi, C., Sousa, D.A., Dias, S.C., and Franco, O.L. (2010) Antimicrobial peptides from marine invertebrates as a new frontier for microbial infection control. *FASEB Journal*, **24**, 1320–1334.

85 Sperstad, S.V., Haug, T., Blencke, H.M., Styrvold, O.B., Li, C., and Stensvag, K. (2011) Antimicrobial peptides from marine invertebrates: challenges and perspectives in marine antimicrobial peptide discovery. *Biotechnology Advances*, **29**, 519–530.

86 Ghosh, J., Lun, C.M., Majeske, A.J., Sacchi, S., Schrankel, C.S., and Smith, L.C. (2011) Invertebrate immune diversity. *Developmental & Comparative Immunology*, **35**, 959–974.

87 Hancock, R.E.W., Brown, K.L., and Mookherjee, N. (2006) Host defence peptides from invertebrates – emerging antimicrobial strategies. *Immunobiology*, **211**, 315–322.

88 Lamont, R.J. (2004) *Bacterial Invasion of Host Cells*, Cambridge University Press, Cambridge.

89 Gläser, R., Harder, J., and Schröder, J.-M. (2008) Antimicrobial peptides as first-line effector molecules of the human innate immune system, in *Innate Immunity of Plants, Animals, and Humans* (ed. H. Heine), Springer, Berlin, pp. 187–218.

90 Bevins, C.L. and Salzman, N.H. (2011) Paneth cells, antimicrobial peptides and maintenance of intestinal homeostasis. *Nature Reviews Microbiology*, **9**, 356–368.

91 Masuda, K., Nakamura, K., Yoshioka, S., Fukaya, R., Sakai, N., and Ayabe, T. (2011) Regulation of microbiota by antimicrobial peptides in the gut. *Advances in Oto-Rhino-Laryngology*, **72**, 97–99.

92 Santaolalla, R., Fukata, M., and Abreu, M.T. (2011) Innate immunity in the small intestine. *Current Opinion in Gastroenterology*, **27**, 125–131.

93 Rollins-Smith, L.A. (2009) The role of amphibian antimicrobial peptides in protection of amphibians from pathogens linked to global amphibian declines. *Biochimica et Biophysica Acta – Biomembranes*, **1788**, 1593–1599.

94 Conlon, J.M. (2011) Structural diversity and species distribution of host-defense peptides in frog skin secretions. *Cellular and Molecular Life Sciences*, **68**, 2303–2315.

95 Gomes, A., Giri, B., Saha, A., Mishra, R., Dasgupta, S.C., Debnath, A., and Gomes, A. (2007) Bioactive molecules from amphibian skin: their biological activities with reference to therapeutic potentials for possible drug development. *Indian Journal of Experimental Biology*, **45**, 579–593.

96 Harrington, J.M. (2011) Antimicrobial peptide killing of African trypanosomes. *Parasite Immunology*, **33**, 461–469.

97 Rivas, L., Luque-Ortega, J.R., and Andreu, D. (2009) Amphibian antimicrobial peptides and Protozoa: lessons from parasites. *Biochimica et Biophysica Acta – Biomembranes*, **1788**, 1570–1581.

98 Bell, A. (2011) Antimalarial peptides: the long and the short of it. *Current Pharmaceutical Design*, **17**, 2719–2731.

99 Fusetani, N. (2010) Antifungal peptides in marine invertebrates. *Invertebrate Survival Journal*, **7**, 53–66.

100 Matejuk, A., Leng, Q., Begum, M.D., Woodle, M.C., Scaria, P., Chou, S.T., and Mixson, A.J. (2010) Peptide-based antifungal therapies against emerging infections. *Drugs of the Future*, **35**, 197–217.

101 De Brucker, K., Cammue, B.P.A., and Thevissen, K. (2011) Apoptosis-inducing antifungal peptides and proteins. *Biochemical Society Transactions*, **39**, 1527–1532.

102 Slocinska, M., Marciniak, P., and Rosinski, G. (2008) Insects antiviral and anticancer peptides: new leads for the future? *Protein and Peptide Letters*, **15**, 578–585.

103 Jenssen, H. (2009) Therapeutic approaches using host defence peptides to tackle herpes virus infections. *Viruses*, **1**, 939–964.

104 Barlow, P.G., Svoboda, P., Mackellar, A., Nash, A.A., York, I.A., Pohl, J., Davidson, D.J., and Donis, R.O. (2011) Antiviral activity and increased host defense against influenza infection elicited by the human cathelicidin LL-37. *PLoS ONE*, **6**, e25333.

105 Laverty, G., Gorman, S.P., and Gilmore, B.F. (2011) The potential of antimicrobial peptides as biocides. *International Journal of Molecular Sciences*, **12**, 6566–6596.

106 Park, S.C., Park, Y., and Hahm, K.S. (2011) The role of antimicrobial peptides in preventing multidrug-resistant bacterial infections and biofilm formation. *International Journal of Molecular Sciences*, **12**, 5971–5992.

107 Hoskin, D.W. and Ramamoorthy, A. (2008) Studies on anticancer activities of antimicrobial peptides. *Biochimica et Biophysica Acta – Biomembranes*, **1778**, 357–375.

108 Harris, F., Dennison, S.R., Singh, J., and Phoenix, D.A. (2011) On the selectivity and efficacy of defense peptides with respect to cancer cells. *Medicinal Research Reviews*, doi: 10.1002/med.20252.

109 Afacan, N.J., Yeung, A.T.Y., Pena, O.M., and Hancock, R.E.W. (2012) Therapeutic potential of host defense peptides in antibiotic-resistant infections. *Current Pharmaceutical Design*, **18**, 807–819.

110 Hale, J.D.F. (2012) *Bacterial Resistance to Antimicrobial Peptides*, Caister Academic Press, Wymondham.

111 Peschel, A. and Sahl, H.-G. (2006) The co-evolution of host cationic antimicrobial peptides and microbial resistance. *Nature Reviews Microbiology*, **4**, 529–536.

112 Kang, S.J., Kim, D.H., Mishig-Ochir, T., and Lee, B.J. (2012) Antimicrobial peptides: their physicochemical properties and therapeutic application. *Archives of Pharmacal Research*, **35**, 409–413.

113 Baltzer, S.A. and Brown, M.H. (2011) Antimicrobial peptides – promising alternatives to conventional antibiotics. *Journal of Molecular Microbiology and Biotechnology*, **20**, 228–235.

114 Peters, B.M., Shirtliff, M.E., and Jabra-Rizk, M.A. (2010) Antimicrobial peptides: primeval molecules or future drugs? *PLoS Pathogens*, **6**, e1001067.

115 Zhang, L. and Falla, T.J. (2010) Potential therapeutic application of host defense peptides, in *Antimicrobial Peptides: Methods and Protocols* (ed. A. Giuliani and A.C. Rinaldi), Humana Press, Totowa, NJ, pp. 303–327.

116 Huang, Y., Huang, J., and Chen, Y. (2010) Alpha-helical cationic antimicrobial peptides: relationships of structure and function. *Protein & Cell*, **1**, 143–152.

117 Simmaco, M., Kreil, G., and Barra, D. (2009) Bombinins, antimicrobial peptides from *Bombina* species. *Biochimica et Biophysica Acta – Biomembranes*, **1788**, 1551–1555.

118 Koh, J.M.S., Bansal, P.S., Torres, A.M., and Kuchel, P.W. (2009) Platypus venom: source of novel compounds. *Australian Journal of Zoology*, **57**, 203–210.

119 Torres, A.M., Tsampazi, M., Tsampazi, C., Kennett, E.C., Belov, K., Geraghty, D.P., Bansal, P.S., Alewood, P.F., and Kuchel, P.W. (2006) Mammalian L-to-D-amino-acid-residue isomerase from platypus venom. *FEBS Letters*, **580**, 1587–1591.

120 Bansal, P.S., Torres, A.M., Crossett, B., Wong, K.K.Y., Koh, J.M.S., Geraghty, D.P., Vandenberg, J.I., and Kuchel, P.W. (2008) Substrate specificity of platypus venom L-to-D-peptide isomerase. *Journal of Biological Chemistry*, **283**, 8969–8975.

121 Jilek, A., Mollay, C., Tippelt, C., Grassi, J., Mignogna, G., Mullegger, J., Sander, V., Fehrer, C., Barra, D., and Kreil, G. (2005) Biosynthesis of a D-amino acid in peptide linkage by an enzyme from frog skin secretions. *Proceedings of the National Academy of Sciences of the United States of America*, **102**, 4235–4239.

122 Jilek, A., Mollay, C., Lohner, K., and Kreil, G. (2012) Substrate specificity of a peptidyl-aminoacyl-L/D-isomerase from frog skin. *Amino Acids*, **42**, 1757–1764.

123 Koh, J.M.S., Haynes, L., Belov, K., and Kuchel, P.W. (2010) L-to-D-peptide isomerase in male echidna venom. *Australian Journal of Zoology*, **58**, 284–288.

124 Koh, J.M.S., Chow, S.J.P., Crossett, B., and Kuchel, P.W. (2010) Mammalian peptide isomerase: platypus-type activity is present in mouse heart. *Chemistry & Biodiversity*, **7**, 1603–1611.

125 Bittner, S., Scherzer, R., and Harlev, E. (2007) The five bromotryptophans. *Amino Acids*, **33**, 19–42.

126 Lehrer, R.I., Tincu, J.A., Taylor, S.W., Menzel, L.P., and Waring, A.J. (2003) Natural peptide antibiotics from tunicates: structures, functions and potential uses. *Integrative and Comparative Biology*, **43**, 313–322.

127 Taylor, S.W., Craig, A.G., Fischer, W.H., Park, M., and Lehrer, R.I. (2000) Styelin D, an extensively modified antimicrobial peptide from ascidian hemocytes. *Journal of Biological Chemistry*, **275**, 38417–38426.

128 Hicks, R.P., and Russell, A.L. (2012) Application of unnatural amino acids to the de novo design of selective antibiotic peptides, in *Unnatural Amino Acids: Methods and Protocols* (ed. L.S.S. Pollegioni), Springer, Berlin, pp. 135–167.

129 Russell, A.L., Kennedy, A.M., Spuches, A.M., Gibson, W.S., Venugopal, D., Klapper, D., Srouji, A.H., Bhonsle, J.B., and Hicks, R.P. (2011) Determining the effect of the incorporation of unnatural amino acids into antimicrobial peptides on the interactions with zwitterionic and anionic membrane model systems. *Chemistry and Physics of Lipids*, **164**, 740–758.

130 Russell, A.L., Williams, B.C., Spuches, A., Klapper, D., Srouji, A.H., and Hicks, R.P. (2012) The effect of the length and flexibility of the side chain of basic amino acids on the binding of antimicrobial peptides to zwitterionic and anionic membrane model systems. *Bioorganic & Medicinal Chemistry*, **20**, 1723–1739.

131 Tossi, A., Scocchi, M., Zahariev, S., and Gennaro, R. (2012) Use of unnatural amino acids to probe structure–activity relationships and mode-of-action of antimicrobial peptides, in *Unnatural Amino Acids: Methods and Protocols* (ed. L.S.S. Pollegioni), Springer, Berlin, pp. 169–183.

132 Brogden, K.A. (2005) Antimicrobial peptides: pore formers or metabolic inhibitors in bacteria? *Nature Reviews Microbiology*, **3**, 238–250.

133 Dennison, S.R., Wallace, J., Harris, F., and Phoenix, D.A. (2005) Amphiphilic alpha-helical antimicrobial peptides and their structure–function relationships. *Protein and Peptide Letters*, **12**, 31–39.

134 Boland, M.P. and Separovic, F. (2006) Membrane interactions of antimicrobial peptides from Australian tree frogs. *Biochimica et Biophysica Acta – Biomembranes*, **1758**, 1178–1183.

135 Fernandez, D.I., Gehman, J.D., and Separovic, F. (2009) Membrane interactions of antimicrobial peptides from Australian frogs. *Biochimica et Biophysica Acta – Biomembranes*, **1788**, 1630–1638.

136 Rozek, T., Bowie, J.H., Wallace, J.C., and Tyler, M.J. (2000) The antibiotic and anticancer active aurein peptides from the Australian Bell Frogs *Litoria aurea* and *Litoria raniformis*. Part 2. Sequence determination using electrospray mass spectrometry. *Rapid Communications in Mass Spectrometry*, **14**, 2002–2011.

137 Rozek, T., Wegener, K.L., Bowie, J.H., Olver, I.N., Carver, J.A., Wallace, J.C., and Tyler, M.J. (2000) The antibiotic and anticancer active aurein peptides from the Australian Bell Frogs *Litoria aurea* and *Litoria raniformis* – the solution structure of aurein 1.2. *European Journal of Biochemistry*, **267**, 5330–5341.

138 Teixeira, V., Feio, M.J., and Bastos, M. (2012) Role of lipids in the interaction of antimicrobial peptides with membranes. *Progress in Lipid Research*, **51**, 149–177.

139 Wimley, W.C. and Hristova, K. (2011) Antimicrobial peptides: successes, challenges and unanswered questions. *Journal of Membrane Biology*, **239**, 27–34.

140 Stromstedt, A.A., Ringstad, L., Schmidtchen, A., and Malmsten, M. (2010) Interaction between amphiphilic peptides and phospholipid membranes. *Current Opinion in Colloid & Interface Science*, **15**, 467–478.

141 Hwang, P.M. and Vogel, H.J. (1998) Structure–function relationships of antimicrobial peptides. *Biochemistry and Cell Biology/Biochimie et Biologie Cellulaire*, **76**, 235–246.

142 Nguyen, L.T., Haney, E.F., and Vogel, H.J. (2011) The expanding scope of antimicrobial peptide structures and their modes of action. *Trends in Biotechnology*, **29**, 464–472.

143 Zasloff, M. (1987) Magainins, a class of antimicrobial peptides from *Xenopus* skin: isolation, characterization of two active forms, and partial cDNA sequence of a precursor. *Proceedings of the National Academy of Sciences of the United States of America*, **84**, 5449–5453.

144 Bolintineanu, D.S. and Kaznessis, Y.N. (2011) Computational studies of protegrin antimicrobial peptides: a review. *Peptides*, **32**, 188–201.

145 Scocchi, M., Tossi, A., and Gennaro, R. (2011) Proline-rich antimicrobial peptides: converging to a non-lytic mechanism of action. *Cellular and Molecular Life Sciences*, **68**, 2317–2330.

146 Almeida, P.F., Ladokhin, A.S., and White, S.H. (2012) Hydrogen-bond energetics drive helix formation in membrane interfaces. *Biochimica et Biophysica Acta – Biomembranes*, **1818**, 178–182.

147 Wieprecht, T., Beyermann, M., and Seelig, J. (2002) Thermodynamics of the coil-alpha-helix transition of amphipathic peptides in a membrane environment: the role of vesicle curvature. *Biophysical Chemistry*, **96**, 191–201.

148 Harris, F., Daman, A., Wallace, J., Dennison, S.R., and Phoenix, D.A. (2006) Oblique orientated alpha-helices and their prediction. *Current Protein & Peptide Science*, **7**, 529–537.

149 Harris, F., Dennison, S., and Phoenix, D.A. (2006) The prediction of hydrophobicity gradients within membrane interactive protein alpha-helices using a novel graphical technique. *Protein and Peptide Letters*, **13**, 595–600.

150 Phoenix, D.A. and Harris, F. (2002) The hydrophobic moment and its use in the

classification of amphiphilic structures [review]. *Molecular Membrane Biology*, **19**, 1–10.

151 Phoenix, D.A., Harris, F., Daman, O.A., and Wallace, J. (2002) The prediction of amphiphilic alpha-helices. *Current Protein & Peptide Science*, **3**, 201–221.

152 Dennison, S.R., Whittaker, M., Harris, F., and Phoenix, D.A. (2006) Anticancer alpha-helical peptides and structure–function relationships underpinning their interactions with tumour cell membranes. *Current Protein & Peptide Science*, **7**, 487–499.

153 Harris, F., Wallace, J., and Phoenix, D.A. (2000) Use of hydrophobic moment plot methodology to aid the identification of oblique orientated alpha-helices. *Molecular Membrane Biology*, **17**, 201–207.

154 Jin, Y., Hammer, J., Pate, M., Zhang, Y., Zhu, F., Zmuda, E., and Blazyk, J. (2005) Antimicrobial activities and structures of two linear cationic peptide families with various amphipathic beta-sheet and alpha-helical potentials. *Antimicrobial Agents and Chemotherapy*, **49**, 4957–4964.

155 Yeaman, M.R. and Yount, N.Y. (2003) Mechanisms of antimicrobial peptide action and resistance. *Pharmacological Reviews*, **55**, 27–55.

156 Zelezetsky, I. and Tossi, A. (2006) Alpha-helical antimicrobial peptides – using a sequence template to guide structure–activity relationship studies. *Biochimica et Biophysica Acta – Biomembranes*, **1758**, 1436–1449.

157 Dathe, M., Wieprecht, T., Nikolenko, H., Handel, L., Maloy, W.L., MacDonald, D.L., Beyermann, M., and Bienert, M. (1997) Hydrophobicity, hydrophobic moment and angle subtended by charged residues modulate antibacterial and haemolytic activity of amphipathic helical peptides. *FEBS letters*, **403**, 208–212.

158 Fjell, C.D., Hiss, J.A., Hancock, R.E.W., and Schneider, G. (2012) Designing antimicrobial peptides: form follows function. *Nature Reviews Drug Discovery*, **11**, 37–51.

159 Dennison, S.R., Harris, F., and Phoenix, D.A. (2003) Factors determining the efficacy of alpha-helical antimicrobial peptides. *Protein and Peptide Letters*, **10**, 497–502.

160 Tossi, A., Sandri, L., and Giangaspero, A. (2000) Amphipathic, alpha-helical antimicrobial peptides. *Biopolymers*, **55**, 4–30.

161 McCaldon, P. and Argos, P. (1988) Oligopeptide biases in protein sequences and their use in predicting protein coding regions in nucleotide sequences. *Proteins*, **4**, 99–122.

162 Rivas, L., Roman Luque-Ortega, J., Fernandez-Reyes, M., and Andreu, D. (2010) Membrane-active peptides as anti-infectious agents. *Journal of Applied Biomedicine*, **8**, 159–167.

163 Bowman, S.M. and Free, S.J. (2006) The structure and synthesis of the fungal cell wall. *BioEssays*, **28**, 799–808.

164 Yount, N.Y., Bayer, A.S., Xiong, Y.Q., and Yeaman, M.R. (2006) Advances in antimicrobial peptide immunobiology. *Biopolymers*, **84**, 435–458.

165 Harris, F., Dennison, S.R., and Phoenix, D.A. (2009) Anionic antimicrobial peptides from eukaryotic organisms. *Current Protein & Peptide Science*, **10**, 585–606.

166 Xiao, J., Zhang, H., Niu, L., and Wang, X. (2011) Efficient screening of a novel antimicrobial peptide from *Jatropha curcas* by cell membrane affinity chromatography. *Journal of Agricultural and Food Chemistry*, **59**, 1145–1151.

167 Wu, L.-P., Wu, Z.-J., Lin, D., Fang, F., Lin, Q.-Y., and Xie, L.-H. (2008) Characterization and amino acid sequence of y3, an antiviral protein from mushroom *Coprinus comatus*. *Chinese Journal of Biochemistry and Molecular Biology*, **24**, 597–603.

168 Deslouches, B., Phadke, S.M., Lazarevic, V., Cascio, M., Islam, K., Montelaro, R.C., and Mietzner, T.A. (2005) De novo generation of cationic antimicrobial peptides: influence of length and tryptophan substitution on antimicrobial activity. *Antimicrobial Agents and Chemotherapy*, **49**, 316–322.

169 Koszalka, P., Kamysz, E., Wejda, M., Kamysz, W., and Bigda, J. (2011) Antitumor activity of antimicrobial peptides against U937 histiocytic cell

line. *Acta Biochimica Polonica*, **58**, 111–117.

170 Gallo, R.L., Murakami, M., Ohtake, T., and Zaiou, M. (2002) Biology and clinical relevance of naturally occurring antimicrobial peptides. *The Journal of Allergy and Clinical Immunology*, **110**, 823–831.

171 Findlay, B., Zhanel, G.G., and Schweizer, F. (2010) Cationic amphiphiles, a new generation of antimicrobials inspired by the natural antimicrobial peptide scaffold. *Antimicrobial Agents and Chemotherapy*, **54**, 4049–4058.

172 Rosenfeld, Y., Lev, N., and Shai, Y. (2010) Effect of the hydrophobicity to net positive charge ratio on antibacterial and anti-endotoxin activities of structurally similar antimicrobial peptides. *Biochemistry*, **49**, 853–861.

173 Yin, L.M., Edwards, M.A., Li, J., Yip, C.M., and Deber, C.M. (2012) Roles of hydrophobicity and charge distribution of cationic antimicrobial peptides in peptide–membrane interactions. *Journal of Biological Chemistry*, **287**, 7738–7745.

174 Papanastasiou, E.A., Hua, Q., Sandouk, A., Son, U.H., Christenson, A.J., Van Hoek, M.L., and Bishop, B.M. (2009) Role of acetylation and charge in antimicrobial peptides based on human beta-defensin-3. *APMIS: Acta Pathologica, Microbiologica, et Immunologica Scandinavica*, **117**, 492–499.

175 Jiang, Z., Vasil, A.I., Hale, J., Hancock, R.E.W., Vasil, M.L., and Hodges, R.S. (2009) Effects of net charge and the number of positively charged residues on the biological activity of amphipathic alpha-helical cationic antimicrobial peptides, in *Peptides for Youth* (ed. S. DelValle, E. Escher, and W.D. Lubell), Springer, Berlin, pp. 561–562.

176 Jiang, Z., Vasil, A.I., Hale, J.D., Hancock, R.E.W., Vasil, M.L., and Hodges, R.S. (2008) Effects of net charge and the number of positively charged residues on the biological activity of amphipathic alpha-helical cationic antimicrobial peptides. *Biopolymers*, **90**, 369–383.

177 Ahn, H.S., Cho, W., Kang, S.H., Ko, S.S., Park, M.S., Cho, H., and Lee, K.H. (2006) Design and synthesis of novel antimicrobial peptides on the basis of alpha helical domain of Tenecin 1, an insect defensin protein, and structure–activity relationship study. *Peptides*, **27**, 640–648.

178 Dathe, M., Nikolenko, H., Meyer, J., Beyermann, M., and Bienert, M. (2001) Optimization of the antimicrobial activity of magainin peptides by modification of charge. *FEBS Letters*, **501**, 146–150.

179 Yang, Y.X., Feng, Y., Wang, B.Y., and Wu, Q. (2004) PCR-based site-specific mutagenesis of peptide antibiotics FALL-39 and its biologic activities. *Acta Pharmacologica Sinica*, **25**, 239–245.

180 Jiang, Z., Vasil, A.I., Gera, L., Vasil, M.L., and Hodges, R.S. (2011) Rational design of α-helical antimicrobial peptides to target Gram-negative pathogens, *Acinetobacter baumannii* and *Pseudomonas aeruginosa*: utilization of charge, "specificity determinants," total hydrophobicity, hydrophobe type and location as design parameters to improve the therapeutic ratio. *Chemical Biology & Drug Design*, **77**, 225–240.

181 Muhle, S.A. and Tam, J.P. (2001) Design of Gram-negative selective antimicrobial peptides. *Biochemistry*, **40**, 5777–5785.

182 Tokunaga, Y., Niidome, T., Hatakeyama, T., and Aoyagi, H. (2001) Antibacterial activity of bactenecin 5 fragments and their interaction with phospholipid membranes. *Journal of Peptide Science*, **7**, 297–304.

183 Brogden, K.A., Ackermann, M., McCray, P.B., and Tack, B.F. (2003) Antimicrobial peptides in animals and their role in host defences. *International Journal of Antimicrobial Agents*, **22**, 465–478.

184 Tomita, M., Wakabayashi, H., Shin, K., Yamauchi, K., Yaeshima, T., and Iwatsuki, K. (2009) Twenty-five years of research on bovine lactoferrin applications. *Biochimie*, **91**, 52–57.

185 Chen, J., Falla, T.J., Liu, H., Hurst, M.A., Fujii, C.A., Mosca, D.A., Embree, J.R., Loury, D.J., Radel, P.A., Cheng Chang, C., Gu, L., and Fiddes, J.C. (2000)

Development of protegrins for the treatment and prevention of oral mucositis: structure–activity relationships of synthetic protegrin analogues. *Biopolymers*, **55**, 88–98.
186 Chan, D.I., Prenner, E.J., and Vogel, H.J. (2006) Tryptophan- and arginine-rich antimicrobial peptides: structures and mechanisms of action. *Biochimica et Biophysica Acta*, **1758**, 1184–1202.
187 Yang, S.T., Shin, S.Y., Lee, C.W., Kim, Y.C., Hahm, K.S., and Kim, J.I. (2003) Selective cytotoxicity following Arg-to-Lys substitution in tritrpticin adopting a unique amphipathic turn structure. *FEBS Letters*, **540**, 229–233.
188 Koo, Y.S., Kim, J.M., Park, I.Y., Yu, B.J., Jang, S.A., Kim, K.-S., Park, C.B., Cho, J.H., and Kim, S.C. (2008) Structure–activity relations of parasin I, a histone H2A-derived antimicrobial peptide. *Peptides*, **29**, 1102–1108.
189 Dougherty, D.A. (2007) Cation–pi interactions involving aromatic amino acids. *Journal of Nutrition*, **137**, 1504S–1508S.
190 Chen, Y.X., Guarnieri, M.T., Vasil, A.I., Vasil, M.L., Mant, C.T., and Hodges, R.S. (2007) Role of peptide hydrophobicity in the mechanism of action of alpha-helical antimicrobial peptides. *Antimicrobial Agents and Chemotherapy*, **51**, 1398–1406.
191 Dennison, S.R., Wallace, J., Harris, F., and Phoenix, D.A. (2005) Relationships between the physiochemical properties, microbial specificity and antimicrobial activity of alpha-helical antimicrobial peptides: a statistical investigation. *Current Topics in Peptide and Protein Research*, **5**, 52–63.
192 Jiang, Z., Kullberg, B.J., van der Lee, H., Vasil, A.I., Hale, J.D., Mant, C.T., Hancock, R.E.W., Vasil, M.L., Netea, M.G., and Hodges, R.S. (2008) Effects of hydrophobicity on the antifungal activity of alpha-helical antimicrobial peptides. *Chemical Biology & Drug Design*, **72**, 483–495.
193 Dennison, S.R., Morton, L.H.G., Brandenburg, K., Harris, F., and Phoenix, D.A. (2006) Investigations into the ability of an oblique alpha-helical template to provide the basis for design of an antimicrobial anionic amphiphilic peptide. *FEBS Journal*, **273**, 3792–3803.
194 Chen, Y., Mant, C.T., Farmer, S.W., Hancock, R.E.W., Vasil, M.L., and Hodges, R.S. (2005) Rational design of α-helical antimicrobial peptides with enhanced activities and specificity/therapeutic index. *Journal of Biological Chemistry*, **280**, 12316–12329.
195 Eisenberg, D., Weiss, R.M., and Terwilliger, T.C. (1982) The helical hydrophobic moment: a measure of the amphiphilicity of a helix. *Nature*, **299**, 371–374.
196 Piotto, S.P., Sessa, L., Concilio, S., and Iannelli, P. (2012) YADAMP: yet another database of antimicrobial peptides. *International Journal of Antimicrobial Agents*, **39**, 346–351.
197 Wallace, J., Harris, F., and Phoenix, D. (2003) A statistical investigation of amphiphilic properties of C-terminally anchored peptidases. *European Biophysics Journal*, **32**, 589–598.
198 Dathe, M. and Wieprecht, T. (1999) Structural features of helical antimicrobial peptides: their potential to modulate activity on model membranes and biological cells. *Biochimica et Biophysica Acta*, **1462**, 71–87.
199 Raj, P.A. and Dentino, A.R. (2002) Current status of defensins and their role in innate and adaptive immunity. *FEMS Microbiology Letters*, **206**, 9–18.
200 Chou, H.-T., Kuo, T.-Y., Chiang, J.-C., Pei, M.-J., Yang, W.-T., Yu, H.-C., Lin, S.-B., and Chen, W.-J. (2008) Design and synthesis of cationic antimicrobial peptides with improved activity and selectivity against *Vibrio* spp. *International Journal of Antimicrobial Agents*, **32**, 130–138.
201 Thevissen, K., Ferket, K.K., Francois, I.E., and Cammue, B.P. (2003) Interactions of antifungal plant defensins with fungal membrane components. *Peptides*, **24**, 1705–1712.
202 Briolat, J., Wu, S.D., Mahata, S.K., Gonthier, B., Bagnard, D., Chasserot-Golaz, S., Helle, K.B., Aunis, D., and Metz-Boutigue, M.H. (2005) New antimicrobial activity for the catecholamine release-inhibitory peptide from chromogranin A.

Cellular and Molecular Life Sciences, **62**, 377–385.

203 Jean-Francois, F., Desbat, B., and Dufourc, E.J. (2009) Selectivity of cateslytin for fungi: the role of acidic lipid-ergosterol membrane fluidity in antimicrobial action. *FASEB Journal*, **23**, 3692–3701.

204 Varkey, J., Singh, S., and Nagaraj, R. (2006) Antibacterial activity of linear peptides spanning the carboxy-terminal beta-sheet domain of arthropod defensins. *Peptides*, **27**, 2614–2623.

205 Hong, J., Oren, Z., and Shai, Y. (1999) Structure and organization of hemolytic and nonhemolytic diastereomers of antimicrobial peptides in membranes. *Biochemistry*, **38**, 16963–16973.

206 Oren, Z., Hong, J., and Shai, Y. (1999) A comparative study on the structure and function of a cytolytic alpha-helical peptide and its antimicrobial beta-sheet diastereomer. *European Journal of Biochemistry*, **259**, 360–369.

207 Castano, S., Desbat, B., and Dufourcq, J. (2000) Ideally amphipathic beta-sheeted peptides at interfaces: structure, orientation, affinities for lipids and hemolytic activity of (KL)(m)K peptides. *Biochimica et Biophysica Acta–Biomembranes*, **1463**, 65–80.

208 Michael Conlon, J., Mechkarska, M., and King, J.D. (2012) Host-defense peptides in skin secretions of African clawed frogs (Xenopodinae, Pipidae). *General and Comparative Endocrinology*, **176**, 513–518.

209 Blazyk, J., Wiegand, R., Klein, J., Hammer, J., Epand, R.M., Epand, R.F., Maloy, W.L., and Kari, U.P. (2001) A novel linear amphipathic beta-sheet cationic antimicrobial peptide with enhanced selectivity for bacterial lipids. *Journal of Biological Chemistry*, **276**, 27899–27906.

210 Bellm, L., Lehrer, R.I., and Ganz, T. (2000) Protegrins: new antibiotics of mammalian origin. *Expert Opinion on Investigational Drugs*, **9**, 1731–1742.

211 Ganz, T. (2001) Antimicrobial peptides: from host defense to therapeutics. *Aids*, **15**, S57–S57.

212 Lehrer, R.I. and Ganz, T. (2002) Cathelicidins: a family of endogenous antimicrobial peptides. *Current Opinion in Hematology*, **9**, 18–22.

213 Capone, R., Mustata, M., Jang, H., Arce, F.T., Nussinov, R., and Lal, R. (2010) Antimicrobial protegrin-1 forms ion channels: molecular dynamic simulation, atomic force microscopy, and electrical conductance studies. *Biophysical Journal*, **98**, 2644–2652.

214 Lam, K.L.H., Wang, H., Siaw, T.A., Chapman, M.R., Waring, A.J., Kindt, J.T., and Lee, K.Y.C. (2012) Mechanism of structural transformations induced by antimicrobial peptides in lipid membranes. *Biochimica et Biophysica Acta–Biomembranes*, **1818**, 194–204.

215 Su, Y., Waring, A.J., Ruchala, P., and Hong, M. (2011) Structures of beta-hairpin antimicrobial protegrin peptides in lipopolysaccharide membranes: mechanism of gram selectivity obtained from solid-state nuclear magnetic resonance. *Biochemistry*, **50**, 2072–2083.

216 Bolintineanu, D., Hazrati, E., Davis, H.T., Lehrer, R.I., and Kaznessis, Y.N. (2010) Antimicrobial mechanism of pore-forming protegrin peptides: 100 pores to kill *E. coli*. *Peptides*, **31**, 1–8.

217 Kagan, B.L., Jang, H., Capone, R., Arce, F.T., Ramachandran, S., Lal, R., and Nussinov, R. (2012) Antimicrobial properties of amyloid peptides. *Molecular Pharmaceutics*, **9**, 708–717.

218 Jang, H., Arce, F.T., Mustata, M., Ramachandran, S., Capone, R., Nussinov, R., and Lal, R. (2011) Antimicrobial protegrin-1 forms amyloid-like fibrils with rapid kinetics suggesting a functional link. *Biophysical Journal*, **100**, 1775–1783.

219 Harris, F., Dennison, S.R., and Phoenix, D.A. (2012) Aberrant action of amyloidogenic host defense peptides: a new paradigm to investigate neurodegenerative disorders? *FASEB Journal*, **26**, 1776–1781.

220 Wong, J.H., Xia, L.X., and Ng, T.B. (2007) A review of defensins of diverse origins. *Current Protein & Peptide Science*, **8**, 446–459.

221 Yamauchi, H., Maehara, N., Takanashi, T., and Nakashima, T. (2010) Defensins as host defense molecules against

microbes: the characteristics of the defensins from arthropods, mollusks and fungi. *Bulletin of the Forestry and Forest Products Research Institute*, **9**, 1–18.
222 Carvalho, A.D.O. and Gomes, V.M. (2011) Plant defensins and defensin-like peptides – biological activities and biotechnological applications. *Current Pharmaceutical Design*, **17**, 4270–4293.
223 Sharma, M. (2011) Plant defensins: novel antimicrobial peptides. *Vegetos*, **24**, 126–135.
224 Wilmes, M., Cammue, B.P.A., Sahl, H.-G., and Thevissen, K. (2011) Antibiotic activities of host defense peptides: more to it than lipid bilayer perturbation. *Natural Product Reports*, **28**, 1350–1358.
225 Thevissen, K., Kristensen, H.-H., Thomma, B.P.H.J., Cammue, B.P.A., and Francois, I.E.J.A. (2007) Therapeutic potential of antifungal plant and insect defensins. *Drug Discovery Today*, **12**, 966–971.
226 van Dijk, A., Veldhuizen, E.J.A., and Haagsman, H.P. (2008) Avian defensins. *Veterinary Immunology and Immunopathology*, **124**, 1–18.
227 Penberthy, W.T., Chari, S., Cole, A.L., and Cole, A.M. (2011) Retrocyclins and their activity against HIV-1. *Cellular and Molecular Life Sciences*, **68**, 2231–2242.
228 Arnett, E. and Seveau, S. (2011) The multifaceted activities of mammalian defensins. *Current Pharmaceutical Design*, **17**, 4254–4269.
229 Ganz, T. (2005) Defensins and other antimicrobial peptides: a historical perspective and an update. *Combinatorial Chemistry & High Throughput Screening*, **8**, 209–217.
230 Selsted, M.E. and Ouellette, A.J. (2005) Mammalian defensins in the antimicrobial immune response. *Nature Immunology*, **6**, 551–557.
231 Bruhn, O., Paul, S., Tetens, J., and Thaller, G. (2009) The repertoire of equine intestinal alpha-defensins. *BMC Genomics*, **10**, 631.
232 Hazlett, L. and Wu, M.H. (2011) Defensins in innate immunity. *Cell and Tissue Research*, **343**, 175–188.

233 Ganz, T. (2004) Defensins: antimicrobial peptides of vertebrates. *Comptes Rendus Biologies*, **327**, 539–549.
234 Pazgier, M., Hoover, D., Yang, D., Lu, W., and Lubkowski, J. (2006) Human β-defensins. *Cellular and Molecular Life Sciences*, **63**, 1294–1313.
235 Semple, F. and Dorin, J.R. (2012) Beta-defensins: multifunctional modulators of infection, inflammation and more. *Journal of Innate Immunity*, **4**, 337–348.
236 Taylor, K., Barran, P.E., and Dorin, J.R. (2008) Review: structure–activity relationships in beta-defensin peptides. *Biopolymers*, **90**, 1–7.
237 Lehrer, R.I. and Lu, W. (2012) Alpha-defensins in human innate immunity. *Immunological Reviews*, **245**, 84–112.
238 Bevins, C.L. (2005) Events at the host-microbial Interface of the gastrointestinal tract – V. Paneth cell alpha-defensins in intestinal host defense. *American Journal of Physiology*, **289**, G173–G176.
239 Ganz, T. (2003) Defensins: antimicrobial peptides of innate immunity. *Nature Reviews Immunology*, **3**, 710–720.
240 Morrison, G., Kilanowski, F., Davidson, D., and Dorin, J. (2002) Characterization of the mouse beta defensin 1, Defb1, mutant mouse model. *Infection and Immunity*, **70**, 3053–3060.
241 Moser, C., Weiner, D.J., Lysenko, E., Bals, R., Weiser, J.N., and Wilson, J.M. (2002) Beta-defensin 1 contributes to pulmonary innate immunity in mice. *Infection and Immunity*, **70**, 3068–3072.
242 Harder, J., Glaser, R., and Schroder, J.M. (2007) Human antimicrobial proteins – effectors of innate immunity. *Journal of Endotoxin Research*, **13**, 317–338.
243 Gran, L. (1970) An oxytocic principle found in *Oldenlandia affinis* DC. *Meddelelser fra Norsk Farmaceutisk Selskap*, **32**, 173–180.
244 Gran, L. (1973) On the effect of a polypeptide isolated from "Kalata-Kalata" (*Oldenlandia affinis* DC) on the oestrogen dominated uterus. *Acta Pharmacologica et Toxicologica*, **33**, 400–408.

245 Gruber, C.W. and O'Brien, M. (2011) Uterotonic plants and their bioactive constituents. *Planta Medica*, **77**, 207–220.

246 Gran, L. (1973) Oxytocic principles of *Oldenlandia affinis*. *Lloydia –The Journal of Natural Products*, **36**, 174–178.

247 Gran, L., Sandberg, F., and Sletten, K. (2000) *Oldenlandia affinis* (R&S) DC – a plant containing uteroactive peptides used in African traditional medicine. *Journal of Ethnopharmacology*, **70**, 197–203.

248 Gran, L., Sletten, K., and Skjeldal, L. (2008) Cyclic peptides from *Oldenlandia affinis* DC. Molecular and biological properties. *Chemistry & Biodiversity*, **5**, 2014–2022.

249 Gruber, C.W., Elliott, A.G., Ireland, D.C., Delprete, P.G., Dessein, S., Goransson, U., Trabi, M., Wang, C.K., Kinghorn, A.B., Robbrecht, E., and Craik, D. (2008) Distribution and evolution of circular miniproteins in flowering plants. *Plant Cell*, **20**, 2471–2483.

250 Daly, N.L., Rosengren, K.J., and Craik, D.J. (2009) Discovery, structure and biological activities of cyclotides. *Advanced Drug Delivery Reviews*, **61**, 918–930.

251 Tang, J. and Tan, N.H. (2010) Progress of cyclotides in plants. *Progress in Chemistry*, **22**, 677–683.

252 Camarero, J.A. (2011) Legume cyclotides shed light on the genetic origin of knotted circular proteins. *Proceedings of the National Academy of Sciences of the United States of America*, **108**, 10025–10026.

253 Giang, K.T.N., Zhang, S., Ngan, T.K.N., Phuong, Q.T.N., Chiu, M.S., Hardjojo, A., and Tam, J.P. (2011) Discovery and characterization of novel cyclotides originated from chimeric precursors consisting of albumin-1 chain a and cyclotide domains in the fabaceae family. *Journal of Biological Chemistry*, **286**, 24275–24287.

254 Poth, A.G., Colgrave, M.L., Lyons, R.E., Daly, N.L., and Craik, D.J. (2011) Discovery of an unusual biosynthetic origin for circular proteins in legumes. *Proceedings of the National Academy of Sciences of the United States of America*, **108**, 10127–10132.

255 Poth, A.G., Colgrave, M.L., Philip, R., Kerenga, B., Daly, N.L., Anderson, M.A., and Craik, D.J. (2010) Discovery of cyclotides in the fabaceae plant family provides new insights into the cyclization, evolution, and distribution of circular proteins. *ACS Chemical Biology*, **6**, 345–355.

256 Wang, C.K.L., Kaas, Q., Chiche, L., and Craik, D.J. (2008) CyBase: a database of cyclic protein sequences and structures, with applications in protein discovery and engineering. *Nucleic Acids Research*, **36**, D206–D210.

257 Mulvenna, J.P., Wang, C., and Craik, D.J. (2006) CyBase: a database of cyclic protein sequence and structure. *Nucleic Acids Research*, **34**, D192–D194.

258 Craik, D.J. (2010) Discovery and applications of the plant cyclotides. *Toxicon*, **56**, 1092–1102.

259 Jones, J.D.G. and Dangl, J.L. (2006) The plant immune system. *Nature*, **444**, 323–329.

260 Mylne, J.S., Wang, C.K., van der Weerden, N.L., and Craik, D.J. (2010) Cyclotides are a component of the innate defense of *Oldenlandia affinis*. *Biopolymers*, **94**, 635–646.

261 Pranting, M., Loov, C., Burman, R., Goransson, U., and Andersson, D.I. (2010) The cyclotide cycloviolacin O2 from *Viola odorata* has potent bactericidal activity against Gram-negative bacteria. *Journal of Antimicrobial Chemotherapy*, **65**, 1964–1971.

262 Pinto, M.F.S., Almeida, R.G., Porto, W.F., Fensterseifer, I.C.M., Lima, L.A., Dias, S.C., and Franco, O.L. (2012) Cyclotides: from gene structure to promiscuous multifunctionality. *Journal of Evidence-Based Complementary & Alternative Medicine*, **17**, 40–53.

263 Nair, S.S., Romanuka, J., Billeter, M., Skjeldal, L., Emmett, M.R., Nilsson, C.L., and Marshall, A.G. (2006) Structural characterization of an unusually stable cyclic peptide, kalata B2 from *Oldenlandia affinis*. *Biochim Biophys Acta*, **1764**, 1568–1576.

264 Jennings, C.V., Rosengren, K.J., Daly, N.L., Plan, M., Stevens, J., Scanlon, M.J., Waine, C., Norman, D.G., Anderson, M.A., and Craik, D.J. (2005) Isolation, solution structure, and insecticidal activity of kalata B2, a circular protein with a twist: do Mobius strips exist in nature? *Biochemistry*, **44**, 851–860.

265 Colgrave, M.L., Kotze, A.C., Huang, Y.H., O'Grady, J., Simonsen, S.M., and Craik, D.J. (2008) Cyclotides: natural, circular plant peptides that possess significant activity against gastrointestinal nematode parasites of sheep. *Biochemistry*, **47**, 5581–5589.

266 Plan, M.R., Saska, I., Cagauan, A.G., and Craik, D.J. (2008) Backbone cyclised peptides from plants show molluscicidal activity against the rice pest *Pomacea canaliculata* (golden apple snail). *J Agric Food Chem*, **56**, 5237–5241.

267 Kaas, Q., Westermann, J.-C., Troeira Henriques, S., and Craik, D.J. (2010) Antimicrobial peptides in plants, in *Antimicrobial Peptides, Discovery, Design and Novel Therapeutic Strategies* (ed. G. Wang), CABI, Wallingford, pp. 40–70.

268 Pinto, M.F.S., Fensterseifer, I.C.M., and Franco, O.L. (2012) Plant cyclotides: an unusual protein family with multiple functions, in *Plant Defence: Biological Control* (eds J.M. Merillon and K.G. Ramawat), Springer, Dordrecht, pp. 333–344.

269 Craik, D.J. (2012) Host-defense activities of cyclotides. *Toxins*, **4**, 139–156.

270 Wang, C.K., Colgrave, M.L., Gustafson, K.R., Ireland, D.C., Goransson, U., and Craik, D.J. (2008) Anti-HIV cyclotides from the Chinese medicinal herb *Viola yedoensis*. *Journal of Natural Products*, **71**, 47–52.

271 Ireland, D.C., Wang, C.K., Wilson, J.A., Gustafson, K.R., and Craik, D.J. (2008) Cyclotides as natural anti-HIV agents. *Biopolymers*, **90**, 51–60.

272 Burman, R., Strömstedt, A.A., Malmsten, M., and Göransson, U. (2011) Cyclotide–membrane interactions: defining factors of membrane binding, depletion and disruption. *Biochimica et Biophysica Acta*, **1808**, 2665–2673.

273 Wang, C.K., Colgrave, M.L., Ireland, D.C., Kaas, Q., and Craik, D.J. (2009) Despite a conserved cystine knot motif, different cyclotides have different membrane binding modes. *Biophysical Journal*, **97**, 1471–1481.

274 Gould, A., Ji, Y., Aboye, T.L., and Camarero, J.A. (2011) Cyclotides, a novel ultrastable polypeptide scaffold for drug discovery. *Current Pharmaceutical Design*, **17**, 4294–4307.

275 Daly, N.L., Clark, R.J., Plan, M.R., and Craik, D.J. (2006) Kalata B8, a novel antiviral circular protein, exhibits conformational flexibility in the cystine knot motif. *Biochemical Journal*, **393**, 619–626.

276 Henriques, S.T. and Craik, D.J. (2012) Importance of the cell membrane on the mechanism of action of cyclotides. *ACS Chemical Biology*, **7**, 626–636.

277 Henriques, S.T., Huang, Y.-H., Rosengren, K.J., Franquelim, H.G., Carvalho, F.A., Johnson, A., Sonza, S., Tachedjian, G., Castanho, M.A.R.B., Daly, N.L., and Craik, D.J. (2011) Decoding the membrane activity of the cyclotide Kalata B1 the importance of phosphatidylethanolamine phospholipids and lipid organization on hemolytic and anti-HIV activities. *Journal of Biological Chemistry*, **286**, 24231–24241.

278 Sando, L., Henriques, S.T., Foley, F., Simonsen, S.M., Daly, N.L., Hall, K.N., Gustafson, K.R., Aguilar, M.-I., and Craik, D.J. (2011) A synthetic mirror image of Kalata B1 reveals that cyclotide activity is independent of a protein receptor. *ChemBioChem*, **12**, 2456–2462.

279 Wang, C.K.L., Clark, R.J., Harvey, P.J., Johan Rosengren, K., Cemazar, M., and Craik, D.J. (2011) The role of conserved Glu residue on cyclotide stability and activity: a structural and functional study of Kalata B12, a naturally occurring Glu to Asp mutant. *Biochemistry*, **50**, 4077–4086.

280 Shenkarev, Z.O., Nadezhdin, K.D., Lyukmanova, E.N., Sobol, V.A., Skjeldal, L., and Arseniev, A.S. (2008) Divalent cation coordination and mode of membrane interaction in cyclotides:

NMR spatial structure of ternary complex Kalata B7/Mn^{2+}/DPC micelle. *Journal of Inorganic Biochemistry*, **102**, 1246–1256.

281 Shenkarev, Z.O., Nadezhdin, K.D., Sobol, V.A., Sobol, A.G., Skjeldal, L., and Arseniev, A.S. (2006) Conformation and mode of membrane interaction in cyclotides. Spatial structure of kalata B1 bound to a dodecylphosphocoline micelle. *FEBS Journal*, **273**, 2658–2672.

282 Skjeldal, L., Gran, L., Sletten, K., and Volkman, B.F. (2002) Refined structure and metal binding site of the Kalata B1 peptide. *Archives of Biochemistry and Biophysics*, **399**, 142–148.

283 Gerlach, S.L., Burman, R., Bohlin, L., Mondal, D., and Goransson, U. (2010) Isolation, characterization, and bioactivity of cyclotides from the micronesian plant *Psychotria leptothyrsa*. *Journal of Natural Products*, **73**, 1207–1213.

284 Nguyen, G.K.T., Lim, W.H., Nguyen, P.Q.T., and Tam, J.P. (2012) Novel cyclotides and uncyclotides with highly shortened precursors from *Chassalia chartacea* and effects of methionine oxidation on bioactivities. *The Journal of Biological Chemistry*, **287**, 17598–17607.

285 Ireland, D.C., Colgravel, M.L., Nguyencong, P., Daly, N.L., and Craik, D.J. (2006) Discovery and characterization of a linear cyclotide from *Viola odorata*: implications for the processing of circular proteins. *Journal of Molecular Biology*, **357**, 1522–1535.

286 Giang Kien Truc, N., Zhang, S., Wang, W., Wong, C.T.T., Ngan Thi Kim, N., and Tam, J.P. (2011) Discovery of a linear cyclotide from the bracelet subfamily and its disulfide mapping by top-down mass spectrometry. *Journal of Biological Chemistry*, **286**, 44833–44844.

287 Hall, K., Lee, T.-H., Daly, N.L., Craik, D.J., and Aguilar, M.-I. (2012) Gly6 of kalata B1 is critical for the selective binding to phosphatidylethanolamine membranes. *Biochimica et Biophysica Acta – Biomembranes*, **1818**, 2354–2361.

288 Sousa, J.C., Berto, R.F., Gois, E.A., Fontenele-Cardi, N.C., Honorio-Junior, J.E.R., Konno, K., Richardson, M., Rocha, M.F.G., Camargo, A.A.C.M., Pimenta, D.C., Cardi, B.A., and Carvalho, K.M. (2009) Leptoglycin: a new Glycine/Leucine-rich antimicrobial peptide isolated from the skin secretion of the South American frog *Leptodactylus pentadactylus* (Leptodactylidae). *Toxicon*, **54**, 23–32.

289 Sun, D.D., Wu, S.Q., Jing, C.F., Zhang, N., Liang, D., and Xu, A.L. (2012) Identification, synthesis and characterization of a novel antimicrobial peptide HKPLP derived from *Hippocampus kuda* Bleeker. *Journal of Antibiotics*, **65**, 117–121.

290 Park, C.J., Park, C.B., Hong, S.S., Lee, H.S., Lee, S.Y., and Kim, S.C. (2000) Characterization and cDNA cloning of two glycine- and histidine-rich antimicrobial peptides from the roots of shepherd's purse, *Capsella bursa-pastoris*. *Plant Molecular Biology*, **44**, 187–197.

291 Remuzgo, C.R., Lopes, T.R.S., Oewel, T.S., and Miranda, M.T.M. (2010) Glycine and histidine-rich antifungal peptides: on the way to the mode of action of shepherin I. *Journal of Peptide Science*, **16**, 151–152.

292 Liu, H., Feng, Z., Lang, J., Li, Y., He, G., and Chen, Z. (2009) Fusion expression and high-level preparation of a glycine-rich antibacterial peptide (SK66) derived from *Drosophila* in *Escherichia coli*. *African Journal of Biotechnology*, **8**, 4608–4612.

293 Lu, J. and Chen, Z.-W. (2010) Isolation, characterization and anti-cancer activity of SK84, a novel glycine-rich antimicrobial peptide from *Drosophila virilis*. *Peptides*, **31**, 44–50.

294 Pisuttharachai, D., Fagutao, F.F., Yasuike, M., Aono, H., Yano, Y., Murakami, K., Kondo, H., Aoki, T., and Hirono, I. (2009) Characterization of crustin antimicrobial proteins from Japanese spiny lobster *Panulirus japonicus*. *Developmental and Comparative Immunology*, **33**, 1049–1054.

295 Baumann, T., Kaempfer, U., Schuerch, S., Schaller, J., Largiader, C., Nentwig, W., and Kuhn-Nentwig, L. (2010) Ctenidins: antimicrobial glycine-rich peptides from the hemocytes of the spider *Cupiennius salei*. *Cellular and Molecular Life Sciences*, **67**, 2787–2798.

296 Gao, B. and Zhu, S.Y. (2010) Characterization of a hymenoptaecin-like antimicrobial peptide in the parasitic wasp *Nasonia vitripennis*. *Process Biochemistry*, **45**, 139–146.

297 Imjongjirak, C., Amparyup, P., and Tassanakajon, A. (2011) Two novel antimicrobial peptides, arasin-likeSp and GRPSp, from the mud crab *Scylla paramamosain*, exhibit the activity against some crustacean pathogenic bacteria. *Fish & Shellfish Immunology*, **30**, 706–712.

298 Conlon, J.M., Abdel-Wahab, Y.H.A., Flatt, P.R., Leprince, J., Vaudry, H., Jouenne, T., and Condamine, E. (2009) A glycine–leucine-rich peptide structurally related to the plasticins from skin secretions of the frog *Leptodactylus laticeps* (Leptodactylidae). *Peptides*, **30**, 888–892.

299 El Amri, C. and Nicolas, P. (2008) Plasticins: membrane-damaging peptides with "chameleon-like" properties. *Cellular and Molecular Life Sciences*, **65**, 895–909.

300 Nicolas, P. and El Amri, C. (2009) The dermaseptin superfamily: a gene-based combinatorial library of antimicrobial peptides. *Biochimica et Biophysica Acta – Biomembranes*, **1788**, 1537–1550.

301 Calderon-Santiago, M. and Luque de Castro, M.D. (2009) The dual trend in histatins research. *Trends in Analytical Chemistry*, **28**, 1011–1018.

302 Sitaram, N. (2006) Antimicrobial peptides with unusual amino acid compositions and unusual structures. *Current Medicinal Chemistry*, **13**, 679–696.

303 Chernysh, S., Kim, S.I., Bekker, G., Pleskach, V.A., Filatova, N.A., Anikin, V.B., Platonov, V.G., and Bulet, P. (2002) Antiviral and antitumor peptides from insects. *Proceedings of the National Academy of Sciences of the United States of America*, **99**, 12628–12632.

304 Padovan, L., Segat, L., Pontillo, A., Antcheva, N., Tossi, A., and Crovella, S. (2010) Histatins in non-human primates: gene variations and functional effects. *Protein and Peptide Letters*, **17**, 909–918.

305 Fabian, T.K., Hermann, P., Beck, A., Fejerdy, P., and Fabian, G. (2012) Salivary defense proteins: their network and role in innate and acquired oral immunity. *International Journal of Molecular Sciences*, **13**, 4295–4320.

306 Iovino, M., Falconi, M., Marcellini, A., and Desideri, A. (2001) Molecular dynamics simulation of the antimicrobial salivary peptide histatin-5 in water and in trifluoroethanol: a microscopic description of the water destructuring effect. *The Journal of Peptide Research*, **58**, 45–55.

307 Helmerhorst, E.J., Venuleo, C., Beri, A., and Oppenheim, F.G. (2005) *Candida glabrata* is unusual with respect to its resistance to cationic antifungal proteins. *Yeast*, **22**, 705–714.

308 Nikawa, H., Jin, C., Fukushima, H., Makihira, S., and Hamada, T. (2001) Antifungal activity of histatin-5 against non-albicans *Candida* species. *Oral Microbiology and Immunology*, **16**, 250–252.

309 Konopka, K., Dorocka-Bobkowska, B., Gebremedhin, S., and Duzgunes, N. (2010) Susceptibility of *Candida* biofilms to histatin 5 and fluconazole. *Antonie Van Leeuwenhoek International Journal of General and Molecular Microbiology*, **97**, 413–417.

310 Peters, B.M., Zhu, J., Fidel, P.L., Jr, Scheper, M.A., Hackett, W., El Shaye, S., and Jabra-Rizk, M.A. (2010) Protection of the oral mucosa by salivary histatin-5 against *Candida albicans* in an *ex vivo* murine model of oral infection. *FEMS Yeast Research*, **10**, 597–604.

311 Situ, H., Balasubramanian, S.V., and Bobek, L.A. (2000) Role of α-helical conformation of histatin-5 in candidacidal activity examined by proline variants. *Biochimica et Biophysica Acta*, **1475**, 377–382.

312 Helmerhorst, E.J., van't Hof, W., Breeuwer, P., Veerman, E.C.I., Abee, T., Troxler, R.F., Amerongen, A.V.N., and Oppenheim, F.G. (2001) Characterization of histatin 5 with respect to amphipathicity, hydrophobicity, and effects on cell and mitochondrial membrane integrity excludes a candidacidal mechanism of pore formation. *Journal of Biological Chemistry*, **276**, 5643–5649.

313 Mochon, A.B. and Liu, H. (2008) The antimicrobial peptide histatin-5 causes a spatially restricted disruption on the *Candida albicans* surface, allowing rapid entry of the peptide into the cytoplasm. *PLoS Pathogens*, **4**, e1000190.

314 Helmerhorst, E.J., Breeuwer, P., van't Hof, W., Walgreen-Weterings, E., Oomen, L., Veerman, E.C.I., Amerongen, A.V.N., and Abee, T. (1999) The cellular target of histatin 5 on *Candida albicans* is the energized mitochondrion. *Journal of Biological Chemistry*, **274**, 7286–7291.

315 Mayer, F.L., Wilson, D., Jacobsen, I.D., Miramon, P., GroSse, K., and Hube, B. (2012) The novel *Candida albicans* transporter Dur31 Is a multi-stage pathogenicity factor. *PLoS Pathogens*, **8**, e1002592.

316 Sun, J.N., Li, W., Jang, W.S., Nayyar, N., Sutton, M.D., and Edgerton, M. (2008) Uptake of the antifungal cationic peptide Histatin 5 by *Candida albicans* Ssa2p requires binding to non-conventional sites within the ATPase domain. *Molecular Microbiology*, **70**, 1246–1260.

317 Li, X.S., Sun, J.N., Okamoto-Shibayama, K., and Edgerton, M. (2006) *Candida albicans* cell wall Ssa proteins bind and facilitate import of salivary histatin 5 required for toxicity. *Journal of Biological Chemistry*, **281**, 22453–22463.

318 Jang, W.S., Bajwa, J.S., Sun, J.N., and Edgerton, M. (2010) Salivary histatin 5 internalization by translocation, but not endocytosis, is required for fungicidal activity in *Candida albicans*. *Molecular Microbiology*, **77**, 354–370.

319 Kavanagh, K. and Dowd, S. (2004) Histatins: antimicrobial peptides with therapeutic potential. *Journal of Pharmacy and Pharmacology*, **56**, 285–289.

320 Komatsu, T., Salih, E., Helmerhorst, E.J., Offner, G.D., and Oppenheim, F.G. (2011) Influence of histatin 5 on *Candida albicans* mitochondrial protein expression assessed by quantitative mass spectrometry. *Journal of Proteome Research*, **10**, 646–655.

321 De Smet, K. and Contreras, R. (2005) Human antimicrobial peptides: defensins, cathelicidins and histatins. *Biotechnology Letters*, **27**, 1337–1347.

322 Sun, X.L., Salih, E., Oppenheim, F.G., and Helmerhorst, E.J. (2009) Kinetics of histatin proteolysis in whole saliva and the effect on bioactive domains with metal-binding, antifungal, and wound-healing properties. *FASEB Journal*, **23**, 2691–2701.

323 Kulon, K., Valensin, D., Kamysz, W., Valensin, G., Nadolski, P., Porciatti, E., Gaggelli, E., and Koztowski, H. (2008) The His–His sequence of the antimicrobial peptide demegen P-113 makes it very attractive ligand for Cu^{2+}. *Journal of Inorganic Biochemistry*, **102**, 960–972.

324 Grogan, J., McKnight, C.J., Troxler, R.F., and Oppenheim, F.G. (2001) Zinc and copper bind to unique sites of histatin 5. *FEBS Letters*, **491**, 76–80.

325 Tay, W.M., Hanafy, A.I., Angerhofer, A., and Ming, L.J. (2009) A plausible role of salivary copper in antimicrobial activity of histatin-5–metal binding and oxidative activity of its copper complex. *Bioorganic & Medicinal Chemistry Letters*, **19**, 6709–6712.

326 Houghton, E.A. and Nicholas, K.M. (2009) *In vitro* reactive oxygen species production by histatins and copper(I,II). *Journal of Biological Inorganic Chemistry*, **14**, 243–251.

327 Cabras, T., Patamia, M., Melino, S., Inzitari, R., Messana, I., Castagnola, M., and Petruzzelli, R. (2007) Pro-oxidant activity of histatin 5 related Cu(II)-model peptide probed by mass spectrometry. *Biochemical and Biophysical Research Communications*, **358**, 277–284.

328 Melino, S., Rufini, S., Sette, M., Morero, R., Grottesi, A., Paci, M., and Petruzzelli, R. (1999) Zn^{2+} ions selectively induce antimicrobial salivary peptide histatin-5 to fuse negatively charged vesicles. Identification and characterization of a zinc-binding motif present in the functional domain. *Biochemistry*, **38**, 9626–9633.

329 de Planque, M.R.R. and Killian, J.A. (2003) Protein–lipid interactions studied with designed transmembrane peptides: role of hydrophobic matching and interfacial anchoring [review]. *Molecular Membrane Biology*, **20**, 271–284.

330 Norman, K.E. and Nymeyer, H. (2006) Indole localization in lipid membranes revealed by molecular simulation. *Biophysical Journal*, **91**, 2046–2054.

331 Capparelli, R., Amoroso, M.G., Palumbo, D., Iannaccone, M., Faleri, C., and Cresti, M. (2005) Two plant puroindolines colocalize in wheat seed and *in vitro* synergistically fight against pathogens. *Plant Molecular Biology*, **58**, 857–867.

332 Palumbo, D., Iannaccone, M., Porta, A., and Capparelli, R. (2010) Experimental antibacterial therapy with puroindolines, lactoferrin and lysozyme in *Listeria monocytogenes*-infected mice. *Microbes and Infection*, **12**, 538–545.

333 Bhave, M. and Methuku, D.R. (2011) Small cysteine-rich proteins from plants: a rich resource of antimicrobial agents, in *Science against Microbial Pathogens: Communicating Current Research and Technological Advances* (ed. A. Mendez-Vilas), Formatex, Badajos, pp. 1074–1083.

334 Jing, W.G., Demcoe, A.R., and Vogel, H.J. (2003) Conformation of a bactericidal domain of puroindoline a: structure and mechanism of action of a 13-residue antimicrobial peptide. *Journal of Bacteriology*, **185**, 4938–4947.

335 Vogel, H.J., Schibli, D.J., Jing, W.G., Lohmeier-Vogel, E.M., Epand, R.F., and Epand, R.M. (2002) Towards a structure–function analysis of bovine lactoferricin and related tryptophan- and arginine-containing peptides. *Biochemistry and Cell Biology/Biochimie et Biologie Cellulaire*, **80**, 49–63.

336 Strom, M.B., Haug, B.E., Rekdal, O., Skar, M.L., Stensen, W., and Svendsen, J.S. (2002) Important structural features of 15-residue lactoferricin derivatives and methods for improvement of antimicrobial activity. *Biochemistry and Cell Biology/Biochimie et Biologie Cellulaire*, **80**, 65–74.

337 Melo, M.N., Ferre, R., and Castanho, M.A.R.B. (2009) Antimicrobial peptides: linking partition, activity and high membrane-bound concentrations. *Nature Reviews Microbiology*, **7**, 245–250.

338 Jenssen, H., Hamill, P., and Hancock, R.E.W. (2006) Peptide antimicrobial agents. *Clinical Microbiology Reviews*, **19**, 491–511.

339 Schibli, D.J., Epand, R.F., Vogel, H.J., and Epand, R.M. (2002) Tryptophan-rich antimicrobial peptides: comparative properties and membrane interactions. *Biochemistry and Cell Biology/Biochimie et Biologie Cellulaire*, **80**, 667–677.

340 Li, W.-F., Ma, G.-X., and Zhou, X.-X. (2006) Apidaecin-type peptides: biodiversity, structure–function relationships and mode of action. *Peptides*, **27**, 2350–2359.

341 Otvos, L. (2002) The short proline-rich antibacterial peptide family. *Cellular and Molecular Life Sciences*, **59**, 1138–1150.

342 Cuthbertson, B.J., Deterding, L.J., Williams, J.G., Tomer, K.B., Etienne, K., Blackshear, P.J., Büllesbach, E.E., and Gross, P.S. (2008) Diversity in penaeidin antimicrobial peptide form and function. *Developmental & Comparative Immunology*, **32**, 167–181.

343 Gennaro, R., Zanetti, M., Benincasa, M., Podda, E., and Miani, M. (2002) Pro-rich antimicrobial peptides from animals: structure, biological functions and mechanism of action. *Current Pharmaceutical Design*, **8**, 763–778.

344 Li, J.X., Xu, X.Q., Yu, H.N., Yang, H.L., Huang, Z.X., and Lai, R. (2006) Direct antimicrobial activities of PR-bombesin. *Life Sciences*, **78**, 1953–1956.

345 Rolland, J.L., Abdelouahab, M., Dupont, J., Lefevre, F., Bachere, E., and Romestand, B. (2010) Stylicins, a new family of antimicrobial peptides from the Pacific blue shrimp *Litopenaeus stylirostris*. *Molecular Immunology*, **47**, 1269–1277.

346 Gueguen, Y., Romestand, B., Fievet, J., Schmitt, P., Destoumieux-Garzon, D., Vandenbulcke, F., Bulet, P., and Bachere, E. (2009) Oyster hemocytes express a proline-rich peptide displaying synergistic antimicrobial activity with a

defensin. *Molecular Immunology*, **46**, 516–522.

347 Yang, Y., Poncet, J., Garnier, J., Zatylny, C., Bachere, E., and Aumelas, A. (2003) Solution structure of the recombinant penaeidin-3, a shrimp antimicrobial peptide. *The Journal of Biological Chemistry*, **278**, 36859–36867.

348 Anderson, R.C., Hancock, R.E.W., and Yu, P.L. (2004) Antimicrobial activity and bacterial-membrane interaction of ovine-derived cathelicidins. *Antimicrobial Agents and Chemotherapy*, **48**, 673–676.

349 Chan, Y.R., Zanetti, M., Gennaro, R., and Gallo, R.L. (2001) Anti-microbial activity and cell binding are controlled by sequence determinants in the anti-microbial peptide PR-39. *Journal of Investigative Dermatology*, **116**, 230–235.

350 Benincasa, M., Pelillo, C., Zorzet, S., Garrovo, C., Biffi, S., Gennaro, R., and Scocchi, M. (2010) The proline-rich peptide Bac7(1–35) reduces mortality from *Salmonella typhimurium* in a mouse model of infection. *BMC Microbiology*, **10**, 178.

351 Ghiselli, R., Giacometti, A., Cirioni, O., Circo, R., Mocchegiani, F., Skerlavaj, B., D'Amato, G., Scalise, G., Zanetti, M., and Saba, V. (2003) Neutralization of endotoxin *in vitro* and *in vivo* by Bac7(1–35), a proline-rich antibacterial peptide. *Shock*, **19**, 577–581.

352 Rozgonyi, F., Szabo, D., Kocsis, B., Ostorhazi, E., Abbadessa, G., Cassone, M., Wade, J.D., and Otvos, L. (2009) The antibacterial effect of a proline-rich antibacterial peptide A3-APO. *Current Medicinal Chemistry*, **16**, 3996–4002.

353 Knappe, D., Kabankov, N., and Hoffmann, R. (2011) Bactericidal oncocin derivatives with superior serum stabilities. *International Journal of Antimicrobial Agents*, **37**, 166–170.

354 Knappe, D., Mueller, U., Sauer, U., Schiffer, G., Alber, G., and Hoffmann, R. (2011) Antibacterial efficacy of oncocin analogue Onc72 in a mouse model of peritoneal sepsis. *Biopolymers*, **96**, 478–478.

355 Knappe, D., Piantavigna, S., Hansen, A., Mechler, A., Binas, A., Nolte, O., Martin, L.L., and Hoffmann, R. (2010) Oncocin (VDKPPYLPRPRPPRRIYNR-NH$_2$): a novel antibacterial peptide optimized against Gram-negative human pathogens. *Journal of Medicinal Chemistry*, **53**, 5240–5247.

356 Knappe, D., Zahn, M., Sauer, U., Schiffer, G., Straeter, N., and Hoffmann, R. (2011) Rational design of oncocin derivatives with superior protease stabilities and antibacterial activities based on the high-resolution structure of the oncocin–DnaK complex. *ChemBioChem*, **12**, 874–876.

3
Anionic Antimicrobial Peptides

Summary

Anionic antimicrobial peptides (AAMPs) are important components of the innate immune system and here we review recent research into these peptides using human AAMPs as an example. This chapter shows that AAMPs are present in the respiratory tract, the brain, the epidermis, the epididymis, blood components, and the gastrointestinal tract. These peptides use a diverse range of antimicrobial mechanisms, which in some cases, such as respiratory surfactant-associated anionic peptides, involve translocation across the membrane to utilize intracellular sites of antimicrobial action. In other cases, the membrane itself is the major site of action for AAMPs, as is the case for epididymal β-defensins, which induce the disintegration of membranes via carpet-type mechanisms prior to action against intracellular targets. These AAMPs show the potential for development as topical biocides, fertility control agents, therapeutically useful antibiotics, decontaminants, food preservatives, and agents against dental and periodontal diseases.

3.1
Introduction

Anionic antimicrobial peptides (AAMPs) and proteins are used for host defense throughout the prokaryotic kingdom, and are produced by either of two biosynthetic routes: they may be gene-encoded or manufactured by ribosome-independent pathways [1]. Gene-encoded AAMPs from prokaryotes include microcins from Gram-negative bacteria [2, 3], bactericons and lantibiotics from Gram-positive bacteria [4, 5], and halocins from the Archaea [6, 7]. In general, these peptides show relatively narrow ranges of activity against organisms that are closely related to the host, and utilize a diverse array of antimicrobial mechanisms involving specific receptors, membrane components, and intracellular targets. Based on this activity, a number of these AAMPs show promise for development as antimicrobial agents in areas such as food preservation and healthcare [7–9].

Non-ribosomal prokaryotic AAMPs are generally secondary metabolites from fermentations that are assembled by large multifunctional enzyme complexes [10],

---

Antimicrobial Peptides, First Edition. David A. Phoenix, Sarah R. Dennison, and Frederick Harris.
© 2013 Wiley-VCH Verlag GmbH & Co. KGaA. Published 2013 by Wiley-VCH Verlag GmbH & Co. KGaA.

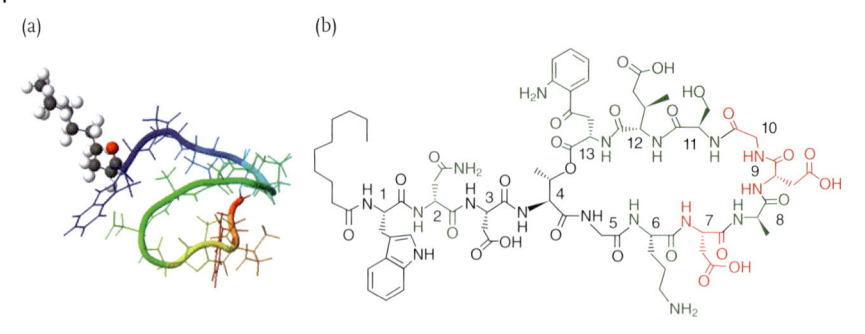

Figure 3.1 Structure of daptomycin: cyclic three-dimensional (a) and two-dimensional (b) structure of daptomycin (adapted from [16]). Non-proteinogenic amino acids are shown in green and residues responsible for the Ca^{2+} binding properties of the peptide are highlighted in red. The relative position of each residue is given, with residues 4–13 forming the forming the 10-member macrolactone ring of the molecule and residues 1–3 forming its exocyclic "tail," which is completed by an n-decanoyl fatty acid moiety attached to residue 1.

and belong to a large class of compounds that includes established antibiotics such as the glycopeptides, vancomycin, and teicoplanin [11, 12]. Among the best characterized of these prokaryotic AAMPs is a family of acidic lipopeptides, which exhibit cyclic peptide core structures that include non-proteinogenic amino acid residues and have exocyclic lipid moieties covalently attached to form tails. However, these lipopeptides show great diversity both in relation to their structures, which vary widely in the length and composition of their lipid tails and peptide chains, and their host bacteria, which include a range of Gram-positive and Gram-negative organisms [13–15]. Daptomycin from *Streptomyces roseosporous* is by far the most studied of these AAMPs and consists of a 10-residue ring system, which includes L-3-methylglutamic acid and L-kynurenine, only known to daptomycin, with an exocyclic tail containing three residues adorned by a decanoyl fatty acid attachment (Figure 3.1) [17]. The peptide has been shown to exhibit activity against a range of Gram-positive bacteria for which the presence of Ca^{2+} is known to be essential; however, the detailed mechanisms underlying this antibacterial activity are far from fully elucidated [18]. It has been proposed that Ca^{2+} mediates the formation of daptomycin oligomers, which serve to deliver the peptide to bacterial membranes in a functional conformation and to create high local concentrations at the membrane itself. The membrane environment then induces the dissociation of these oligomers, permitting the insertion of daptomycin into the bilayer, which is promoted by the peptide's lipid tail [19]. The present consensus view is that these insertion events lead to the formation of bilayer pores, which depolarize the membrane through potassium efflux and disrupt membrane-associated processes, ultimately leading to cell death [16]. However, it has been proposed that the antibacterial action of daptomycin may also involve other targets [20]. Consistent with this proposal, a recent study has suggested that although the membrane is the primary site of action for the peptide, secondary effects from this action lead to the formation of membrane distortions or patches that mediate bacterial cell death. Essentially, these membrane patches are able to redirect the localization of proteins involved in cell division and cell wall synthesis, thereby precipitating wholesale damage to the cell wall and membrane, resulting in membrane permeabilization and cell death [21].

Daptomycin has been shown to have potent activity against a number of medically relevant bacterial pathogens and is increasingly being used to treat infections due to bacteria that have developed resistance to conventional antibiotics. For example, the lipopeptide was recently shown to have an efficacy greater than rifampicin, a derivative of rifamycin, when directed against biofilms of *Staphylococcus epidermidis*, which are becoming increasingly problematic as infections of indwelling medical devices [22]. Rifampicin is one of the most potent known antibiotics against *S. epidermidis* biofilms and it was suggested that daptomycin serve as an alternative in cases where rifampicin resistance becomes prevalent [23]. Several studies have shown that the lipopeptide is also active against Staphylococci and Enterococci with resistance to glycopeptide antibiotics such as linezolid, teicoplanin, and quinupristin–dalfopristin [24, 25]. However, daptomycin is probably best known for its efficacy against pathogenic Enterococci and Staphylococci, including methicillin-resistant *Staphylococcus aureus* (MRSA), which have acquired resistance to vancomycin [20, 26], and is generally regarded as the last line of defense against these pathogens [27].

Given the clear potential of prokaryotic AAMPs for use in the medical arena, their eukaryotic counterparts have received far less research attention. However, these eukaryotic peptides have been identified across vertebrates, invertebrates, and plants, with activity against bacteria, fungi, viruses, nematodes, and insects [28, 29]. In response, this chapter gives an overview of the structure–function relationships of these peptides, and reviews current research into their antimicrobial mechanisms and therapeutic potential. To illustrate this research, we use AAMPs identified in humans that represent one of the most diverse arrays of these peptides found in eukaryotes.

3.2
AAMPs in the Respiratory Tract

The respiratory tract comprises the nasal cavity, sinuses, and trachea of the upper tract, which, along with airways of the lower tract, is lined by the respiratory epithelium [30]. This mucosal surface is regularly exposed to potentially harmful substances and microbes, and in response, the innate immune system of the respiratory tract has developed a variety of defense mechanisms, including the production of several short AAMPs rich in aspartic acid [31]. First identified in the respiratory

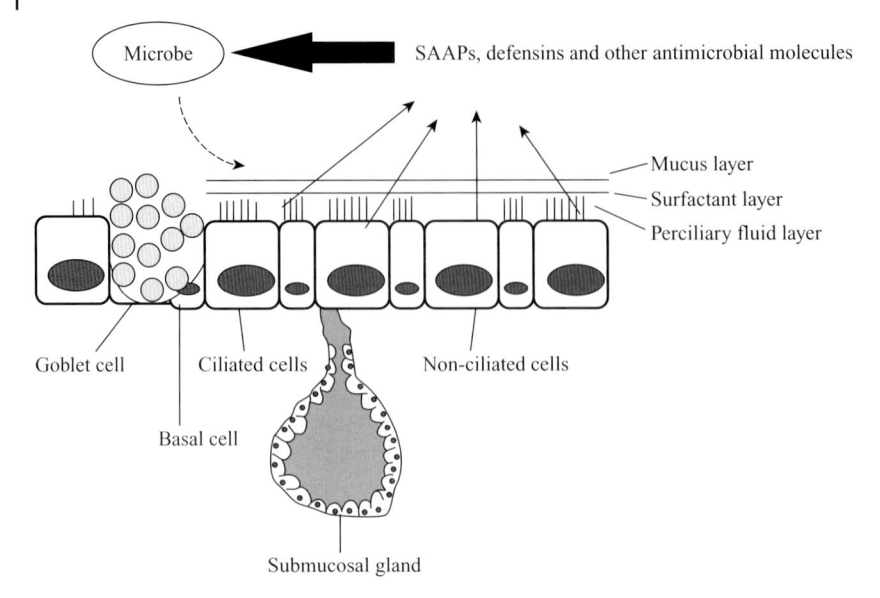

Figure 3.2 Schematic representation of the components of the respiratory epithelium, which lines the upper respiratory tract, including the nasal cavity, sinuses, and trachea, and larger airways of the lower respiratory tract. This lining is composed of a pseudostratified, ciliated epithelium with associated submucosal glands and other epithelial cells such as mucus-secreting goblet cells. The respiratory epithelium is a primary site for the deposition of microbes that are acquired during inspiration and as a contribution to the innate immune system of the respiratory tract; respiratory epithelia secrete SAAPs into the airway surface liquid along with CAMPs such as defensins to eliminate these pathogens.

tract of humans and ruminants, these AAMPs (net charge −4 to −7) were designated surfactant-associated anionic peptides (SAAPs), and are believed to be constitutively expressed in the epithelial cells of the trachea and airways of the lower tract (Figure 3.2) [31, 32]. SAAPs appear to have several biological functions [28, 29], and have been shown to exhibit weak activity against a variety of Gram-positive and Gram-negative bacteria that is enhanced by the presence of Zn^{2+}. Characterization of this activity suggested that Zn^{2+} may form a cationic salt bridge between these peptides and the anionic bacterial cell surface, thereby facilitating their translocation across the membrane. SAAPs internalized to the bacterial cell cytoplasm were found to induce intracellular damage and the flocculation of cellular components, which led to the suggestion that they may attack targets such as ribosomes or DNA [33]. To gain insight into the defense function of SAAPs, a number of investigations characterized the expression of these peptides in ruminant pulmonary diseases, which suggested that they may play an important role in the resolution of infections near epithelial surfaces of the respiratory tract [34]. In response, investigations were undertaken to characterize the expression of SAAPs in individuals with cystic fibrosis (CF) [35], which is a human disorder characterized by bacterial infection of the respiratory tract, arising from impaired innate immune functions [36, 37]. The distributions of these AAMPs in the respiratory tracts of healthy people and patients with CF were compared and it was found there were significant differences between these two groups of individuals in relation to both the concentration and location of SAAPs. Based on these observations, it was suggested that in the case of patients with CF, a block in the release of SAAPs from their original sites of synthesis may give rise to a physiological deficiency in these peptides that predisposes a person with the disease to respiratory infections [35]. Consistent with this suggestion, there is evidence to suggest that viscous secretions associated with CF may obstruct the submucosal gland ducts of the respiratory tract, thereby inhibiting the secretion of antimicrobial molecules into the epithelium and promoting colonization of the respiratory epithelium by bacterial pathogens (Figure 3.1) [37]. No further research on the activity of SAAPs in CF-related infections appears to have been conducted, but it has been suggested that these peptides should be further investigated for potential development as novel AAMPs in the treatment of these infections [28]. It has been shown that SAAPs possess activity against *Pseudomonas aeruginosa* [38, 39], which is the major pathogen associated with CF infections [40]. Moreover, recent studies have shown that AAMPs are effective against sessile bacteria [41–43] and the multidrug-resistant (MDR) biofilms of *P. aeruginosa* associated with chronic CF infections are the major current hurdle in the development of novel therapies for these infections [44].

3.3
AAMPs in the Brain

The human brain was long considered to be an immunologically privileged organ based on the observation that antigenic material, including bacteria and viruses, failed to elicit a systemic, T-cell-mediated immunological response [45, 46]. However, it is now known that that this immune privilege does not relate to the total absence of immunological components and that the brain possesses its own innate immune system, which acts with the blood–brain barrier and the meninges to form a coordinated defense network directed against microbial infection [47, 48]. Both infiltrating immune cells and cells resident in the brain such as neurons, astrocytes, and microglia contribute to the function of this network by producing a diverse array of defense molecules, and identified within these molecular species are a number of AAMPs [49]. Several peptides known to function as AAMPs, including dermcidin (DCD) and thymosin-β_4, have been detected in the brain, but their antimicrobial activity has also been identified and characterized within the context of other bodily locations, and therefore these peptides have been described below [49, 50]. The most studied AAMPs in the brain are peptide B and its truncated variant, enkelytin (net charge −6 and −7, respectively), which are cleaved from proenkephalin A (PEA), a precursor of the opioid peptides [51], but with no known neuropeptide function [52]. In addition to humans, these AAMPs have been identified within the sequence of PEA across a range of vertebrates and invertebrates [28] and have been shown by number of studies to exhibit potent activity (minimum inhibitory concentration (MIC) < 3 µM) against Gram-positive bacteria, including *Micrococcus luteus*, *S. aureus*, and *Bacillus megaterium* [53–55]. To gain insight into the antibacterial action of peptide B/enkelytin, computer modeling studies were undertaken, which predicted that the active form of these peptides was an L-shaped structure formed by two linear α-helical arms angularly juxtaposed through a proline-induced bend (Figure 3.3). These studies further predicted that in this conformation, glutamic acid residues on the short arm of

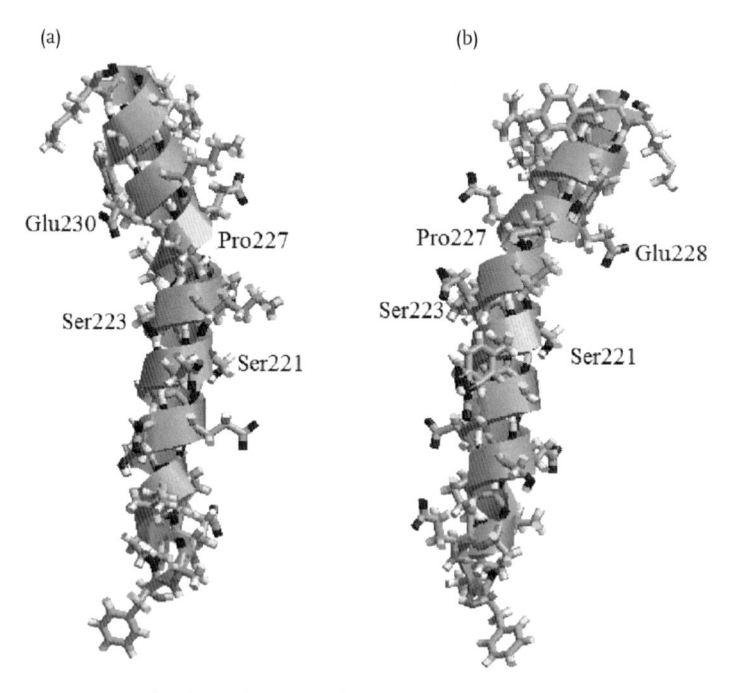

Figure 3.3 Molecular architecture of enkelytin in the absence of post-translational phosphorylation (adapted from [53]). The annotated letters represent the single-letter code for amino acid residues while the numbers refer to the position of associated residues within the primary structure of proenkephalin A. In these structures, proline at position 227 is in either a *cis* (a) or a *trans* (b) conformation. In both cases, the presence of this proline residue leads to a kink or bend that breaks the α-helical structure of these peptides to form juxtaposed α-helical arms that bring a glutamic acid residue and a serine residue into proximity. These residues are serine at position 223 and glutamic acid at position 230 for molecules with proline in a *cis* conformation (a), and serine at position 221 and glutamic acid at position 228 for molecules with proline in a *trans* conformation (b). It was predicted that phosphorylation of these serine residues would lead to repulsive electrostatic interactions with neighboring glutamic acid residues, thereby promoting opening of the kinked α-helical structure and the antibacterial activity of enkelytin/peptide B.

the structure would be brought proximal to phosphorylated serine residues on the longer arm, resulting in repulsive electrostatic interactions that would promote opening of the L-shaped structure. This action would then facilitate the antibacterial action of peptide B/enkelytin by allowing the C-terminal amphiphilic α-helix on the short arm of the molecule to interact with the membranes of the target bacteria. Based on comparisons to other AAMPs, it was also speculated that the antibacterial action of peptide B/enkelytin may require interaction between their sites of phosphorylation and divalent metal ions to promote adoption of the active conformation assumed by these peptides [53, 54].

Despite insights gained into the antibacterial mechanisms used by peptide B/enkelytin, detailed descriptions of these mechanisms currently remain lacking. However, given that these peptides are coproduced with opioid peptides, it has been postulated that these AAMPs help facilitate a unified neuroimmune response to immediate threat such as bacterial challenge, stress, or other stimuli [56, 57]. In this response (Figure 3.4), the processing of PEA leads to the liberation of opioid peptides such as Met-Enk, which would participate in the activation of immunocytes and provide a chemotaxic signal to further stimulate immunocyte recruitment. During the time required for this induction and mobilization of the adaptive immune system, peptide B/enkelytin would be released as factors of the innate immune system, providing an immediate counter to invading bacteria, or as a precautionary measure [57–59]. Although generally beyond the scope of this chapter, very recent studies on the brains of several toads have revealed the presence of multiple homologous AAMPs, some of which appear to have dual function, also serving as neuropeptides [60]. This ability provides a clear contrast between these peptides and peptide B/enkelytin, and suggests that at least in some amphibians, localizing AAMP and neuropeptide capability within a single peptide may be important to the functioning of a unified neuroimmune response. Indeed, it is interesting to note that a number of established neuropeptides have recently been shown to possess direct antimicrobial activity [52, 61–63], and to date many anionic neuropeptides that have been identified within frogs and toads appear to be untested for antimicrobial potential [64].

Most recently, evidence has been presented to suggest that amyloidogenic AAMPs may be present in the human brain in the shape of the peptides, Aβ40 and Aβ42 [65]. These peptides are best known for their neurotoxicity and role in the pathogenesis of Alzheimer's disease (AD), which is believed to involve the ability of Aβ40 and Aβ42 to form amyloid-associated structures that permeabilize the neuronal membrane [66]. However, despite extensive investigation, the normal physiological functions of Aβ40 and Aβ42 are unknown [67], which led Soscia et al. [65] to investigate the hypothesis that these peptides may play a role in host defense of the brain. This hypothesis was based on similarities between the membrane-permeabilizing mechanisms involved in the neurotoxicity of these latter peptides [68, 69] and those utilized by the antimicrobial activity of known amyloidogenic AMPs (Chapter 5). In their investigations, Soscia et al. [65] took brain tissues from AD patients and showed that these tissues exhibited

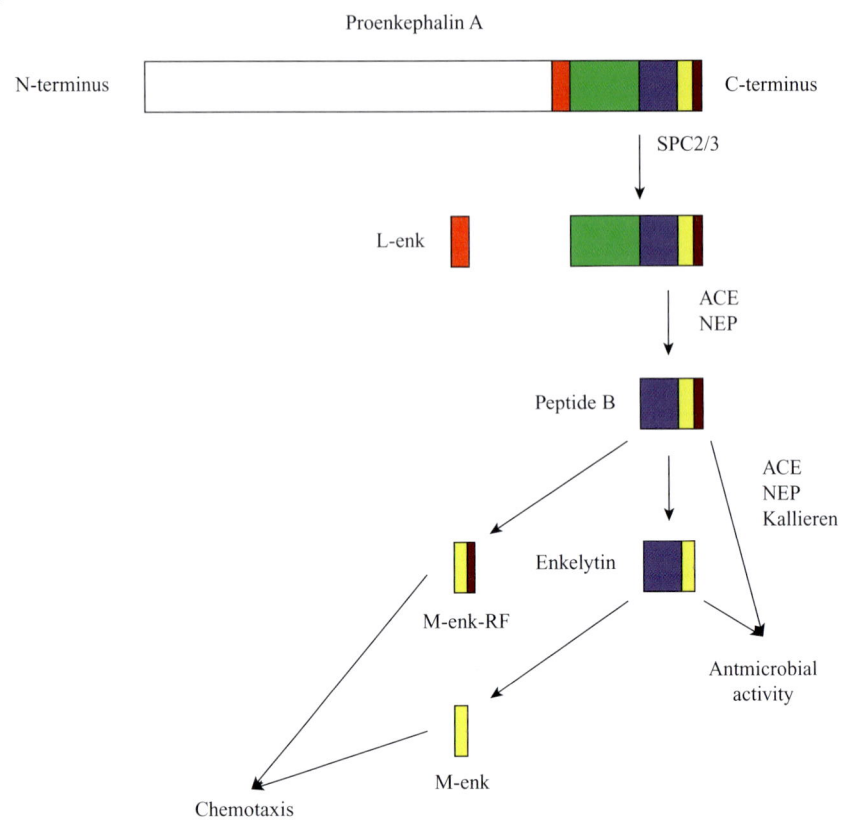

Figure 3.4 Putative model for the processing of human PEA in response to microbial challenge [58]. According to this model, the action of prohormone convertase 2 or 3 (SPC2/3) on the C-terminal region of PEA yields a segment containing peptide B. This segment is then cleaved to peptide B, which is then further fragmented by angiotensin-converting enzyme (ACE) and neutral endopeptidase (NEP) to either enkelytin or M-enk-RF (methionine-enkephalin-arginine-phenylalanine). Enkelytin is also degraded by ACE and NEP to M-enk (methionine-enkephalin). M-enk and M-enk-RF are opioid peptides, and induce chemotaxis; peptide B and enkelytin engage in antibacterial activity.

antimicrobial activity at levels significantly above those of corresponding tissues from healthy individuals. Moreover, it was established that in these brain tissues there was a correlation between antimicrobial activity and the levels of Aβ40 and Aβ42 present [65]. Based on these studies, it was suggested that these peptides may function as AAMPs, and consistent with this suggestion, both Aβ40 and Aβ42 exhibited potent activity against a range of clinically relevant organisms such as *Candida albicans* [65], which is a major cause of neurocandidiasis [70, 71]. The studies of Soscia et al. [65] further showed that Aβ40 and Aβ42 bound to bacterial membranes, which, taken with data from other studies, led to the proposal of a model for the antimicrobial activity of these peptides [72]. According to this model, anionic lipid in the microbial membrane promotes conformational change in Aβ40 and Aβ42, which leads to adoption of amphiphilic β-sheet structures by these peptides. These β-sheet structures interact with the microbial membrane via electrostatic contributions, which involve associations between cationic residues in Aβ42 and Aβ40, and anionic moieties in the membrane lipid head-group region. Most recently, these peptides were shown to have a strong affinity for phosphatidylethanol [73], which is a common component of bacterial membranes (Chapter 5). Accompanying these electrostatic associations is a hydrophobic contribution that involves interactions between the C-terminal region of Aβ42 and Aβ40 and acyl chains in the membrane lipid core region. In combination, these peptide–membrane interactions promote amyloidogenesis by Aβ42 and Aβ40, which then mediates the formation of ion-conducting channels in the bilayer and thereby inactivation of the target microbe [72]. Taking the results of these various studies overall, there would seem to be compelling evidence to suggest that the normal physiological function of Aβ42 and Aβ40 is to serve as AAMPs in protecting the human brain against microbial infection. Indeed, this paradigm of host defense could help explain the fact that AD appears to be frequently associated with microbial infections [74, 75] and suggests several possible mechanisms by which these peptides may contribute to the genesis of the disease [65]. For example, it was postulated by these latter authors [65] that a transient brain infection by microbial pathogens could trigger a self-perpetuating innate immune response, and thereby excessive production of Aβ42 and Aβ40, resulting in the deposition of these peptides to form the extracellular senile plaques associated with AD. A clear implication from this paradigm of host defense is that AD could be an instance of a disease that results of the inappropriate accumulation of amyloidic defense peptides [72] as has been reported in a number of other cases [76–78]. This observation could prove to be of great medical significance by giving insights into the pathogenesis of AD and hence into the search for channel-blocking drugs that may serve as candidate therapeutics for the disease [79–82].

3.4 AAMPs in the Epidermis

The epidermis is a stratified squamous epithelium composed of proliferating basal and differentiated suprabasal keratinocytes, which forms a protective shield to prevent invasion by microbes [83]. However, in addition to acting as a passive mechanical barrier, epidermal cells also actively combat microbial invasion by secreting a variety of antimicrobial molecules, which include the sweat-borne protein, DCD [84]. As described above, DCD is present in the brain and has also been identified in tears [85], but it is generally regarded as an epidermal protein. It is well established that DCD is constitutively expressed in the dark mucous cells of the eccrine sweat glands, secreted into sweat, and transported to the epidermal surface [86, 87]. In sweat, DCD is processed by cathepsin D and other proteases to yield a spectrum of AAMPs and other DCD-derived AMPs that varied with the individual. It was suggested for a given individual, these peptides synergize to produce a spectrum of antimicrobial action, thereby maximizing the protection of the host from microbial infection [87–91]. These peptides show potent broad range antimicrobial activity (MICs in the low micromolar range) that is unaffected by the low pH and elevated salt levels that are characteristic of human sweat [86, 87]. Based on these results, it was suggested that these AAMPs play a role in the innate immune responses of the skin [87, 92] and strong support for this suggestion came from studies on atopic dermatitis (ATD) [93, 94], which is an inflammatory skin disorder characterized by recurrent bacterial or viral skin infections [95]. These studies found that when compared to healthy individuals, sufferers of ATD had reduced amounts of DCD-derived AAMPs in their sweat and an impaired ability to kill bacteria on the skin surface [93, 94]. It has also been suggested that in addition to their antimicrobial function, DCD-derived peptides may play other roles in the innate immune system of the skin [96]. In support of this suggestion, a recent *in vitro* study demonstrated that DCD-1 and DCD-1L, which are cleaved from the C-terminal region of DCD [28], were able to stimulate the production of cytokines and other molecules involved in innate immune signaling from keratinocytes [97]. This result was surprising as sweat is extruded onto the surface of the skin, and under normal circumstances there is no direct contact between these peptides and keratinocytes [89, 98]. However, there is evidence to suggest that disease-related dysfunction of the sweat delivery system can cause DCD-derived peptides to accumulate in dermal tissues [99] and it has been proposed that in disorders where epithelial integrity is lost, these peptides may be able to access and activate keratinocytes [97].

Efforts to better understand the role of DCD-derived AAMPs in the innate immune system of the skin have focused on the antimicrobial capability of DCD-1 and DCD-1L [28]. These efforts have shown that in conditions mimetic of sweat, both peptides are able to kill skin associated bacteria such as *Staphylococcal epidermis* as well as a range of nosocomial pathogens, including *P. aeruginosa*, MRSA, MDR *Mycobacterium tuberculosis*, *Salmonella typhimurium*, and *Candida albicans* [87, 88, 94, 100–104]. A few cases of bacterial resistance to DCD-derived AAMPs have been described [105–107], but in general, as with other AMPs, the incidence of such cases is low [108]. To gain insight into the antimicrobial mechanisms used by DCD-derived AAMPs, a number of structure–function studies have been undertaken. The DCD pre-protein was found to be predominantly random coil in aqueous solution but was able to adopt α-helical architecture in the presence of membrane mimics, and it was predicted that this ability may be relevant to biological functions of the protein and its derivative peptides [109]. Consistent with this prediction, it was found that DCD-1L, DCD-1, and other DCD-derived AAMPs adopted an α-helical structure in the presence of membrane mimics, although no clear correlation between the levels of this secondary structure, net charge, and antimicrobial activity of these peptides was apparent [110]. More recent studies showed that DCD-1L and DCD-1 adopted an α-helical architecture, which was essentially formed by a series of α-helices (I–IV, Figure 3.5a) with a helix–hinge–helix motif [102, 111]. Two short α-helices (I and II) were located in the highly

3.4 AAMPs in the Epidermis

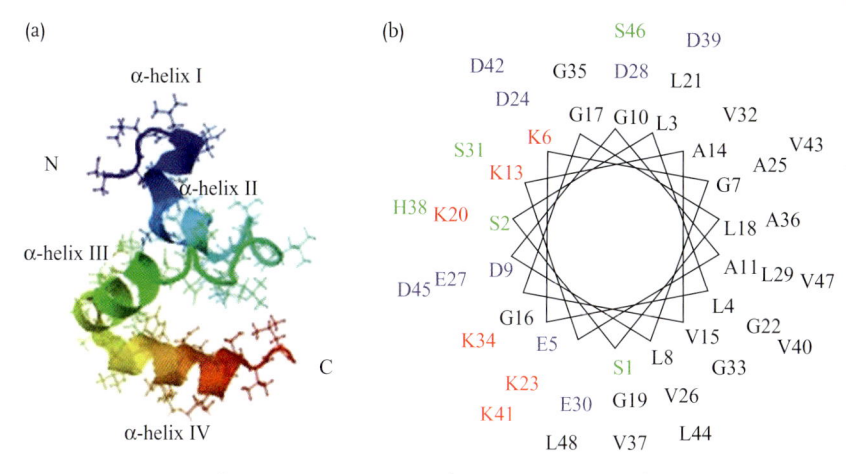

Figure 3.5 Structure of DCD-1L. (a) Molecular architecture of DCD-1L. The overall topology of the peptide consists of four α-helices and a number of β-turns. The α-helices I and II are relatively short, and are located in N-terminal half of DCD-1L, which was found to be highly flexible due to the presence of multiple glycine residues. This region also contains a cationic segment that has been shown to be essential for the antimicrobial activity of the peptide. (b) DCD-1L represented as a two-dimensional axial projection where anionic residues are in blue, cationic residues are in red, polar residues are in green, and hydrophobic residues are in gray. In this conformation, the peptide shows a clear segregation of hydrophilic and hydrophobic residues, indicating a strong potential to adopt an amphiphilic α-helical structure and interact with membranes.

flexible N-terminal region of this architecture [111] and previous studies have shown that this region includes a cationic segment, which appears to be essential for the antimicrobial activity of these AAMPs [110]. Several studies showed that the α-helical architecture adopted by DCD-1L and DCD-1 was strongly amphiphilic (Figure 3.5b) [109, 111], indicating the potential for membrane interaction (Chapter 5) [112]. Confirming this potential, these AAMPs have been shown to bind lipid mimics of bacterial membranes [111], naturally occurring membranes of Gram-positive and Gram-negative bacteria [113, 114], and various components of the membrane systems possessed by both these latter bacterial classes [115]. Based on these results, it is generally accepted that the antimicrobial action of DCD-1L and DCD-1 depends upon their ability to bind the membranes of target organisms, and several lines of evidence have suggested that this action may involve specific lipids or membrane receptors [113, 114]. Several studies have shown that DCD-1L and DCD-1 are able to self-associate [101, 109, 113], which taken with the results of a recent study led to a proposed model for the antimicrobial activity of DCD-1L (Figure 3.6) [116]. According to this model, the unstructured monomeric peptide is secreted in human sweat (Figure 3.6a), targets the anionic bacterial membrane via its cationic N-terminal region, and binds to this membrane with the concomitant adoption of amphiphilic α-helical structure (Figure 3.6b). Interaction with the bacterial membrane leads to oligomerization of the peptide and two possible

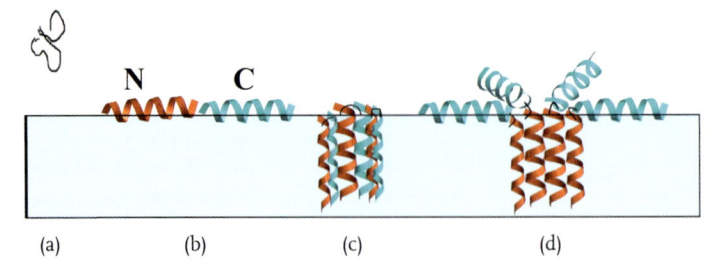

Figure 3.6 Putative model for the membrane interaction of DCD-1L (adapted from [116]). Using this model, unstructured, monomeric DCD-1L (a) targets the anionic bacterial membrane via its cationic N-terminal region and binds to this membrane with the concomitant adoption of an amphiphilic α-helical structure (b). Interaction with the bacterial membrane leads to oligomerization of the peptide and two possible mechanisms of membrane invasion. Using the first of these mechanisms, the cationic N-terminus of DCD-1L folds back onto its anionic C-terminal region to form an intramolecular hairpin and these hairpin structures assemble to form a pore (c). In the second of these mechanisms, the cationic N-terminal region of DCD-1L forms a toroidal pore across the lipid bilayer while the amphiphilic C-terminal region remains associated with the membrane surface (d).

mechanisms of membrane invasion. Using the first of these mechanisms, the cationic N-terminus of DCD-1L folds back onto its anionic C-terminal region to form an intramolecular hairpin and these hairpin structures assemble to form a pore (Figure 3.6c). In the second of these mechanisms, the cationic N-terminal region of the peptide forms a toroidal pore (Chapter 5) across the lipid bilayer while the amphiphilic C-terminal region remains associated with the membrane surface (Figure 3.6d) [116]. However, other studies have suggested that the binding of DCD-1L and DCD-1 to bacterial membranes leads to the inactivation of these organisms via mechanisms that do involve bilayer perturbation or permeation, contrasting to other epidermal AMPs such as LL-37, which functions by membranolytic antimicrobial mechanisms [113–115]. It has also been speculated that the antimicrobial action of these peptides may involve intracellular microbial targets [28, 113] as previously described for several other human AAMPs [29, 33, 117]. Currently, the detailed mechanisms underlying the antimicrobial activity of DCD-1L and DCD-1 await elucidation; however, based the efficacy of these AAMPs against a broad range of bacterial pathogens with resistance to conventional antibiotics, it has been suggested that they have strong potential for development as antimicrobial agents in the clinical setting [102].

3.5
AAMPs in the Epididymis

The epididymis ("De Graaf's thread") is a major organ of the male reproductive tract, where immature sperm released from the testis are concentrated and undergo sequential maturation to acquire forward motility and fertilizing ability

[118]. Analogous to the brain, the epididymis is immunologically privileged and is protected by the blood–epididymal barrier, which provides a highly specialized, microenvironment where sperm remain isolated from other compartments of the body [119]. A spectrum of proteins, including AMPs, released into the lumen of the epididymis bind sperm and are believed to play an important role in epididymal innate immunity as well as serving their function in sperm maturation [120, 121]. Efforts to identify novel AMPs involved in epididymal immunity led to the description of a large family of genes that encode β-defensins, and are primarily clustered on human chromosomes 6, 8, and 20 [117]. Two of these genes, on chromosomes 8 and 20, respectively, were found to encode anionic β-defensins (net charge −1), which were designated HE2C [122] and DEFB118 [123]. It was found that HE2C is secreted into the epididymal fluid, becoming part of the ejaculate, and possesses a strong ability to bind the surface of *Escherichia coli* and *S. aureus*. However, the peptide showed no significant activity against these organisms and, currently, its biological function is unclear, although it was observed that it may be active against other microbes [124]. In contrast, DEFB118 was found to be restricted to the epididymis [123] and several lines of evidence suggested that the peptide may serve as an epididymal AAMP [117, 125]. This suggestion was strongly supported when DEFB118 was shown to exhibit potent activity against *E. coli* in the presence of the high salt levels that are associated with seminal plasma [126]. The mechanisms underpinning this antibacterial action were investigated and it was found that the peptide was able to inhibit the biosynthesis of RNA, DNA, and proteins in *E. coli*. These observations clearly suggested that intracellular targets were involved in the antibacterial mechanisms of DEFB118 and, therefore, that the peptide was able to translocate across the bacterial membrane. Consistent with this suggestion, it was found that DEFB118 possessed a strong ability to permeabilize the *E. coli* membrane system, which appeared to be maximal at the dividing septa of the organism [126]. Similar observations have been reported for the antibacterial action of other β-defensins specific to the epididymis, which led to the suggestion that regions of septum division formed during *E. coli* cell division may be especially vulnerable to the action of this family of peptides and other AMPs [127]. Structural characterization of DEFB118 was undertaken, and showed that the core cationic β-sheet domain possessed by the peptide was essential for its antibacterial and membrane permeabilizing ability [126]. Based on these combined results, it was proposed that the antimicrobial action of DEFB118 may be described by the Shai–Huang–Matasaki (SHM) model (Chapter 5) [29, 117]. According to this model, monomeric DEFB118 "carpets" the outer surface of the bacterial membrane via electrostatic interactions involving its cationic β-sheet domain and anionic components of the target bacterial membrane. These interactions lead to incorporation of the peptide into the bilayer, accompanied by thinning of the outer leaflet and an increase in its surface area relative to the inner leaflet, resulting in strain within the bilayer. This action results in the appearance of membrane lesions and transient pores, which facilitate the translocation of DEFB118 to the inner leaflet. The diffusion of DEFB118 to intracellular targets may now occur along with permeabilization of the bacterial cell membrane and the loss of cytoplasmic material [29, 117]. Currently, the

detailed mechanisms by which DEFB118 permeabilizes and translocates the membrane are unknown, but it has been observed that peptide's antimicrobial mechanism could be of medical benefit in light of recent studies [28]. These studies have clearly shown the potential of epididymal β-defensins to treat reproductive tract infections, including those associated with sexually transmitted diseases [128, 129], and there is increasing interest in using these peptides as fertility control targets [120].

3.6
AAMPs in Blood Components

Blood is a part of the cardiovascular system and serves a number of functions, such as transport, regulation, and protection, and a growing body of research has shown that a part of this protective function is to exert antimicrobial activity. An important mechanism that contributes to this antimicrobial activity is the blood-borne mobilization of cationic AMPs (CAMPs) and AAMPs to sites of microbial infection [130]. These molecules are carried either in the cytosolic granules of circulating blood cells, such as leukocytes [131] and platelets [132], or in the plasma of blood where they derive either from cellular degranulation or secretions of the liver [133]. Several AAMPs were reported in blood plasma that corresponded to segments encrypted within the α-chain of the blood coagulant, fibrinogen, and the lipid transport protein, apolipoprotein CIII (net charges −2, respectively). Both peptides were reported to exhibit antibacterial activity, but in neither case does there appear to have been any further investigation into this activity or other properties of these AAMPs [134]. Peptide B/enkelytin have been detected in the secretions of leukocytes (polymorphonuclear neutrophils) [53] and in plasma where the levels of these AAMPs were low they were seen to quickly rise in response to skin incision due to coronary artery bypass grafting [55]. This latter study also showed that peptide B/enkelytin were metabolized *in vivo* to opioid peptides with granulocyte chemotactic activity as described in Figure 3.3. The results of this study appeared to be among the first to demonstrate the presence of AAMPs with immunocyte-activating properties in plasma and their release into the bloodstream as a result of stress [55]. It was suggested by these latter authors that this innate response may be an important initial event in limiting the spread of microbial infection after trauma, further supporting the view that the nervous and immune systems present a unified response to immediate threat [57].

Platelets contain a variety of cytoplasmic granules including α-granules from which AAMPs and other antimicrobial molecules are released in response to stimulation by microbes or host-derived agonists such as thrombin [135]. Two of these platelet AAMPs, fibrinopeptide A (FPA; net charge −3) and fibrinopeptide B (FPB; net charge −2) were shown to correspond to segments of the fibrinogen α- and β-chains, respectively [136]. Based on the work of these latter authors [136], it was suggested that the thrombin-mediated stimulation of platelets may lead to the generation of FPA and FPB via the proteolytic cleavage of fibrinogen secreted by these blood cells [132]. Characterization of these AAMPs showed them to exhibit antimicrobial activity in the low micromolar range when directed against a range of pathogens that have a propensity to enter the bloodstream. These pathogens included *S. aureus*, *E. coli*, and the fungus, *Cryptococcus neoformans* [136]. Later studies showed that FPA was ineffective against *Lactobacillus* spp. in a rat model of experimental endocarditis, but no further characterization of either this peptide or FPB appear to have been undertaken [137]. A further AAMP found in the α-granules of platelets was thymosin-$β_4$ (net charge −2), which was ineffective against fungi, but in the low micromolar range showed activity against *S. aureus* and *E. coli* that was enhanced by alkaline conditions [136]. The peptide is known to be ubiquitous in cells and its relatively high abundance in platelets and other blood components is well established [138], but these former studies appear to be one of the first to demonstrate that thymosin-$β_4$ possesses antimicrobial activity [136]. It has been suggested that the antimicrobial activities of thymosin-$β_4$, FPA, and FPB may serve to complement and synergize those of CAMPs identified in platelets, thereby providing these cells with broad-range efficacy against blood-borne pathogens [132]. Nonetheless, the more recent finding that ocular thymosin-$β_4$ from humans was only weakly active against physiologically relevant bacterial pathogens [139] raised questions as to whether antimicrobial activity is generally a primary function of the peptide [140]. In response, it has been suggested that the ocular antimicrobial activity of thymosin-$β_4$ may serve an important support role with its primary function involving activities of promoting anti-inflammatory and wound-healing effects [50, 139, 141]. Promotion of these effects by the peptide are well described in the skin, heart, and neural system; however, as yet, investigations into the occurrence of associated antimicrobial activity does not appear to have been undertaken [142].

3.7
AAMPs in the Gastrointestinal Tract and Food Proteins

It is clear that endogenous AMPs play a major defense role in the innate immune system, but there is much evidence to suggest that this role is supported by peptides generated through the proteolytic digestion of food proteins. Food proteins can be therefore, regarded not only as sources of nutrition but also as a resource to increase the natural defense of the host organism against invading pathogens [143–147]. Major food proteins include ovalbumin (OA), which comprises between 60 and 65% of the total protein in avian egg white [148, 149] along with the well-characterized bovine milk-borne proteins casein [150], α-lactalbumin (α-LA) [151], and β-lactoglobulin (β-LG) [152]. Encrypted within the primary structures of these proteins are numerous AMPs that are released *in situ* by the action of gastrointestinal enzymes, such as trypsin, α-chymotrypsin, pepsin, chymosin, and pancreatin [153]. These liberated AMPs show activity against a broad range of microbes and it is becoming increasingly clear that included within these peptides are many AAMPs [154], although in most cases these latter peptides have yet to be fully

Table 3.1 Anionic AMPs.

AAMPs	Sequence of AAMPs	Source proteins	Key references
ODT3	YPILPEYLQ	Residues 111–119 of OA	[155]
ODT4	ELINSW	Residues 143–148 of OA	[155]
ODC1	AEERYPILPEYL	Residues 143–148 of OA	[155]
ODC3	TSSNVMEER	Residues 143–148 of OA	[155]
LDT2	GYGGVSLPEWVCTTF ALCSEK	Residues 17–31 of α-LA linked by a disulfide bridge to residues 109–114 of the protein	[156]
LDC	CKDDQNPH ISCDKF	Residues 61–68 of α-LA linked by a disulfide bridge to residues 175–80 of the protein	[156]
LGDT3	VLVLDTDYK	Residues 92–100 of β-LG	[157]
LGDT4	AASDISLLDAQSAPLR	Residues 25–40 of β-LG	[157]
LGDT5	WENGECAQK	Residues 61–69 of β-LG	[157]
LGDT6	LSFNPTQLEEQCHI	Residues 149–162 of β-LG	[157]
Caseicin C	SDIPNPIGSENSEK	Residues 195–208 of bovine $α_{s1}$-casein	[158]
Kappacin (A)	MAIPPKKNQDKTEIPTINTIAS GEPTSTPTTEAVESTVATLEDS PEVIESPPEINTVQVTSTAV	Residues 106–169 of κ-casein	[41]
Kappacin (B)	MAIPPKKNQDKTEIPTINTIAS GEPTSTPTIEAVESTVATLEAS PEVIESPPEINTVQVTSTAV	Residues 106–169 of variant κ-casein	[41]

characterized (Table 3.1). The major encrypted AAMPs to be identified in bovine milk are the kappacins, which appear to be cleaved from κ-casein by digestion with chymosin in the stomach [154]. The best characterized of these AAMPs are kappacins A and B (net charges −7 and −6, respectively), which exhibit potent activity against a range of Gram-positive and Gram-negative bacteria. In particular, these AAMPs were found to be highly effective against oral bacteria, including *Streptococcus mutans*, *Porphyromonas gingivalis*, and *Actinomycesnaes lundii*, which is a major component of supragingival dental plaque [41, 159].

The active region of kappacins has been localized to a segment corresponding to residues 138–158 of κ-casein, which includes a phosphorylated serine residue that is essential for antibacterial activity (Table 3.1). At present, the active conformation of these AAMPs is unknown, although based on architectural similarities to peptide B/enkelytin (Figure 3.3) it has been suggested that they may form a proline-kinked amphiphilic α-helix [42, 159]. The antimicrobial mechanisms used by kappacins are also currently unclear, although there is evidence to suggest that these mechanisms may involve the pH-dependent lysis of microbial membranes, which is enhanced by low pH [159]. Interestingly, kappacin A shows significantly higher antimicrobial activity than kappacin B under low pH conditions, which is primarily due to a single residue difference in the active regions of these AAMPs. Kappacin A possesses an aspartic acid residue at position 148, whereas kappacin B displays an alanine residue in the corresponding location [41, 42]. The functional significance of this difference in sequences is not known, although one possibility is that the additional negative charge possessed by kappacin A may enhance its ability to bind metal ions and thereby the efficacy of its antibacterial action [154]. It has been established that in a membrane mimetic environment, the presence of Ca^{2+} and Zn^{2+} ions leads to conformational changes in kappacins and enhanced levels of antibacterial activity. The conformational changes observed in the peptide could not be clearly assigned to any particular secondary structures, but based on these data it was suggested that metal ions may form a cationic salt bridge between kappacins and anionic components of the bacterial membrane cell, thereby facilitating membrane binding and antibacterial action [42].

3.8
AAMPs and Their Structure–Function Relationships

The prevalent view is that membrane interaction is a key step in the antimicrobial and anticancer action of AMPs, and the contribution of membrane lipid to these interactions is discussed in Chapters 5 and 6. In relation to the contribution made by AMPs, quantitative structure–function analyses of CAMPs have shown that net charge residue composition, sequence length, secondary structure, hydrophobicity, and amphiphilicity all contribute to the ability of these peptides to interact with membranes and kill microbes [160] (Chapter 2). However, similar analyses in the case of AAMPs are hindered by the fact that only around 100 of these peptides are known and many are poorly characterized [28, 29], but some insight into the antimicrobial structure–function relationships of these peptides can be gleaned.

As with CAMPs, human AAMPs clearly exhibit a diverse range of antimicrobial mechanisms and it would appear that the membrane interactions associated with these mechanisms can be described in broad terms by the models proposed for AMPs in general, which are reviewed in Chapter 2. For example, there is evidence to suggest that specific bacterial lipids may be involved in the membrane targeting ability of Aβ40, Aβ42 [73], DCD-1, and DCD-1L [119, 120], which appears to also be reported for a small number of other AAMPs and CAMPs from vertebrates, invertebrates, plants, fungi [28, 162, 163], and bacteria [164, 165]. SAAPs utilize mechanisms of antimicrobial action that are non-membranolytic and involve the internalization of these peptides to attack intracellular targets [33], and similar non-membranolytic antimicrobial mechanisms have been proposed for a number of other AMPs [166–168], such as the synthetic peptide, WRWYCR, which targets bacterial DNA repair mechanisms [169]. Membranolytic antimicrobial mechanisms involving pore formation are employed by kappacins [42] and DCD-1L [116] while membranolytic mechanisms based on amyloidogenesis have been suggested for Aβ40 and Aβ42, which would be the first case of human AAMPs using this mechanism [170]. Amyloid-mediated mechanisms of antimicrobial action have been reported for a variety of other AMPs (Chapter 4) [171–177] such as protegrin-1 from pigs [69].

In relation to residue distribution, glutamic acid and aspartic acid are the major contributors to the net negative charge of human AAMPs, possessing carboxylated side-chains that have pK_a values in the region of 4, which means that these residues remain fully charged across the physiological range of pH [178]. The net charge of human AAMPs ranges from −1 to −7, which would include the bulk of known AAMPs [28], although some have been identified with net charges of up to −20 [179]. In addition to their charge contributions, glutamic acid and aspartic acid residues appear to serve other roles in the function of human AAMPs and other AMPs of this structural class. Both residues have a high propensity for α-helix formation [180], and it has been suggested that they may serve a structural role in α-helical AAMPs such as DCD-1 and DCD-1L along with CAMPs (Chapter 2). When these residues are present in α-helical peptides, they tend to occupy positions in the α-helix that are $i \pm 3$ or $i \pm 4$ relative to cationic residues, and it has been suggested that this structural positioning may promote helix formation via salt bridging and may be a strategy for improving the rigidity of α-helical residue arrangements and hence changing efficacy [181, 182]. Glutamic acid and aspartic acid also play a more direct role in facilitating the membrane interactions of some human AAMPs such as SAAPs, which use these residues to bind metal ions and form cationic salt bridges with negatively charged components of microbial membranes [33]. A similar metal-binding function appears to be served by serine residues, rendered anionic by phosphorylation, in the membrane interactions of peptide B/enkelytin and kappacins. This use of metal ions to directly facilitate membrane interactions has also been reported for AAMPs from non-human sources [28, 29] and other groups of AMPs such as the human antifungal CAMPs, histatins (Chapter 2) [183, 184]. In most cases, the roles played by individual residues in the membrane interactions and antimicrobial mechanisms of AAMPs would seem to be similar those played by these residues in other groups of AMPs (Chapter 6). For example, the ability of proline residues to induce a "kink" or bend in α-helical structure promotes the membrane interactive conformations of peptide B/enkelytin and kappacins and serves a corresponding function in the antimicrobial actions of a number of CAMPs [185–187], such as fowlicidin-2 from chickens [188]. Interestingly, DEFB118, Aβ40, and Aβ42 appear to use regions rich in cationic residues to bind directly to anionic components of the target membrane and clearly these residues serve a similar function in CAMPs. This observation suggests that DEFB118, Aβ40, and Aβ42 effectively function as CAMPs, and based on a similar use of cationic residues, it has been proposed that this may be the case for a variety of other AAMPs such as cyclotides [28, 29]. This would seem to indicate that in contrast to CAMPs, net charge is not necessarily a determining factor in the membrane interactions of AAMPs, although clearly this would not appear to be case for peptides such as SAAPs and some plant AAMPs that only possess anionic charged residues [33, 161].

The human AAMPs reviewed here range in sequence length from six to 100 residues, which compares to the range of five to around 70 residues that is typical of these peptides in general [28], although sequence lengths of up to 300 residues have been described [28, 189, 190]. In general, sequence length does not appear to be a determinant in the antimicrobial and anticancer activity of AMPs; however, as described above, it seems able to influence the choice of membrane interactive mechanism used by DCD-1L [116] and it has been suggested that this may also be the case for some anticancer peptides (Chapter 6) [191], and therefore could be the case for other peptides. In relation to secondary structure, human AAMPs exhibit a range of architectures that are typical of AMPs in general [192, 193], including extended peptides such as SAAPs [33], cysteine-stabilized peptides such as DEFB118 [117], Aβ40, and Aβ42 [170], and α-helical peptides such as kappacins [41], enkelyin/peptide B [53] along with those described above. As also found for AMPs in general [192, 193], the α-helical secondary structure is that most commonly found in human AAMPs, with cysteine-stabilized architectures forming the second largest group. Interestingly, in plant AAMPs, this rank order of secondary structures appears to be reversed [192, 193] and many of these latter peptides form cyclic cysteine-stabilized structures [194]. The reasons for this structural preference are unclear but, to date, AAMPs with cyclic cysteine-stabilized structures do not appear to have been reported for humans or other eukaryotic non-plant sources [194, 195], although cyclic CAMPs have been identified in these hosts, including amatoxins in fungi [196], θ-defensins in primates [194], and retrocyclins from humans [197]. Regardless of their secondary structures, most human AAMPs have molecular architectures that are amphiphilic, which is also generally found across all other AAMPs and AMPs [192, 193]. An example of amphiphilicity in human AAMPs is provided by DCD-1L (Figure 3.5b) where there is a clear segregation between hydrophobic residues and anionic and other hydrophilic residues. This segregation of residues indicates the importance of both amphiphilicity and hydrophobicity for the action of these AAMPs: amphiphilicity facilitates the ability of these peptides to target and interact with microbial membranes. However, the ability of AAMPs to partition into these membranes is facilitated by hydrophobicity-driven interactions with the bilayer's apolar core region. A fuller discussion of the role of these structural characteristics in the interactions of AMPs with microbial membranes is given in Chapter 5 and with cancer cell membranes in Chapter 6.

3.9
Discussion

This chapter has shown that AAMPs are produced in locations across the human body, from the brain to the epididymis and in circulating blood cells. AAMPs such as SAAPs, HE2C, and DEFB118 are secreted at their sites of antimicrobial action while those such as DCD-1, DCD-1L, kappacins, and other gastrointestinal AAMPs are cleaved *in situ* from precursor proteins. These latter cleaved AAMPs are coproduced with CAMPs, reinforcing the view that the former peptides may have arisen to synergize with the antimicrobial activity of the latter peptides in host defense. In contrast, peptide B/enkelytin are cleaved from precursor opioid proteins and are coproduced with neuropeptides to facilitate a unified neuroimmune response to microbial challenge – an indication that some human AAMPs can serve more than one biological function. Interestingly, many of the AAMPs described here, including peptide B/enkelytin, Aβ40, Aβ42, DCD, thymosin-β_4, HE2C, and DEFB118, have been detected in immunologically privileged locations, and whether the anionicity of these peptides is functionally relevant to immune privilege is unclear.

The AAMPs described here exhibit potent activity against a spectrum of bacterial and fungal pathogens, particularly those relevant to their location of action, and only a few instances of microbial resistance to these peptides have been reported. This is the case for most AAMPs and has been ascribed to the relatively nonspecific nature of the membrane-based antimicrobial mechanisms used by these peptides coupled to the improbability of microbes acquiring the multitude of cumulative genetic changes potentially needed to inhibit these mechanisms [28]. A number of the AAMPs described here are also able to exhibit potent activity against a range of microbes that show multiple resistance to conventional antibiotics. This ability appears to reflect the fact that many of these peptides utilize the membrane as their major site of action, rendering it unlikely that their activity would be affected by the efflux pumps used by microbial pathogens to expel conventional antimicrobial agents [198]. Based on their antimicrobial properties, it has been suggested that human AAMPs have the potential for development in a number of capacities, including topical biocides, fertility control agents, anti-invectives against dental and periodontal diseases, and therapeutically useful antibiotics in conditions such as CF and ATD. However, the most important contribution made by AAMPs described in this chapter to the well-being of humans is the insight provided by Aβ40 and Aβ42, which led to a search for channel-blocking drugs with the potential to treat AD. Currently, this disorder is the major global contributor to dementia and it has been predicted that in the absence of improved therapies it will affect around 42 million people worldwide by 2020 [199].

References

1 Hammami, R., Zouhir, A., Le Lay, C., Ben Hamida, J., and Fliss, I. (2010) BACTIBASE second release: a database and tool platform for bacteriocin characterization. *BMC Microbiology*, **10**, 22.

2 Duquesne, S., Destoumieux-Garzon, D., Peduzzi, J., and Rebuffat, S. (2007) Microcins, gene-encoded antibacterial peptides from enterobacteria. *Natural Product Reports*, **24**, 708–734.

3 Rebuffat, S. (2011) *Microcins from Enterobacteria: On the Edge Between Gram-Positive Bacteriocins and Colicins*, Springer, New York.

4 Heng, N.C.K., Wescombe, P.A., Burton, J.P., Jack, R.W., and Tagg, J.R. (2007) The diversity of bacteriocins produced

by Gram-positive bacteria, in *Bacteriocins – Ecology and Evolution* (eds M.A. Riley and M.A. Chavan), Springer, Berlin, pp. 45–92.

5 Rea, M.C., Ross, R.P., Cotter, P.D., and Hill, C. (2011) Classification of bacteriocins from Gram-positive bacteria, in *Prokaryotic Antimicrobial Peptides: From Genes to Applications* (eds S. Drider and S. Rebuffat), Springer, Berlin, pp. 29–53.

6 Shand, R.F. and Leyva, K.J. (2007) Peptide and protein antibiotics from the domain archaea: halocins and sulfolobicins, in *Bacteriocins: Ecology and Evolution* (eds M.A. Riley and M.A. Chavan), Springer, Berlin, pp. 93–110.

7 Shand, R.F. and Leyva, K.J. (2008) *Archaeal Antimicrobials: An Undiscovered Country*, Caister Academic Press, Wymondham.

8 Galvez, A., Abriouel, H., Ben Omar, N., and Lucas, R. (2011) Food applications and regulation, in *Prokaryotic Antimicrobial Peptides: From Genes to Applications* (eds S. Drider and S. Rebuffat), Springer, Berlin, pp. 353–390.

9 Dicks, L.M.T., Heunis, T.D.J., van Staden, D.A., Brand, A., Noll, K.S., and Chikindas, M.L. (2011) *Medical and Personal Care Applications of Bacteriocins Produced by Lactic Acid Bacteria*, Springer, New York.

10 Strieker, M., Tanovic, A., and Marahiel, M.A. (2010) Nonribosomal peptide synthetases: structures and dynamics. *Current Opinion in Structural Biology*, **20**, 234–240.

11 Stegmann, E., Frasch, H.-J., and Wohlleben, W. (2010) Glycopeptide biosynthesis in the context of basic cellular functions. *Current Opinion in Microbiology*, **13**, 595–602.

12 Nailor, M.D. and Sobel, J.D. (2011) Antibiotics for Gram-positive bacterial infection: vancomycin, teicoplanin, quinupristin/dalfopristin, oxazolidinones, daptomycin, telavancin, and ceftaroline. *Medical Clinics of North America*, **95**, 723–742.

13 Jacques, P. (2011) Surfactin and other lipopeptides from *Bacillus* spp, in *Biosurfactants: From Genes to Applications* (ed. G. SoberonChavez), Springer, New York, pp. 57–91.

14 Strieker, M. and Marahiel, M.A. (2009) The structural diversity of acidic lipopeptide antibiotics. *ChemBioChem*, **10**, 607–616.

15 Raaijmakers, J.M., de Bruijn, I., Nybroe, O., and Ongena, M. (2010) Natural functions of lipopeptides from *Bacillus* and *Pseudomonas*: more than surfactants and antibiotics. *FEMS Microbiology Reviews*, **34**, 1037–1062.

16 Robbel, L. and Marahiel, M.A. (2010) Daptomycin, a bacterial lipopeptide synthesized by a nonribosomal machinery. *The Journal of Biological Chemistry*, **285**, 27501–27508.

17 Baltz, R.H. (2009) Biosynthesis and genetic engineering of lipopeptides in *Streptomyces roseosporus*. *Methods in Enzymology*, **458**, 511–531.

18 Baltz, R.H., Miao, V., and Wrigley, S.K. (2005) Natural products to drugs: daptomycin and related lipopeptide antibiotics. *Natural Product Reports*, **22**, 717–741.

19 Straus, S.K. and Hancock, R.E.W. (2006) Mode of action of the new antibiotic for Gram-positive pathogens daptomycin: comparison with cationic antimicrobial peptides and lipopeptides. *Biochimica et Biophysica Acta – Biomembranes*, **1758**, 1215–1223.

20 Cottagnoud, P. (2008) Daptomycin: a new treatment for insidious infections due to Gram-positive pathogens. *Swiss Medical Weekly*, **138**, 93–99.

21 Pogliano, J., Pogliano, N., and Silverman, J. (2012) Daptomycin mediated reorganization of membrane architecture causes mislocalization of essential cell division proteins. *Journal of Bacteriology*, doi: 10.1128/JB.00011-12.

22 Mack, D., Davies, A.P., Harris, L.G., Knobloch, J.K.M., and Rohde, H. (2009) *Staphylococcus epidermidis* biofilms: functional molecules, relation to virulence, and vaccine potential. *Topics in Current Chemistry*, **288**, 157–182.

23 Leite, B., Gomes, F., Teixeira, P., Souza, C., Pizzolitto, E., and Oliveira, R. (2011) *In vitro* activity of daptomycin, linezolid and rifampicin on *Staphylococcus*

24 Picazo, J.J., Betriu, C., Rodríguez-Avial, I., Culebras, E., López-Fabal, F., and Gómez, M. (2011) Comparative activities of daptomycin and several agents against staphylococcal blood isolates. Glycopeptide tolerance. *Diagnostic Microbiology and Infectious Disease*, **70**, 373–379.

25 Johnson, A.P., Mushtaq, S., Warner, M., and Livermore, D.M. (2004) Activity of daptomycin against multi-resistant Gram-positive bacteria including enterococci and *Staphylococcus aureus* resistant to linezolid. *International Journal of Antimicrobial Agents*, **24**, 315–319.

26 Moise, P.A., North, D., Steenbergen, J.N., and Sakoulas, G. (2009) Susceptibility relationship between vancomycin and daptomycin in *Staphylococcus aureus*: facts and assumptions. *The Lancet Infectious Diseases*, **9**, 617–624.

27 Gould, I.M. (2011) Clinical activity of anti-Gram-positive agents against methicillin-resistant *Staphylococcus aureus*. *The Journal of Antimicrobial Chemotherapy*, **66**, IV17–IV21.

28 Harris, F., Dennison, S.R., and Phoenix, D.A. (2009) Anionic antimicrobial peptides from eukaryotic organisms. *Current Protein & Peptide Science*, **10**, 585–606.

29 Harris, F., Dennison, S.R., and Phoenix, D.A. (2011) Anionic antimicrobial peptides from eukaryotic organisms and their mechanisms of action. *Current Chemical Biology*, **5**, 142–153.

30 Lopez, A. (2007) Respiratory system, thoracic cavity and pleura, in *Pathologic Basis of Veterinary Disease* (eds M.D. McGavin and J.F. Zachary), Elsevier, Amsterdam, pp. 463–558.

31 Grubor, B., Meyerholz, D.K., and Ackermann, M.R. (2006) Collectins and cationic antimicrobial peptides of the respiratory epithelia. *Veterinary Pathology*, **43**, 595–612.

32 Anderson, R.C., Wilkinson, B., and Yu, P.L. (2004) Ovine antimicrobial peptides: new products from an age-old industry. *Australian Journal of Agricultural Research*, **55**, 69–75.

33 Brogden, K.A., Ackermann, M., McCray, P.B., and Tack, B.F. (2003) Antimicrobial peptides in animals and their role in host defences. *International Journal of Antimicrobial Agents*, **22**, 465–478.

34 Meyerholz, D.K. and Ackermann, M.R. (2005) Antimicrobial peptides and surfactant proteins in ruminant respiratory tract disease. *Veterinary Immunology and Immunopathology*, **108**, 91–96.

35 Brogden, K.A., Ackermann, M.R., McCray, P.B., and Huttner, K.M. (1999) Differences in the concentrations of small, anionic, antimicrobial peptides in bronchoalveolar lavage fluid and in respiratory epithelia of patients with and without cystic fibrosis. *Infection and Immunity*, **67**, 4256–4259.

36 Yang, L., Jelsbak, L., and Molin, S. (2011) Microbial ecology and adaptation in cystic fibrosis airways. *Environmental Microbiology*, **13**, 1682–1689.

37 Dooring, G. and Gulbins, E. (2009) Cystic fibrosis and innate immunity: how chloride channel mutations provoke lung disease. *Cellular Microbiology*, **11**, 208–216.

38 Kalfa, V.C. and Brogden, K.A. (1999) Anionic antimicrobial peptide–lysozyme interactions in innate pulmonary immunity. *International Journal of Antimicrobial Agents*, **13**, 47–51.

39 Brogden, K.A., Ackermann, M., and Huttner, K.M. (1997) Small, anionic, and charge-neutralizing propeptide fragments of zymogens are antimicrobial. *Antimicrobial Agents and Chemotherapy*, **41**, 1615–1617.

40 Bals, R., Hubert, D., and Tuemmler, B. (2011) Antibiotic treatment of CF lung disease: from bench to bedside. *Journal of Cystic Fibrosis*, **10**, S146–S151.

41 Dashper, S.G., Liu, S.W., and Reynolds, E.C. (2007) Antimicrobial peptides and their potential as oral therapeutic agents. *International Journal of Peptide Research and Therapeutics*, **13**, 505–516.

42 Dashper, S.G., O'Brien-Simpson, N.M., Cross, K.J., Paolini, R.A., Hoffmann, B., Catmull, D.V., Malkoski, M., and Reynolds, E.C. (2005) Divalent metal cations increase the activity of the antimicrobial peptide kappacin. *Antimicrobial Agents and Chemotherapy*, **49**, 2322–2328.

43 Stewart, P.S., Davison, W.M., and Steenbergen, J.N. (2009) Daptomycin rapidly penetrates a *Staphylococcus epidermidis* biofilm. *Antimicrobial Agents and Chemotherapy*, **53**, 3505–3507.

44 Rybtke, M.T., Jensen, P.O., Hoiby, N., Givskov, M., Tolker-Nielsen, T., and Bjarnsholt, T. (2011) The implication of *Pseudomonas* aeruginosa biofilms in infections. *Inflammation & Allergy Drug Targets*, **10**, 141–157.

45 Galea, I., Bechmann, I., and Perry, V.H. (2007) What is immune privilege (not)? *Trends in Immunology*, **28**, 12–18.

46 Bechmann, I., Galea, I., and Perry, V.H. (2007) What is the blood–brain barrier (not)? *Trends in Immunology*, **28**, 5–11.

47 Rivest, S. (2009) Regulation of innate immune responses in the brain. *Nature Reviews Immunology*, **9**, 429–439.

48 Nicholas, M.K. and Lukas, R. (2010) Immunologic privilege and the brain. *NeuroImmune Biology*, **9**, 169–181.

49 Su, Y., Zhang, K., and Schluesener, H.J. (2010) Antimicrobial peptides in the brain. *Archivum Immunologiae et Therapiae Experimentalis*, **58**, 365–377.

50 Sosne, G., Qiu, P., Goldstein, A.L., and Wheater, M. (2010) Biological activities of thymosin beta$_4$ defined by active sites in short peptide sequences. *FASEB Journal*, **24**, 2144–2151.

51 Metz-Boutigue, M.H., Kieffer, A.E., Goumon, Y., and Aunis, D. (2003) Innate immunity: involvement of new neuropeptides. *Trends in Microbiology*, **11**, 585–592.

52 Brogden, K.A., Guthmiller, J.M., Salzet, M., and Zasloff, M. (2005) The nervous system and innate immunity: the neuropeptide connection. *Nature Immunology*, **6**, 558–564.

53 Goumon, Y., Lugardon, K., Kieffer, B., Lefevre, J.F., Van Dorsselaer, A., Aunis, D., and Metz-Boutigue, M.H. (1998) Characterization of antibacterial COOH-terminal proenkephalin-A-derived peptides (PEAP) in infectious fluids. Importance of enkelytin, the antibacterial PEAP209–237 secreted by stimulated chromaffin cells. *The Journal of Biological Chemistry*, **273**, 29847–29856.

54 Goumon, Y., Strub, J.M., Moniatte, M., Nullans, G., Poteur, L., Hubert, P., Van Dorsselaer, A., Aunis, D., and Metz-Boutigue, M.H. (1996) The C-terminal bisphosphorylated proenkephalin-A-(209–237)-peptide from adrenal medullary chromaffin granules possesses antibacterial activity. *European Journal of Biochemistry*, **235**, 516–525.

55 Tasiemski, A., Salzet, M., Benson, H., Fricchione, G.L., Bilfinger, T.V., Goumon, Y., Metz-Boutigue, M.H., Aunis, D., and Stefano, G.B. (2000) The presence of antibacterial and opioid peptides in human plasma during coronary artery bypass surgery. *Journal of Neuroimmunology*, **109**, 228–235.

56 Stefano, G.B., Fricchione, G.L., Goumon, Y., and Esch, T. (2005) Pain, immunity, opiate and opioid compounds and health. *Medical Science Monitor*, **11**, Ms47–Ms53.

57 Salzet, M., Vieau, D., and Day, R. (2000) Crosstalk between nervous and immune systems through the animal kingdom: focus on opioids. *Trends in Neurosciences*, **23**, 550–555.

58 Salzet, M. and Tasiemski, A. (2001) Involvement of pro-enkephalin-derived peptides in immunity. *Developmental and Comparative Immunology*, **25**, 177–185.

59 Salzet, M. (2001) Invertebrates to human. *Neuro Endocrinology Letters*, **22**, 467–474.

60 Liu, R., Liu, H., Ma, Y., Wu, J., Yang, H., Ye, H., and Lai, R. (2011) There are abundant antimicrobial peptides in brains of two kinds of *Bombina* toads. *Journal of Proteome Research*, **10**, 1806–1815.

61 El Karim, I.A., Linden, G.J., Orr, D.F., and Lundy, F.T. (2008) Antimicrobial activity of neuropeptides against a range

of micro-organisms from skin, oral, respiratory and gastrointestinal tract sites. *Journal of Neuroimmunology*, **200**, 11–16.

62 Delgado, M. and Ganea, D. (2008) Anti-inflammatory neuropeptides: a new class of endogenous immunoregulatory agents. *Brain, Behavior, and Immunity*, **22**, 1146–1151.

63 Gonzalez-Rey, E., Ganea, D., and Delgado, M. (2010) Neuropeptides: keeping the balance between pathogen immunity and immune tolerance. *Current Opinion in Pharmacology*, **10**, 473–481.

64 Pukala, T.L., Bowie, J.H., Maselli, V.M., Musgrave, I.F., and Tyler, M.J. (2006) Host-defence peptides from the glandular secretions of amphibians: structure and activity. *Natural Product Reports*, **23**, 368–393.

65 Soscia, S.J., Kirby, J.E., Washicosky, K.J., Tucker, S.M., Ingelsson, M., Hyman, B., Burton, M.A., Goldstein, L.E., Duong, S., Tanzi, R.E., and Moir, R.D. (2010) The Alzheimer's disease-associated amyloid beta-protein is an antimicrobial peptide. *PLoS ONE*, **5**, e9505.

66 Kawahara, M., Ohtsuka, I., Yokoyama, S., Kato-Negishi, M., and Sadakane, Y. (2011) Membrane incorporation, channel formation, and disruption of calcium homeostasis by Alzheimer's beta-amyloid protein. *International Journal of Alzheimer's Disease*, **2011**, 304583.

67 Querfurth, H.W. and LaFerla, F.M. (2010) Mechanisms of disease Alzheimer's disease. *The New England Journal of Medicine*, **362**, 329–344.

68 Butterfield, S.M. and Lashuel, H.A. (2010) Amyloidogenic protein membrane interactions: mechanistic insight from model systems. *Angewandte Chemie International Edition*, **49**, 5628–5654.

69 Jang, H.J.H., Arce, F.T., Mustata, M., Ramachandran, S., Capone, R., Nussinov, R., and Lal, R. (2011) Antimicrobial protegrin-1 forms amyloid-like fibrils with rapid kinetics suggesting a functional link. *Biophysical Journal*, **100**, 1775–1783.

70 Scully, E.P., Baden, L.R., and Katz, J.T. (2008) Fungal brain infections. *Current Opinion in Neurology*, **21**, 347–352.

71 Sánchez-Portocarrero, J., Pérez-Cecilia, E., Corral, O., Romero-Vivas, J., and Picazo, J.J. (2000) The central nervous system and infection by *Candida* species. *Diagnostic Microbiology and Infectious Disease*, **37**, 169–179.

72 Harris, F., Dennison, S.R., and Phoenix, D.A. (2012) Aberrant action of amyloidogenic host defense peptides: a new paradigm to investigate neurodegenerative disorders? *FASEB Journal*, **26**, 1776–1781.

73 Cazzaniga, E., Bulbarelli, A., Lonati, E., Orlando, A., Re, F., Gregori, M., and Masserini, M. (2011) Abeta peptide toxicity is reduced after treatments decreasing phosphatidylethanolamine content in differentiated neuroblastoma cells. *Neurochemical Research*, **36**, 863–869.

74 Shima, K., Kuhlenbaeumer, G., and Rupp, J. (2010) *Chlamydia pneumoniae* infection and Alzheimer's disease: a connection to remember? *Medical Microbiology and Immunology*, **199**, 283–289.

75 Honjo, K., van Reekum, R., and Verhoeff, N.P.L.G. (2009) Alzheimer's disease and infection: do infectious agents contribute to progression of Alzheimer's disease? *Alzheimers & Dementia*, **5**, 348–360.

76 Torrent, M., Odorizzi, F., Nogues, M.V., and Boix, E. (2010) Eosinophil cationic protein aggregation: identification of an N-terminus amyloid prone region. *Biomacromolecules*, **11**, 1983–1990.

77 Brorsson, A.C., Kumita, J.R., MacLeod, I., Bolognesi, B., Speretta, E., Luheshi, L.M., Knowles, T.P.J., Lomas, D.A., Dobson, C.M., and Crowther, D.C. (2010) Methods and models in neurodegenerative and systemic protein aggregation diseases. *Frontiers in Bioscience*, **15**, 373–396.

78 Araki-Sasaki, K., Osakabe, Y., Miyata, K., Amano, S., Yamada, M., Kitagawa, K., Hirano, K., and Kinoshita, S. (2009) What is this thing called "amyloidosis"? *Cornea*, **28**, S80–S83.

79 Diaz, J.C., Simakova, O., Jacobson, K.A., Arispe, N., and Pollard, H.B. (2009) Small molecule blockers of the Alzheimer A beta calcium channel potently protect neurons from A beta cytotoxicity. *Proceedings of the National Academy of Sciences of the United States of America*, **106**, 3348–3353.

80 Kawahara, M. (2010) Neurotoxicity of beta-amyloid protein: oligomerization, channel formation and calcium dyshomeostasis. *Current Pharmaceutical Design*, **16**, 2779–2789.

81 Kawahara, M., Ohtsuka, I., Yokoyama, S., Kato-Negishi, M., and Sadakane, Y. (2011) Membrane incorporation, channel formation, and disruption of calcium homeostasis by Alzheimer's β-amyloid protein. *International Journal of Alzheimer's Disease*, **2011**, 304583.

82 Arispe, N., Diaz, J.C., and Simakova, O. (2007) A beta ion channels. Prospects for treating Alzheimer's disease with A beta channel blockers. *Biochimica et Biophysica Acta – Biomembranes*, **1768**, 1952–1965.

83 Bangert, C., Brunner, P.M., and Stingl, G. (2011) Immune functions of the skin. *Clinics in Dermatology*, **29**, 360–376.

84 Wiesner, J. and Vilcinskas, A. (2010) Antimicrobial peptides. The ancient arm of the human immune system. *Virulence*, **1**, 440–464.

85 You, J.J., Fitzgerald, A., Cozzi, P.J., Zhao, Z., Graham, P., Russell, P.J., Walsh, B.J., Willcox, M., Zhong, L., Wasinger, V., and Li, Y. (2010) Post-translation modification of proteins in tears. *Electrophoresis*, **31**, 1853–1861.

86 Rieg, S., Garbe, C., Sauer, B., Kalbacher, H., and Schittek, B. (2004) Dermcidin is constitutively produced by eccrine sweat glands and is not induced in epidermal cells under inflammatory skin conditions. *The British Journal of Dermatology*, **151**, 534–539.

87 Schittek, B., Hipfel, R., Sauer, B., Bauer, J., Kalbacher, H., Stevanovic, S., Schirle, M., Schroeder, K., Blin, N., Meier, F., Rassner, G., and Garbe, C. (2001) Dermcidin: a novel human antibiotic peptide secreted by sweat glands. *Nature Immunology*, **2**, 1133–1137.

88 Baechle, D., Flad, T., Cansier, A., Steffen, H., Schittek, B., Tolson, J., Herrmann, T., Dihazi, H., Beck, A., Mueller, G.A., Mueller, M., Stevanovic, S., Garbe, C., Mueller, C.A., and Kalbacher, H. (2006) Cathepsin D is present in human eccrine sweat and involved in the postsecretory processing of the antimicrobial peptide DCD-1L. *The Journal of Biological Chemistry*, **281**, 5406–5415.

89 Rieg, S., Seeber, S., Steffen, H., Humeny, A., Kalbacher, H., Stevanovic, S., Kimura, A., Garbe, C., and Schittek, B. (2006) Generation of multiple stable dermcidin-derived antimicrobial peptides in sweat of different body sites. *The Journal of Investigative Dermatology*, **126**, 354–365.

90 Sagawa, K., Kimura, A., Saito, Y., Inoue, H., Yasuda, S., Nosaka, M., and Tsuji, T. (2003) Production and characterization of a monoclonal antibody for sweat-specific protein and its application for sweat identification. *International Journal of Legal Medicine*, **117**, 90–95.

91 Flad, T., Bogumil, R., Tolson, J., Schittek, B., Garbe, C., Deeg, M., Mueller, C.A., and Kalbacher, H. (2002) Detection of dermcidin-derived peptides in sweat by ProteinChip technology. *Journal of Immunological Methods*, **270**, 53–62.

92 Rieg, S., Garbe, C., Sauer, B., Kalbacher, H., and Schittek, B. (2004) Dermcidin is constitutively produced by eccrine sweat glands and is not induced in epidermal cells under inflammatory skin conditions. *The British Journal of Dermatology*, **151**, 534–539.

93 Kimata, H. (2007) Increase in dermcidin-derived peptides in sweat of patients with atopic eczema caused by a humorous video. *Journal of Psychosomatic Research*, **62**, 57–59.

94 Rieg, S., Steffen, H., Seeber, S., Humeny, A., Kalbacher, H., Dietz, K., Garbe, C., and Schittek, B. (2005) Deficiency of dermcidin-derived antimicrobial peptides in sweat of patients with atopic dermatitis correlates with an impaired innate defense of human skin *in vivo*. *Journal of Immunology*, **174**, 8003–8010.

95 Schroeder, J.-M. (2011) Antimicrobial peptides in healthy skin and atopic dermatitis. *Allergology International*, **60**, 17–24.

96 Metz-Boutigue, M.-H., Shooshtarizadeh, P., Prevost, G., Haikel, Y., and Chich, J.-F. (2010) Antimicrobial peptides present in mammalian skin and gut are multifunctional defence molecules. *Current Pharmaceutical Design*, **16**, 1024–1039.

97 Niyonsaba, F., Suzuki, A., Ushio, H., Nagaoka, I., Ogawa, H., and Okumura, K. (2009) The human antimicrobial peptide dermcidin activates normal human keratinocytes. *The British Journal of Dermatology*, **160**, 243–249.

98 Rieg, S., Garbe, C., Sauer, B., Kalbacher, H., and Schittek, B. (2004) The antimicrobial peptide Dermcidin is not induced in epidermal keratinocytes under inflammatory skin conditions. *Archives of Dermatological Research*, **295**, 342.

99 Shiohara, T., Doi, T., and Hayakawa, J. (2011) Defective sweating responses in atopic dermatitis, in *Pathogenesis and Management of Atopic Dermatitis* (ed. T. Shiohara), Karger, Basel, pp. 68–79.

100 Senyurek, I., Paulmann, M., Sinnberg, T., Kalbacher, H., Deeg, M., Gutsmann, T., Hermes, M., Kohler, T., Gotz, F., Wolz, C., Peschel, A., and Schittek, B. (2009) Dermcidin-derived peptides show a different mode of action against *Staphylococcus aureus* than the Cathelicidin LL-37. *Antimicrobial Agents and Chemotherapy*, **53**, 2499–2509.

101 Cipakova, I., Gasperik, J., and Hostinova, E. (2006) Expression and purification of human antimicrobial peptide, dermcidin, in *Escherichia coli*. *Protein Expression and Purification*, **45**, 269–274.

102 Lai, Y.P., Peng, Y.F., Zuo, Y., Li, J., Huang, J., Wang, L.F., and Wu, Z.R. (2005) Functional and structural characterization of recombinant dermcidin-1L, a human antimicrobial peptide. *Biochemical and Biophysical Research Communications*, **328**, 243–250.

103 Murakami, M., Ohtake, T., Dorschner, R.A., Schittek, B., Garbe, C., and Gallo, R.L. (2002) Cathelicidin anti-microbial peptide expression in sweat, an innate defense system for the skin. *The Journal of Investigative Dermatology*, **119**, 1090–1095.

104 Hong, I., Kim, Y.S., and Choi, S.G. (2010) Simple purification of the human antimicrobial peptide dermcidin (MDCD-1L) by intein-mediated expression in *E. coli*. *Journal of Microbiology and Biotechnology*, **20**, 350–355.

105 Lai, Y., Villaruz, A.E., Li, M., Cha, D.J., Sturdevant, D.E., and Otto, M. (2007) The human anionic antimicrobial peptide dermcidin induces proteolytic defence mechanisms in staphylococci. *Molecular Microbiology*, **63**, 497–506.

106 Vuong, C., Voyich, J.M., Fischer, E.R., Braughton, K.R., Whitney, A.R., DeLeo, F.R., and Otto, M. (2004) Polysaccharide intercellular adhesin (PIA) protects *Staphylococcus epidermidis* against major components of the human innate immune system. *Cellular Microbiology*, **6**, 269–275.

107 Senyuerek, I., Doering, G., Kalbacher, H., Deeg, M., Peschel, A., Wolz, C., and Schittek, B. (2009) Resistance to dermcidin-derived peptides is independent of bacterial protease activity. *International Journal of Antimicrobial Agents*, **34**, 86–90.

108 Peschel, A. and Sahl, H.-G. (2006) The co-evolution of host cationic antimicrobial peptides and microbial resistance. *Nature Reviews Microbiology*, **4**, 529–536.

109 Majczak, G., Lilla, S., Garay-Malpartida, M., Markovic, J., Medrano, F.J., de Nucci, G., and Belizario, J.E. (2007) Prediction and biochemical characterization of intrinsic disorder in the structure of proteolysis-inducing factor/dermcidin. *Genetics and Molecular Research*, **6**, 1000–1011.

110 Steffen, H., Rieg, S., Wiedemann, I., Kalbacher, H., Deeg, M., Sahl, H.G., Peschel, A., Gotz, F., Garbe, C., and Schittek, B. (2006) Naturally processed dermcidin-derived peptides do not

permeabilize bacterial membranes and kill microorganisms irrespective of their charge. *Antimicrobial Agents and Chemotherapy*, **50**, 2608–2620.

111 Jung, H.H., Yang, S.-T., Sim, J.-Y., Lee, S., Lee, J.Y., Kim, H.H., Shin, S.Y., and Kim, J.I. (2010) Analysis of the solution structure of the human antibiotic peptide dermcidin and its interaction with phospholipid vesicles. *BMB Reports*, **43**, 362–368.

112 Phoenix, D.A. and Harris, F. (2002) The hydrophobic moment and its use in the classification of amphiphilic structures [review]. *Molecular Membrane Biology*, **19**, 1–10.

113 Steffen, H., Rieg, S., Wiedemann, I., Kalbacher, H., Deeg, A., Sahl, H.G., Peschel, A., Goetz, F., Garbe, C., and Schittek, B. (2006) Naturally processed dermcidin-derived peptides do not permeabilize bacterial membranes and kill microorganisms irrespective of their charge. *Antimicrobial Agents and Chemotherapy*, **50**, 2608–2620.

114 Li, M., Rigby, K., Lai, Y., Nair, V., Peschel, A., Schittek, B., and Otto, M. (2009) *Staphylococcus aureus* mutant screen reveals interaction of the human antimicrobial peptide dermcidin with membrane phospholipids. *Antimicrobial Agents and Chemotherapy*, **53**, 4200–4210.

115 Senyuerek, I., Paulmann, M., Sinnberg, T., Kalbacher, H., Deeg, M., Gutsmann, T., Hermes, M., Kohler, T., Goetz, F., Wolz, C., Peschel, A., and Schittek, B. (2009) Dermcidin-derived peptides show a different mode of action than the Cathelicidin LL-37 against *Staphylococcus aureus*. *Antimicrobial Agents and Chemotherapy*, **53**, 2499–2509.

116 Paulmann, M., Arnold, T., Linke, D., Özdirekcan, S., Kopp, A., Gutsmann, T., Kalbacher, H., Wanke, I., Schuenemann, V.J., Habeck, M., Bürck, J., Ulrich, A.S., and Schittek, B. (2012) Structure–activity analysis of the dermcidin-derived peptide DCD 1L, an anionic antimicrobial peptide present in human sweat. *The Journal of Biological Chemistry*, **287**, 8434–8443.

117 Hall, S.H., Yenugu, S., Radhakrishnan, Y., Avellar, M.C.W., Petrusz, P., and French, F.S. (2007) Characterization and functions of beta defensins in the epididymis. *Asian Journal of Andrology*, **9**, 453–462.

118 Turner, T.T. (2008) De Graaf's thread: the human epididymis. *Journal of Andrology*, **29**, 237–250.

119 Mital, P., Hinton, B.T., and Dufour, J.M. (2011) The blood–testis and blood–epididymis barriers are more than just their tight junctions. *Biology of Reproduction*, **84**, 851–858.

120 Sipila, P., Jalkanen, J., Huhtaniemi, I.T., and Poutanen, M. (2009) Novel epididymal proteins as targets for the development of post-testicular male contraception. *Reproduction*, **137**, 379–389.

121 Hall, S.H., Hamil, K.G., and French, F.S. (2002) Host defense proteins of the male reproductive tract. *Journal of Andrology*, **23**, 585–597.

122 von Horsten, H.H., Derr, P., and Kirchhoff, C. (2002) Novel antimicrobial peptide of human epididymal duct origin. *Biology of Reproduction*, **67**, 804–813.

123 Liu, Q., Hamil, K.G., Sivashanmugam, P., Grossman, G., Soundararajan, R., Rao, A.J., Richardson, R.T., Zhang, Y.L., O'Rand, M.G., Petrusz, P., French, F.S., and Hall, S.H. (2001) Primate epididymis-specific proteins: characterization of ESC42, a novel protein containing a trefoil-like motif in monkey and human. *Endocrinology*, **142**, 4529–4539.

124 von Horsten, H.H., Schafer, B., and Kirchhoff, C. (2004) SPAG11/isoform HEX, an atypical anionic beta-defensin-like peptide. *Peptides*, **25**, 1223–1233.

125 Hollox, E.J. and Armour, J.A.L. (2008) Directional and balancing selection in human beta-defensins. *BMC Evolutionary Biology*, **8**, 113.

126 Yenugu, S., Hamil, K.G., Radhakrishnan, Y., French, F.S., and Hall, S.H. (2004) The androgen-regulated epididymal sperm-binding protein, human beta-defensin 118 (DEFB118) (formerly ESC42), is an antimicrobial beta-defensin. *Endocrinology*, **145**, 3165–3173.

127 Yenugu, S., Hamil, K.G., French, F.S., and Hall, S.H. (2004) Antimicrobial actions of the human epididymis 2 (HE2) protein isoforms, HE2alpha, HE2beta1 and HE2beta2. *Reproductive Biology and Endocrinology*, **2**, 61.

128 Yenugu, S. and Narmadha, G. (2010) The human male reproductive tract antimicrobial peptides of the HE2 family exhibit potent synergy with standard antibiotics. *Journal of Peptide Science*, **16**, 337–341.

129 Liao, M., Ruddock, P.S., Rizvi, A.S., Hall, S.H., French, F.S., and Dillon, J.R. (2005) Cationic peptide of the male reproductive tract, HE2 alpha, displays antimicrobial activity against *Neisseria gonorrhoeae*, *Staphylococcus aureus* and *Enterococcus faecalis*. *The Journal of Antimicrobial Chemotherapy*, **56**, 957–961.

130 Gudmundsson, G.H. and Agerberth, B. (2005) Antimicrobial peptides in human blood, in *Antimicrobial Peptides in Human Health and Disease* (ed. R.L. Gallo), Horizon Scientific Press/Caister Academic Press, Hethersett, pp. 193–228.

131 Risso, A. (2000) Leukocyte antimicrobial peptides: multifunctional effector molecules of innate immunity. *Journal of Leukocyte Biology*, **68**, 785–792.

132 Yeaman, M.R. and Bayer, A.S. (2007) Antimicrobial host defense, in *Platelets* (ed. A.D. Michelson), 2nd edn, Academic Press, New York, pp. 727–755.

133 Levy, O. (2000) Antimicrobial proteins and peptides of blood: templates for novel antimicrobial agents. *Blood*, **96**, 2564–2572.

134 Salzet, M. (2005) Neuropeptide-derived antimicrobial peptides from invertebrates for biomedical applications. *Current Medicinal Chemistry*, **12**, 3055–3061.

135 Blair, P. and Flaumenhaft, R. (2009) Platelet alpha-granules: basic biology and clinical correlates. *Blood Reviews*, **23**, 177–189.

136 Tang, Y.Q., Yeaman, M.R., and Selsted, M.E. (2002) Antimicrobial peptides from human platelets. *Infection and Immunity*, **70**, 6524–6533.

137 Vankerckhoven, V., Moreillon, P., Piu, S., Giddey, M., Huys, G., Vancanneyt, M., Goossens, H., and Entenza, J.M. (2007) Infectivity of *Lactobacillus rhamnosus* and *Lactobacillus paracasei* isolates in a rat model of experimental endocarditis. *Journal of Medical Microbiology*, **56**, 1017–1024.

138 Goldstein, A.L., Hannappel, E., and Kleinman, H.K. (2005) Thymosin [beta]4: actin-sequestering protein moonlights to repair injured tissues. *Trends in Molecular Medicine*, **11**, 421–429.

139 Huang, L.C., Jean, D., Proske, R.J., Reins, R.Y., and McDermott, A.M. (2007) Ocular surface expression and *in vitro* activity of antimicrobial peptides. *Current Eye Research*, **32**, 595–609.

140 Kolar, S.S. and McDermott, A.M. (2011) Role of host-defence peptides in eye diseases. *Cellular and Molecular Life Sciences*, **68**, 2201–2213.

141 Sosne, G., Qiu, P., Kurpakus-Wheater, M., and Matthew, H. (2010) Thymosin beta 4 and corneal wound healing: visions of the future, in *Thymosins in Health and Disease* (eds E. Garaci and A.L. Goldstein), Wiley, New York, pp. 190–198.

142 Crockford, D., Turjman, N., Allan, C., and Angel, J. (2010) Thymosin beta 4: structure, function, and biological properties supporting current and future clinical applications, in *Thymosins in Health and Disease* (eds E. Garaci and A.L. Goldstein), Wiley, New York, pp. 179–189.

143 Ryan, J.T., Ross, R.P., Bolton, D., Fitzgerald, G.F., and Stanton, C. (2011) Bioactive peptides from muscle sources: meat and fish. *Nutrients*, **3**, 765–791.

144 Atanasova, J. and Ivanova, I. (2010) Antibacterial peptides from goat and sheep milk proteins. *Biotechnology & Biotechnological Equipment*, **24**, 1799–1803.

145 Szwajkowska, M., Wolanciuk, A., Barlowska, J., Krol, J., and Litwinczuk, Z. (2011) Bovine milk proteins as the source of bioactive peptides influencing the consumers' immune system – a

review. *Animal Science Papers and Reports*, **29**, 269–280.

146 Urista, C.M., Fernandez, R.A., Rodriguez, F.R., Cuenca, A.A., and Jurado, A.T. (2011) Review: production and functionality of active peptides from milk. *Food Science and Technology International*, **17**, 293–317.

147 Pellegrini, A. (2003) Antimicrobial peptides from food proteins. *Current Pharmaceutical Design*, **9**, 1225–1238.

148 Mine, Y. and Kovacs-Nolan, J. (2006) New insights in biologically active proteins and peptides derived from hen egg. *Worlds Poultry Science Journal*, **62**, 87–95.

149 Huntington, J.A. and Stein, P.E. (2001) Structure and properties of ovalbumin. *Journal of Chromatography B*, **756**, 189–198.

150 Phelan, M., Aherne, A., FitzGerald, R.J., and O'Brien, N.M. (2009) Casein-derived bioactive peptides: biological effects, industrial uses, safety aspects and regulatory status. *International Dairy Journal*, **19**, 643–654.

151 Kamau, S.M., Cheison, S.C., Chen, W., Liu, X.M., and Lu, R.R. (2010) Alpha-lactalbumin: its production technologies and bioactive peptides. *Comprehensive Reviews in Food Science and Food Safety*, **9**, 197–212.

152 Hernandez-Ledesma, B., Recio, I., and Amigo, L. (2008) Beta-lactoglobulin as source of bioactive peptides. *Amino Acids*, **35**, 257–265.

153 Biziulevičius, G.A., Kislukhina, O.V., Kazlauskaite, J., and Žukaite, V. (2006) Food-protein enzymatic hydrolysates possess both antimicrobial and immunostimulatory activities: a "cause and effect" theory of bifunctionality. *FEMS Immunology & Medical Microbiology*, **46**, 131–138.

154 Benkerroum, N. (2010) Antimicrobial peptides generated from milk proteins: a survey and prospects for application in the food industry. A review. *International Journal of Dairy Technology*, **63**, 320–338.

155 Pellegrini, A., Hülsmeier, A.J., Hunziker, P., and Thomas, U. (2004) Proteolytic fragments of ovalbumin display antimicrobial activity. *Biochimica et Biophysica Acta – General Subjects*, **1672**, 76–85.

156 Pellegrini, A., Thomas, U., Bramaz, N., Hunziker, P., and von Fellenberg, R. (1999) Isolation and identification of three bactericidal domains in the bovine alpha-lactalbumin molecule. *Biochimica et Biophysica Acta – General Subjects*, **1426**, 439–448.

157 Pellegrini, A., Dettling, C., Thomas, U., and Hunziker, P. (2001) Isolation and characterization of four bactericidal domains in the bovine beta-lactoglobulin. *Biochimica et Biophysica Acta – General Subjects*, **1526**, 131–140.

158 Hayes, A., Ross, R.P., Fitzgerald, G.F., Hill, C., and Stanton, C. (2006) Casein-derived antimicrobial peptides generated by *Lactobacillus acidophilus* DPC6026. *Applied and Environmental Microbiology*, **72**, 2260–2264.

159 Malkoski, M., Dashper, S.G., O'Brien-Simpson, N.M., Talbo, G.H., Macris, M., Cross, K.J., and Reynolds, E.C. (2001) Kappacin, a novel antibacterial peptide from bovine milk. *Antimicrobial Agents and Chemotherapy*, **45**, 2309–2315.

160 Yeaman, M.R. and Yount, N.Y. (2003) Mechanisms of antimicrobial peptide action and resistance. *Pharmacological Reviews*, **55**, 27–55.

161 Laverty, G., Gorman, S.P., and Gilmore, B.F. (2011) The potential of antimicrobial peptides as biocides. *International Journal of Molecular Sciences*, **12**, 6566–6596.

162 Wilmes, M., Cammue, B.P.A., Sahl, H.-G., and Thevissen, K. (2011) Antibiotic activities of host defense peptides: more to it than lipid bilayer perturbation. *Natural Product Reports*, **28**, 1350–1358.

163 Burman, R., Stromstedt, A.A., Malmsten, M., and Goransson, U. (2011) Cyclotide–membrane interactions: defining factors of membrane binding, depletion and disruption. *Biochimica et Biophysica Acta – Biomembranes*, **1808**, 2665–2673.

164 Zhao, M. (2011) Lantibiotics as probes for phosphatidylethanolamine. *Amino Acids*, **41**, 1071–1079.

165 Asaduzzaman, S.M. and Sonomoto, K. (2009) Lantibiotics: diverse activities and unique modes of action. *Journal of Bioscience and Bioengineering*, **107**, 475–487.

166 Hao, G., Shi, Y., Tang, Y., and Le, G. (2010) The action mechanism of analogues of the antimicrobial peptide Buforin II with *E. coli* genomic DNA. *Weishengwu Xuebao*, **50**, 328–333.

167 Nicolas, P. (2009) Multifunctional host defense peptides: intracellular-targeting antimicrobial peptides. *FEBS Journal*, **276**, 6483–6496.

168 Henriques, S.T., Melo, M.N., and Castanho, M. (2006) Cell-penetrating peptides and antimicrobial peptides: how different are they? *Biochemical Journal*, **399**, 1–7.

169 Su, L.Y., Willner, D.L., and Segall, A.M. (2010) Antimicrobial peptide that targets DNA repair intermediates *in vitro* inhibits salmonella growth within murine macrophages. *Antimicrobial Agents and Chemotherapy*, **54**, 1888–1899.

170 Harris, F., Dennison, S.R., and Phoenix, D.A. (2012) Aberrant action of amyloidogenic host defense peptides: a new paradigm to investigate neurodegenerative disorders? *FASEB Journal*, **26**, 1776–1781.

171 Kinnunen, P.K. (2009) Amyloid formation on lipid membrane surfaces. *The Open Biology Journal*, **2**, 163–175.

172 Mahalka, A.K. and Kinnunen, P.K.J. (2009) Binding of amphipathic alpha-helical antimicrobial peptides to lipid membranes: lessons from temporins B and L. *Biochimica et Biophysica Acta – Biomembranes*, **1788**, 1600–1609.

173 Sood, R., Domanov, Y., Pietiainen, M., Kontinen, V.P., and Kinnunen, P.K.J. (2008) Binding of LL-37 to model biomembranes: insight into target vs host cell recognition. *Biochimica et Biophysica Acta – Biomembranes*, **1778**, 983–996.

174 Zhao, H., Sood, R., Jutila, A., Bose, S., Fimland, G., Nissen-Meyer, J., and Kinnunen, P.K.J. (2006) Interaction of the antimicrobial peptide pheromone Plantaricin A with model membranes: implications for a novel mechanism of action. *Biochimica et Biophysica Acta – Biomembranes*, **1758**, 1461–1474.

175 Auvynet, C., El Amri, C., Lacombe, C., Bruston, F., Bourdais, J., Nicolas, P., and Rosenstein, Y. (2008) Structural requirements for antimicrobial versus chemoattractant activities for dermaseptin S9. *FEBS Journal*, **275**, 4134–4151.

176 Nicolas, P. and El Amri, C. (2009) The dermaseptin superfamily: a gene-based combinatorial library of antimicrobial peptides. *Biochimica et Biophysica Acta – Biomembranes*, **1788**, 1537–1550.

177 Gossler-Schofberger, R., Hesser, G., Muik, M., Wechselberger, C., and Jilek, A. (2009) An orphan dermaseptin from frog skin reversibly assembles to amyloid-like aggregates in a pH-dependent fashion. *FEBS Journal*, **276**, 5849–5859.

178 Nozaki, Y., Tanford, C., and Hirs, C.H.W. (1967) Examination of titration behavior. *Methods in Enzymology*, **11**, 715–734.

179 Diego-Garcia, E., Batista, C.V., Garcia-Gomez, B.I., Lucas, S., Candido, D.M., Gomez-Lagunas, F., and Possani, L.D. (2005) The Brazilian scorpion *Tityus costatus* Karsch: genes, peptides and function. *Toxicon*, **45**, 273–283.

180 Pace, C.N. and Scholtz, J.M. (1998) A helix propensity scale based on experimental studies of peptides and proteins. *Biophysical Journal*, **75**, 422–427.

181 Tossi, A., Sandri, L., and Giangaspero, A. (2000) Amphipathic, alpha-helical antimicrobial peptides. *Biopolymers*, **55**, 4–30.

182 Dennison, S.R., Wallace, J., Harris, F., and Phoenix, D.A. (2005) Amphiphilic alpha-helical antimicrobial peptides and their structure–function relationships. *Protein and Peptide Letters*, **12**, 31–39.

183 Ming, L.J. (2010) Metallopeptides – from drug discovery to catalysis. *Journal of the Chinese Chemical Society*, **57**, 285–299.

184 Calderon-Santiago, M. and Luque de Castro, M.D. (2009) The dual trend in histatins research. *Trends in Analytical Chemistry*, **28**, 1011–1018.

185 Brogden, K.A. (2005) Antimicrobial peptides: pore formers or metabolic inhibitors in bacteria? *Nature Reviews Microbiology*, **3**, 238–250.

186 Duclohier, H. (2004) Helical kink and channel behaviour: a comparative study with the peptaibols alamethicin, trichotoxin and antiamoebin. *European Biophysics Journal with Biophysics Letters*, **33**, 169–174.

187 Haney, E.F., Hunter, H.N., Matsuzaki, K., and Vogel, H.J. (2009) Solution NMR studies of amphibian antimicrobial peptides: linking structure to function? *Biochimica et Biophysica Acta – Biomembranes*, **1788**, 1639–1655.

188 Xiao, Y.J., Herrera, A.I., Bommineni, Y.R., Soulages, J.L., Prakash, O., and Zhang, G.L. (2009) The central kink region of fowlicidin-2, an alpha-helical host defense peptide, is critically involved in bacterial killing and endotoxin neutralization. *Journal of Innate Immunity*, **1**, 268–280.

189 Bruhn, H., Winkelmann, J., Andersen, C., Andra, J., and Leippe, M. (2006) Dissection of the mechanisms of cytolytic and antibacterial activity of lysenin, a defence protein of the annelid *Eisenia fetida*. *Developmental and Comparative Immunology*, **30**, 597–606.

190 Prochazkova, P., Silerova, M., Felsberg, J., Joskova, R., Beschin, A., De Baetselier, P., and Bilej, M. (2006) Relationship between hemolytic molecules in *Eisenia fetida* earthworms. *Developmental and Comparative Immunology*, **30**, 381–392.

191 Huang, Y., Wang, X., Wang, H., Liu, Y., and Chen, Y. (2011) Studies on mechanism of action of anticancer peptides by modulation of hydrophobicity within a defined structural framework. *Molecular Cancer Therapeutics*, **10**, 416–426.

192 Wang, G., Li, X., and Wang, Z. (2009) APD2: the updated antimicrobial peptide database and its application in peptide design. *Nucleic Acids Research*, **37**, D933–D937.

193 Wang, G., Li, X., and Zasloff, M. (2010) A database view of naturally occurring antimicrobial peptides: nomenclature, classification and amino acid sequence analysis. *Advances in Molecular and Cellular Microbiology*, **18**, 1–21.

194 Conlan, B.F. and Anderson, M.A. (2011) Circular micro-proteins and mechanisms of cyclization. *Current Pharmaceutical Design*, **17**, 4318–4328.

195 Conlan, B.F., Gillon, A.D., Craik, D.J., and Anderson, M.A. (2010) Circular proteins and mechanisms of cyclization. *Peptide Science*, **94**, 573–583.

196 Cascales, L. and Craik, D.J. (2010) Naturally occurring circular proteins: distribution, biosynthesis and evolution. *Organic & Biomolecular Chemistry*, **8**, 5035–5047.

197 Penberthy, W.T., Chari, S., Cole, A.L., and Cole, A.M. (2011) Retrocyclins and their activity against HIV-1. *Cellular and Molecular Life Sciences*, **68**, 2231–2242.

198 Davies, J. and Davies, D. (2010) Origins and evolution of antibiotic resistance. *Microbiology and Molecular Biology Reviews*, **74**, 417–433.

199 Ballard, C., Gauthier, S., Corbett, A., Brayne, C., Aarsland, D., and Jones, E. (2011) Alzheimer's disease. *Lancet*, **377**, 1019–1031.

4
Graphical Techniques to Visualize the Amphiphilic Structures of Antimicrobial Peptides

Summary

A number of sequence-based analyses have been developed to identify protein segments that are able to form membrane-interactive amphiphilic α-helices. Earlier techniques attempted to detect the characteristic periodicity in hydrophobic amino acid residues while recently, more advanced models have utilized weightings of amphiphilicity to enable the introduction of computational procedures based on techniques such as hydropathy analysis, homology modeling, multiple sequence alignments, and neural networks, with prediction accuracies of the order of greater than 80%. Here, we review a number of these predictive techniques and consider problems associated with their use in the prediction of structure–function relationships, using α-helices from lytic peptides and tilted peptides as examples.

4.1
Introduction

Traditionally, the pharmaceutical industry has disregarded peptides as potential therapeutic agents due to a number of factors, such as their production costs, susceptibility to degradation, and low oral availability. However, since the mid-1960s in response to the increasing cost of developing conventional drugs there has been growing interest in peptides and their mimetics as potential agents for therapeutic intervention. This interest has led to strategies to improve both the productivity and pharmacokinetic properties of peptides, which, along with the development of alternative routes of administration, has led to a number of peptide-based drugs being marketed [1–5].

One of the biggest research areas for the development of peptide-based therapeutics is in the field of antimicrobial peptides (AMPs) and numerous studies have established that the ability of these peptides to kill microbes depends upon their ability to interact with the membranes of these organisms [6, 7] (Chapter 5). Structural analysis has revealed that these peptides are capable of adopting a diverse variety of structures ranging from pure α-helix or β-strands to a mixture of α and

Antimicrobial Peptides, First Edition. David A. Phoenix, Sarah R. Dennison, and Frederick Harris.
© 2013 Wiley-VCH Verlag GmbH & Co. KGaA. Published 2013 by Wiley-VCH Verlag GmbH & Co. KGaA.

β structures. However, numerous studies have shown that to accommodate both polar and non-polar environments of the membrane, these structures have the common property of possessing amphiphilic elements, which are characterized by a clear segregation of hydrophobic and hydrophilic amino acid residues [8, 9] (Chapters 2, 3, and 6).

It is now generally accepted that the possession of structural amphiphilicity is fundamental to the antimicrobial activity of the vast majority of AMPs, but those that adopt amphiphilic α-helical architecture (α-AMPs) are by far the most often studied [10]. This choice of peptide is primarily due to the fact that the spatial regularity of their residues has allowed the development of a range of graphical techniques that are able to visualize and quantify not only their amphiphilicity, but also other structural and physiochemical parameters (Chapters 2, 3, and 6) that can be related to their capacity for lipid/membrane interactions. This ability has also led to a number of methodologies that are based on sequence information alone, and attempt to predict and classify membrane-interactive α-AMPs according to their biological function [11, 12]. In this chapter, we describe the most commonly used of these graphical techniques and methodologies after presenting an overview of the different types of amphiphilic structure adopted by AMPs.

4.2
Amphiphilic Structures Adopted by AMPs

It is generally accepted that three major types of amphiphilic structure can be identified in membrane-interactive proteins and peptides, namely those possessing either primary, secondary, or tertiary amphiphilicity [12]. Structures with primary amphiphilicity are exemplified by the transmembrane (TM), α-helices of integral membrane proteins, which possess stretches of consecutive, predominantly hydrophobic amino acid residues that span the apolar membrane core and are flanked on either side by clusters of hydrophilic residues that interact with the lipid head-group regions [13–15]. This form of amphiphilicity is not generally common in AMPs, and, as described in Chapter 2, the major known examples are indolicidin [16] and tritrpticin [17]. These peptides adopt an extended conformation with a core segment that is composed of a short central apolar sequence, flanked at each end by cationic residues (Figure 2.1). This mode of membrane interaction appears to be essential to the antimicrobial action of both peptides, although as described in Chapter 2 detailed descriptions of their mode of action are lacking [8, 18].

Structures with secondary amphiphilicity include those with unconstrained β-sheet architecture, as seen for the membrane-spanning segments of bacterial β-barrel TM porins. For these segments, hydrophobic residues cluster on one face of the β-sheet structure and face into the apolar membrane core region while hydrophilic residues cluster on the other face of this structure and form a hydrophilic membrane pore [19, 20]. Examples of AMPs that utilize this type of amphiphilic β-sheet architecture to facilitate their membrane interaction and

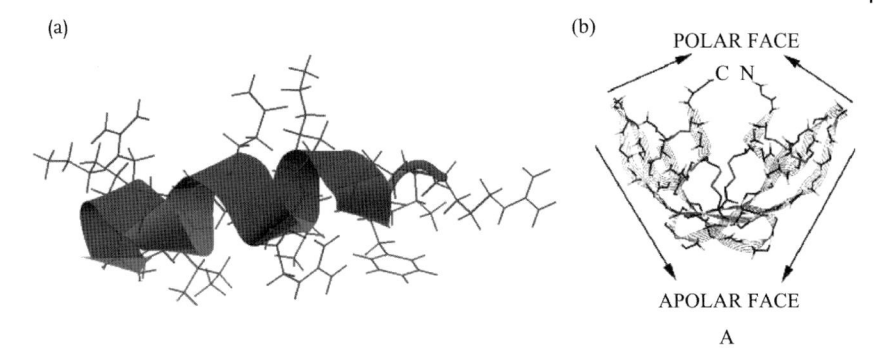

Figure 4.1 Amphiphilicity of AMPs. (a) The secondary amphiphilicity of aurein 1.2, which illustrates the three-dimensional α-helical segregation of polar and apolar residues. (b) The tertiary amphiphilicity possessed by the β-sheet defensin, human neutrophil-3. The side-chains of residues are not shown for clarity, and it can be seen that proximity of N-terminal and C-terminal residues generates the polar face while the folding pattern of the molecule's middle region generates the apolar face.

antimicrobial mechanisms include cateslytin [21, 22] and a number of synthetic AMPs [23–28]. However, the major group of structures with secondary amphiphilicity are those with α-helical architecture, which were first recognized within the molecules of myoglobin and hemoglobin during the 1960s [29]. These amphiphilic α-helices have been detected in a wide variety of biologically active molecules, including apolipoproteins and lipid-binding proteins [30–32], cytotoxic peptides [33–35], and channel-forming peptides [12]. It is well established that α-helices with secondary amphiphilicity show an ordered segregation of polar and apolar amino acid residues about the α-helical long axis, exemplified by AMPs such as aurein 1.2 (Figure 4.1a). This residue arrangement allows membrane-interactive AMPs to adopt an orientation that is approximately parallel to the membrane surface such that their non-polar face interacts with the membrane lipid core and their hydrophilic face engages in electrostatic interactions with the membrane lipid head-group region [8, 9]. Based on this mode of membrane interaction, these α-helices have often been referred to as surface-active or surface-seeking α-helices albeit that their interfacial location is often transient and the first step in a range of membrane-associated functions, including membrane stabilization [36, 37], pore and channel formation [12, 35, 38], and protein–membrane anchoring [39]. More recently, a further, highly specialized group of membrane-interactive peptides generally termed oblique oriented α-helices, were discovered in viral proteins (Chapter 5). Now known to occur in a variety of proteins, these α-helices exhibit fundamental structural differences to other AMPs that allow them to destabilize membranes and promote a range of biological events via membrane penetration at a tilted angle [11, 40].

In contrast to the peptides described above, a major group of AMPs have recently been characterized that possess tertiary amphiphilicity where polar and apolar residues distal in the primary structure of the molecule are brought together in its final three-dimensional structure to form a lipid interactive site [12]. AMPs that possess tertiary amphiphilicity include cyclotides, which are cysteine-stabilized AMPs with a molecular surface characterized by the occurrence of hydrophobic and hydrophilic sites that are formed by the clustering of apolar and polar residues respectively (Chapter 2). It has been shown that possession of these amphiphilic sites are core to the functionality of these peptides, facilitating the invasion of microbial membranes [41–43], along with other biological activities such as toxicity against insects [44, 45], worms [46], and snails [47]. Another major example of cysteine-stabilized AMPs with tertiary amphiphilicity (Chapters 1, 2, and 3) are defensins, which is illustrated for human defensin, neutrophil-3, in Figure 4.1b. It can be seen from Figure 4.1 that the N-terminal and C-terminal residues of this molecule are brought together to form its polar face while the folding pattern of its middle region generates the apolar face [48]. It is well established that these amphiphilic characteristics facilitate the ability of defensins to interact with microbial membranes and kill the host microbe (Chapter 2) [49–54]. Interestingly, some defensins also show chemotactic activity via interaction with various chemokine receptors and it has been suggested that these interactions may involve modification of the membrane microdomain associated with the chemokine receptor (Chapter 1) [55]. This suggestion does not appear to have been verified experimentally, but we speculate that these membrane interactions would be facilitated by the tertiary amphiphilicity of defensins. More generally, tertiary amphiphilicity appears to have been tailored by AMPs to facilitate microbial membrane invasion and the antimicrobial actions of a structurally diverse range of peptides that are appropriate to the defense needs of an equally diverse range of host organisms (Chapters 1, 2, 3, 5, and 6).

However, amphiphilic α-helical segments have the advantage over other amphiphilic protein structures in that the spatial regularity of the residues within these structures has allowed a number of analytical techniques to be developed that are based on sequence analysis, and are able to quantify a range of structural and physiochemical parameters possessed by these α-helices. These techniques form the basis of a number of methodologies designed to predict the structure–function relationships of amphiphilic α-helices [11, 12, 34, 35, 56–59], which we review in this chapter.

4.3
Qualitative Methods for Identifying Amphiphilic Structure

The most commonly used qualitative method for modeling α-helices is the graphical helical wheel methodology of Schiffer and Edmundson [60], which represents helices as a two-dimensional axial projection. This analysis assumes the angular periodicity of 100° for an ideal α-helix and projects the side-chains of the residues forming the α-helix onto a unit circle, which lies in a plane perpendicular to the

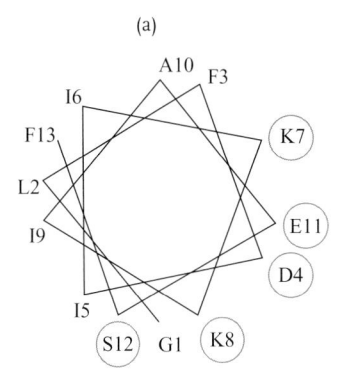

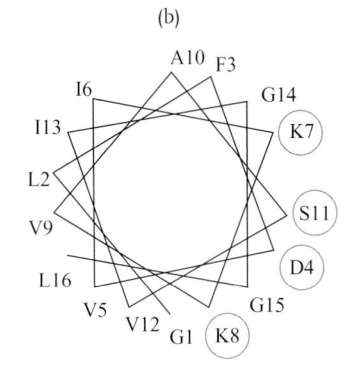

Figure 4.2 Secondary amphiphilicity of AMPs aurein 1.2 (a) and citropin 1.1 (b). Illustrated as a two-dimensional α-helical axial projection, it can be seen that each peptide possesses a narrow hydrophilic face composed of polar and charged residues (circled), which is also rich in glycine residues.

long axis of the helix. This circle is divided into sectors, each representing an amino acid residue in the helical sequence, and as a result, sequences forming secondary amphiphilicity show a clustering of hydrophobic residues on one side of the unit circle and hydrophilic residues on the opposing side, which usually indicates membrane-interactive potential (Figure 4.2).

Graphical techniques are useful tools for predicting amphiphilic α-helices in cases where single short amino acid sequences are being considered and amphiphilic structure is well defined. This is clearly illustrated by the membrane-interactive AMPs shown in Figure 4.2, which shows that there are number of structural similarities between the α-helices formed by aurein 1.2 and citropin 1.1. Each peptide possesses a wide hydrophobic face, which is consistent with their common function of membrane invasion and has been shown to be important to the antimicrobial potency of AMPs [61–63]. The hydrophilic face of these AMPs is also rich in glycine residues, which is a structural feature strongly associated with tilted α-helix formation and can be more rigorously characterized using analytical techniques described below [34, 64].

However, the helical wheel model does not reflect the nature of the longitudinal arrangement of residue side-chains, hence Dunhill [65] introduced the "helical-net" diagram in which the amino acid residues of the α-helix are visualized as a graphical projection on the surface of a cylinder. The cylinder is then cut along a single line parallel to its long axis and flattened into a rectangle to locate hydrophobic clusters and the distribution of charged and polar residues. An example is given in Figure 4.3, which shows a helical net of a synthetic AMP, D1 (K13), along with the corresponding space-filled model of the peptide. These representations have been used produced to help in the recent design of α-AMPs to target Gram-negative pathogens such as *Acinetobacter baumannii* and *Pseudomonas aeruginosa* [66].

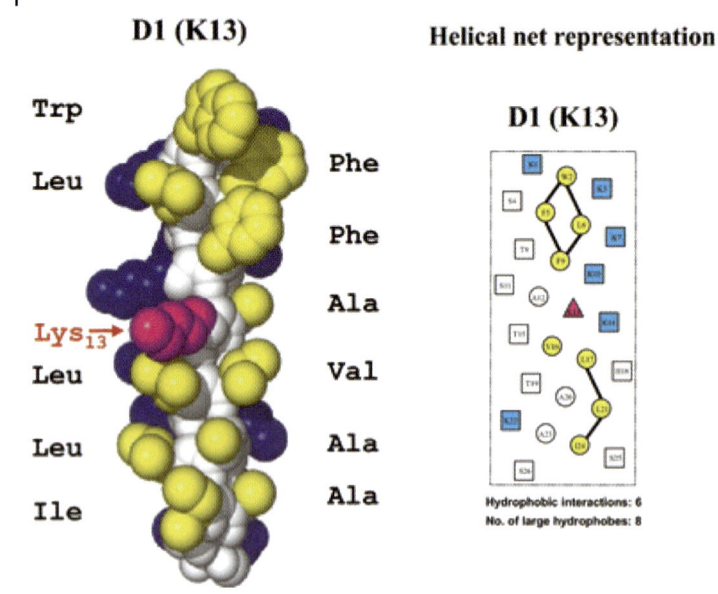

Figure 4.3 Dunhill plot showing a space-filled model and a corresponding helical net representation for the synthetic AMP, D1 (K13) [66]. In the space-filled model, hydrophobic amino acids on the non-polar face are colored yellow; hydrophilic amino acids on the polar face are colored blue; the peptide backbone is colored white. The lysine residue in the center of the non-polar face of peptide D1 (K13) is colored pink. In the helical net representation of the peptide, the lysine residue at position 13 in the center of the non-polar face of the peptide is denoted by a pink triangle. The amino acid residues on the polar face are boxed and the positively charged lysine residues are colored blue. The amino acid residues on the non-polar face are circled and the large hydrophobicities (Trp, Phe, Leu, Val, and Ile) are colored yellow. Hydrophobic interactions between these large hydrophobicities along the helix are shown as black bars.

However, there are limitations to the Dunhill methodology in that, as with the helical wheel, it does not quantify the amphiphilicity of the sequence. Moreover, for longer sequences, graphical techniques are cumbersome to apply and often difficult to interpret. In response, Chou *et al.* [67] introduced the Wenxiang diagram where hydrophobic and hydrophilic residues are considered to be segregated by an interfacial plane, termed the demarcation plane. A two-dimensional diagram is then generated by conical projection of an α-helix perpendicular to its axis as shown in Figure 4.4a. Taken with other data, these projections can be used to yield the inclination angle, ω, between the axis of the helix and the demarcation plane, thereby providing a measure of the degree of amphiphilicity of an α-helix [67]. This facility gives the Wenxiang diagram an advantage over other graphical techniques in that it makes it easier to discern the relative locations of amino acids in an α-helical conformation, which is especially useful in the case of longer α-helices. A Wenxiang diagram representation of aurein 1.2, which is known to be a highly amphiphilic α-helical molecule [64], is shown in Figure 4.4b where

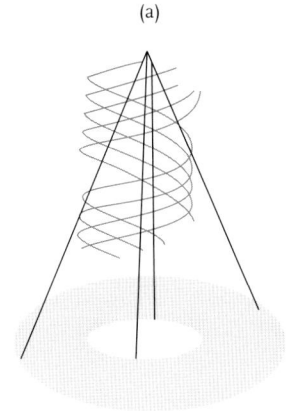

Figure 4.4 Wenxiang analysis. (a) Conical projection of an α-helix H perpendicular to the plane within the ring area R [68]. The lines represent the radiating lines from the apex (A). (b) Two-dimensional Wenxiang diagram for aurein 1.2 with the black circles representing hydrophobic residues and the white circles representing hydrophilic residues.

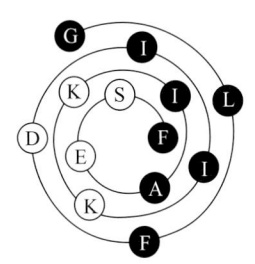

this amphiphilicity is clearly revealed. This diagram was generated using an online program, which currently does not compute ω [68], although previous work has shown that the highly amphiphilic α-helices of membrane-interactive AMPs have ω values around zero. In contrast, globular proteins tend to have values of ω that are less than 25°, showing that these proteins possess moderate amphiphilicity. In cases where ω is high for α-helices, as found in triose phosphate isomerase, this indicates low amphiphilicity with no clear-cut demarcation hydrophobic and hydrophilic residues [67]. More recently, Wenxiang diagrams of a leucine-zipper structure from a coiled-coil protein have been used to identify key residues that play important role in its interaction with another such structure and thereby appears to be of use in characterizing protein–protein interactions in addition to lipid–protein/AMPs interactions [69].

4.4
Quantitative Techniques for Analyzing Amphiphilic Structure

4.4.1
Techniques Based on Hydropathy Plot Analysis

Hydropathy plot analysis [70] was originally used to identify structures with primary amphiphilicity such as the TM α-helices of integral membrane proteins [12]. Essentially, this form of analysis computes a moving mean hydrophobicity (hydropathy index), <H> using an 11-residue window, which is plotted versus the hydrophobicity values of consecutive residues in a protein's primary structure (H_1, H_2, . . . , H_k) to provide a hydrophobicity profile (hydropathy plot) of the sequence [70]. The occurrence of stretches of consecutive hydrophobic residues of 20 residues or more within a protein sequence was taken as indicative of a potential TM α-helix [12]. The methodology has successfully predicted the occurrence of single membrane-spanning α-helices such as that found in receptor tyrosine kinases [71] and glycophorin A [72]. Hydropathy plot analysis also identified bundles of TM α-helices in G-protein-coupled receptors (GPCRs) [73–75] such as bacteriorhodopsin [72]. However, primarily due to the complex topology of their membrane-spanning segments, the methodology was less successful in predicting the TM helices of other GPCRs [76] such as the ClC chloride channel [77] and the KvAP voltage-gated potassium channel [78]. The modeling of GPCRs is now routinely performed using more sophisticated prediction techniques [79–84].

Several authors have endeavored to detect amphiphilic α-helical structures from protein primary structures using variants of hydropathy plot analysis. Jähnig [85] attempted to detect the periodicity of hydrophobic residues associated with such structures by introducing a weighting factor to average hydropathy indices while Lim [86] employed the autocorrelation function, $\Sigma_k H_k H_{k+p}$, where H_k and H_{k+p} are the hydrophobicity values of the kth and $(k+p)$th residues, respectively, $p = 1, \ldots, n-1, k = 1, \ldots, n-p$, for a protein sequence of n residues. This function searches a protein's primary sequence for sequential periodicity in hydrophobicity, p, by seeking values of p for which $\Sigma_k H_k H_{k+p}$ is high. More recently, Harris et al. [58] attempted to adapt hydropathy plot analysis to identify the secondary amphiphilicity associated with tilted α-helices (Chapter 5) using a seven-residue window, which previous studies have shown to give optimal resolution [87, 88]. In what appears to be the first report of AMPs utilizing tilted α-helices to facilitate their antimicrobial action [64], hydropathy plot analysis identified a hydrophobicity gradient within the helical structure of aurein 1.2. This hydrophobicity gradient is illustrated in Figure 4.5 and can be seen to run C → N, coinciding with the localization of strongly hydrophobic residues at the N-terminus of the aurein 1.2 α-helix. Confirming these results, angled membrane penetration by aurein 1.2 has been demonstrated experimentally [89]; however, although this technique has

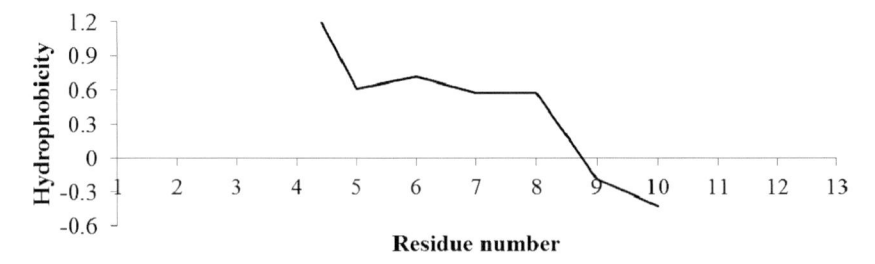

Figure 4.5 Hydropathy plot analyses of aurein 1.2 using the hydrophobicity scale of Kyte and Doolittle [70] and a seven-residue window. A progressive increase in hydrophobicity can be seen along the primary structure of aurein 1.2.

successfully identified the hydrophobicity gradients of some AMPs [34], its potential for the identification of tilted α-helices was found to be limited [58]. Moreover, methodologies based on hydropathy plot analysis do not take into account the nature of the amino acid side-chains or the longitudinal arrangement, which led researchers to develop more rigorous mathematical approaches for the detection of amphiphilic α-helical structure in proteins.

4.4.2
Techniques Based on Fourier Transforms

Essentially, these techniques provide a numerical measure of amphiphilicity by using discrete Fourier transforms to correlate a sequence of hydrophobicities within a protein's primary structure with a test sequence of known periodicity, typically a harmonic sequence, or sinusoidal function [90, 91]. For a given sequence $(H_1, H_2, \ldots, H_n)$, the hydrophobicity values corresponding to residues along a protein sequence of length n, the hydrophobicity of residue H_k is approximated as:

$$H_k = \{A(\theta)\cos(k\theta) + B(\theta)\sin(k\theta)\}, \tag{4.1}$$

where θ is the angular frequency, and $A(\theta)$ and $B(\theta)$ are defined by:

$$A(\theta) = \sum_{k=1}^{n} H_k \cos(k\theta)$$

$$B(\theta) = \sum_{k=1}^{n} H_k \sin(k\theta).$$

The Fourier power spectrum, $P(\theta)$ is therefore defined by:

$$P(\theta) = [A(\theta)]^2 + [B(\theta)]^2,$$

when:

$$P(\theta) = \left[\sum_{k=1}^{n} H_k \cos(k\theta)\right]^2 + \left[\sum_{k=1}^{n} H_k \sin(k\theta)\right]^2. \tag{4.2}$$

In order to detect the periodicity in hydrophobicity, $P(\theta)$, it is examined for the value $\theta = \hat{\theta}$ that maximizes $P(\theta)$. The value of $\hat{\theta}$ is usually interpreted as the characteristic frequency of the series $(H_1, H_2, \ldots, H_n)$, hence the following equation gives the approximation that has the highest non-normalized correlation with $\dot{H_k}$:

$$\hat{H}_k = A(\hat{\theta})\cos(k\hat{\theta}) + B(\hat{\theta})\sin(k\hat{\theta}).$$

If a protein molecule shows a strong regular pattern of hydrophobicity, then the corresponding power spectrum tends to show high peaks at the fundamental period allowing peptide sequences to be visualized by its harmonics, superimposed on a background of random noise. McLachlan and Stewart [92] were among the first to identify, from a background of random noise, a biologically significant 14th-order period for hydrophobic and non-polar amino acid residues in the protein tropomyosin. However, Cornette et al. [91] later developed the Amphipathic Index (AI) based on these methodologies.

4.4.3
Amphipathic Index

Cornett et al. [91] used Fourier transform to derive a method based on a least-squares fit of a harmonic sequence to a sequence of hydrophobicity values, which led to the definition of the AI:

$$AI(\theta) = \frac{\frac{1}{25°}\int_{85°}^{110°} P(\theta)d\theta}{\frac{1}{180°}\int_{0°}^{180°} P(\theta)d\theta},$$

where $P(\theta)$ is defined by Eq. (4.2) for a window size of length n.

The AI is maximized for the hydrophobicity scale chosen to compute the power spectrum and the higher the value of the parameter, the more likely the α-helix is to be amphiphilic [91].

Using the AI and the hydrophobicity scale of Kyte and Doolittle [70], Cornette et al. [91] predicted the occurrence of two amphiphilic α-helices within the structure of citrate synthase. These latter authors [91] then developed an algorithm, AMPHI, based on this statistical version of the AI and successfully predicted 18 out of 23 known antigenic sites of immunodominant helper T-cells to be amphiphilic α-helices. Furthermore, Cornette et al. [91] made a comparison of the two forms of the AI and concluded that when analyzing sequences from a large collection of α-helices greater accuracy for the prediction was seen when they used either the statistically based or Fourier transform-based versions of the AI to detect periodicity in hydrophobicity. However, for smaller single sequences (less than 25 amino acid residues) the statistically based approach was more accurate. Nonetheless, despite the utility of the AI, of all the analytical approaches that are based on Fourier series analysis, the most commonly used to detect amphiphilic α-helices is the hydrophobic moment, which has also formed the basis of a number of related methodologies.

4.4.4
Hydrophobic Moment Analysis

Using Eq. (4.2), Eisenberg et al. [93, 94] derived Eq. (4.3), which defines the mean hydrophobic moment, $<\mu_H>$. Essentially, this parameter quantifies amphiphilicity by treating the hydrophobicity of successive amino acids in α-helical sequences as vectors. These vectors are summed in two dimensions, assuming an amino acid side-chain periodicity of 100°, equivalent to three turns of an α-helix, and the resultant provides the mean hydrophobic moment, <μ_H>, which essentially provides a measure of α-helix amphiphilicity. For a given sequence (L) the mean hydrophobic moment, <μ_H>, is defined as:

$$<\mu_H> = \frac{1}{L}\left\{\left(\sum_{i=1}^{L} H_i \cos(i100°)\right)^2 + \left(\sum_{i=1}^{L} H_i \sin(i100°)\right)^2\right\}^{\frac{1}{2}}, \quad (4.3)$$

where L represents the window size and H_i is the hydrophobicity of the ith residue according to the hydrophobicity scale used, although the most commonly used scale is the Eisenberg consensus scale [93].

To determine which regions of the sequence may form amphiphilic α-helices, a sliding window of length w is used to calculate <μ_H> for each window progressing along the amino acid sequence. For a given sequence (H_1, H_2, ..., H_n), the $\mu_{H(s)}(100°)$ versus H_s defines the amphipathic profile. Visual inspection of the plot enables identification of windows with maximal values of $\mu(100°)$ and these segments are taken as the most likely candidates to form amphiphilic α-helices. Hydrophobic moment analysis has a number of advantages such as the ability to compare sequences of different lengths and the utility of this technique is beyond dispute. Nonetheless, a number of problems with the methodology have been identified in relation to angular frequency, hydrophobicity scale, and window size [59]. In particular, Eisenberg [93–96] assumes the amphiphilic helix would possess a periodicity angle of 100°, but in the case of Cornette et al. [91] the angle was shown to lie closer to 97.5°, whereas for TM α-helices the angle is thought to be closer to 102° [97]. In response, hydrophobic moment profile analysis was extended to include a profile of all values of θ where 180° ≤ θ ≤ 0°. This methodology enables the detection of other periodic secondary structures as well as identification of amphiphilic α-helices [96]. In later work, these latter authors extended the use of <μ_H> analysis and presented hydrophobic moment plot methodology which uses <μ_H> to predict biological function [96]. Essentially, these latter authors computed <μ_H> for each sequence in a dataset of α-helix forming segments, together with their mean hydrophobicity, <H>, which measures the segment's affinity for the membrane interior, and then used these values to construct the hydrophobic moment plot diagram. For a sequence under investigation, <μ_H> and <H> are plotted, and based on the location of the resulting data point, the α-helix is classified as either a surface active α-helix, which is strongly amphiphilic and typically assembles at a membrane interface; a transmembrane α-helix, which is predominantly hydrophobic and possess low amphiphilicity; or a globular α-helix as found in soluble proteins, which possess properties intermediate between those of the former two α-helical classes [96].

4.4.4.1 Choice of Hydrophobicity Scales

Hydrophobicity scales provide a measure of the relative hydrophobicities of amino acid residues and currently over 100 such scales have been presented, which show large variations in the relative values of these hydrophobicities, both within and between scales [98–100]. As by definition the hydrophobic moment is a function of residue hydrophobicity (Eq. 4.3), clearly the choice of hydrophobicity scale used in analyses is of fundamental importance in that it can affect interpretation of the magnitude of hydrophobicity and thereby the significance of the hydrophobic moment [11]. As an example, the definition of the hydrophobic moment shown in Eq. (4.3) assumes that apolar residues will be assigned positive values and polar residues will be assigned negative ones, as is the case with the consensus scale of Eisenberg et al. [94], which is that most commonly used in hydrophobic moment analysis [11]. However, a number of hydrophobicity scales are largely dominated by positive values [91, 98–100]. The use of these scales could lead to residues with large hydrophobicities dominating and largely determining the hydrophobic moment at $\theta = 0°$ and at all subsequent periodic boundaries, with the result that important periodicities are masked [11, 91]. A number of procedures aimed at solving this problem have been presented, but the most common is to apply a correction factor to the hydrophobicity of each residue in a protein sequence to generate an effective hydrophobicity, which then can be used to produce a corresponding effective $<\mu_H>$. This strategy permits the detection of periodicity in residue hydrophobicity around the mean hydrophobicity of a protein segment with the subsequent application of $<\mu_H>$ profile analysis often being used to determine likely amphiphilic α-helical regions [12]. Another problem associated with the dependence of the $<\mu_H>$ on hydrophobicity scales arises in selecting a threshold to decide whether an α-helical segment possesses amphiphilicity. A number of authors have approached this problem by deriving cut-off formula for specific scales to detect particular classes of amphiphilic α-helices [96, 101] such as surface seeking α-helices, which were identified in hydrophobic moment plot methodology by the expression: $\mu(100°)/11 \geq 0.6 + 0.382<H>$, where $<H>$ is the mean hydrophobicity of the putative α-helical segment [96].

4.4.4.2 Effect of Sequence Length on Amphiphilicity

A general problem with most Fourier transform-based measures of amphiphilicity is that they are not comparable for sequences of different lengths. For example, short sequences are more likely to have high hydrophobic moments than longer sequences; hence, a number of statistical approaches such as the least-squares model [91] have been developed to overcome the problem of length variation. To minimize these problems, Fourier transform-based analyses generally utilize fixed-length windows, between 7 and 20 amino acid residues long. More recently, Daman et al. [97] have shown that shorter window sizes give the best discrimination of TM α-helices and varying the window size can lead to differing identification of amphiphilic α-helices. These studies emphasized that optimization of window length is generally crucial to the classification of amphiphilic α-helices and that the values assumed by hydrophobic moment plot analysis for both this parameter and angular periodicity are not optimal for such classification. Furthermore, Daman et al. [97] found that for an 11-residue window, an angular frequency of 102° maximized $<\mu_H>$ for TM sequences. Yet, despite these difficulties, a number of variants of hydrophobic moment-based analysis have been introduced to identify and classify amphiphilic α-helices.

4.4.5
Classification of Amphiphilic α-Helices Using the Approach of Segrest

Segrest et al. [102] assembled a dataset of known and putative α-helix-forming sequences and analyzed these sequences according to conventional hydrophobic moment analysis [94] before modeling them as helical wheel projections [60]. These helical wheel projections enabled Segrest et al. [102] to group amphiphilic α-helices into seven distinct classes (A, C, H, L, G, K, and M) according to their structural characteristics and biological function. Three of these classes, C, G, and K, are involved in protein–protein interactions: the class C α-helices of some coiled-coil protein arrangements, the class G α-helices of globular proteins, and the class K α-helices found in the calmodulin-binding domains of various protein kinases. The remaining major classes of amphiphilic α-helix defined by Segrest et al. [102] are primarily concerned with protein–lipid interactions.

In general, class A, L, and H α-helices orient parallel to the membrane, and are included within the surface active α-helices defined by Eisenberg et al. [96]. The class A α-helices are characterized by high average hydrophobic moment, $<\mu_H> = 0.77$, featuring a unique molecular architecture of positively charged residues at the polar/non-polar interface with negatively charged residues at the center of the 180° polar face (Figure 4.6). Since the original work by Segrest et al. [102], evidence has been presented to suggest that the relative positions of aromatic residue position in the non-polar face of class A α-helices amphipathic may play a role in determining their biological activity [103]. In general, these α-helices appear to stabilize membranes [12] and more recent studies have suggested that they stabilize the edge of discoidal bilayer structures formed in the lipid transport of living organisms [104–107].

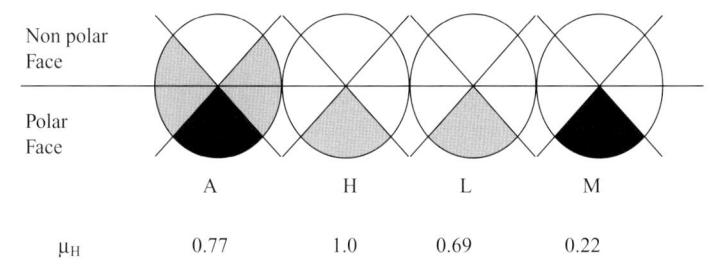

Figure 4.6 Segrest classification of membrane active α-helices [102]. The white areas represent the hydrophobic regions, black areas represent the negatively charged regions, and the gray areas represent positively charged regions.

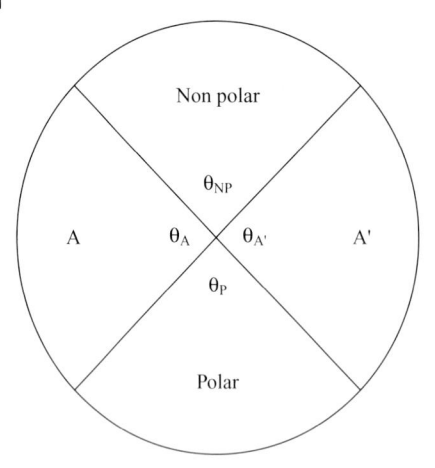

Figure 4.7 Tomich [118] method of identification of lytic channel-forming peptides. The helical wheel is divided into four quadrants: polar, non-polar, and the two inter-helical contact regions A and A′.

In contrast, class L (Figure 4.6), which possess narrow polar faces, are found in a number of lytic peptides [102] due to their role in destabilization of the membrane [108–110] and possess an average hydrophobic moment, $<\mu_H> = 0.69$ of class C, which is comparable to that of class A. The class H (Figure 4.6) are also characterized by narrow polar faces but possess a high average hydrophobic moment, $<\mu_H> = 1.0$. This latter class requires the amphiphilic environment of the membrane surface to catalyze their formation [111, 112] and their high $<\mu_H>$ would appear to be necessary to facilitate the location of hormone receptors [113]. Although the class M (Figure 4.6) are considered as a subclass of TM proteins, they are not generally regarded as amphiphilic helices. These α-helices possess low hydrophobic moments, $<\mu_H> = 0.22$, and low polar face charge densities but high non-polar face hydrophobicity.

The classification scheme provided by Segrest et al. [102] is a useful guide when attempting to assign biological function to potential amphiphilic α-helices, but the taxonomy suffers inherent problems derived from the hydrophobic moment analysis upon which it is founded [12]. Indeed, Segrest et al. [114] observed that the hydrophobic moment algorithm used in their original analysis of amphiphilic α-helices [102] was not a good measure of lipid-associating amphiphilic α-helices. A major criticism of this taxonomy is that many lytic peptides show properties that differ from the class L definition. For example, Harris and Phoenix [115] showed that in contrast to the L class helices, the PBP5 and PBP6 C-terminal α-helical peptides possess wide hydrophilic faces that subtend angles of 120° or greater and yet were highly hemolytic. Furthermore, melittin, which is one of the most potent hemolytic AMPs known [116], shows structural characteristics associated with the class M α-helices. In response Tytler et al. [117] proposed

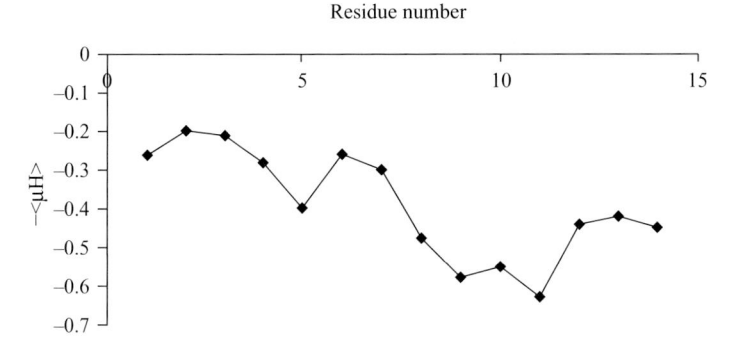

Figure 4.8 Amphiphilic profile analysis. Profile analysis of maximin H5 shows the gradient of the amphiphilicity profile ($-\langle\mu_H\rangle$), which decreases in the N → C direction.

subclassification of the class L amphiphilic α-helices based on the differing abilities of these α-helices to lyse eukaryotic membranes while Tomich [118] proposed an alternative approach to the classification of lytic peptides since many lytic peptides are also channel-forming peptides. For example, paradaxin, an AMP secreted by the Red Sea moses sole, forms ion-conducting pores in lipid bilayers [119]. In response, Tomich [118] proposed that the α-helical faces of lytic peptides should be considered as four quadrants of the helix (Figure 4.7) as follows: a hydrophilic sector and a hydrophobic sector, separated by two sectors, A and A′, which are concerned with α-helix–α-helix interactions. These A and A′ sections are parallel bundles with identical mean hydropathy values. The distribution of amino acids residues responsible for the α-helix–α-helix interactions led to the suggestion of a packing motif in the assembled structures and the prediction that the residues forming the α-helices of some lytic peptides may not only serve the function of membrane destabilization, but may be also involved in ordered self-assembly and channel formation. This packing motif supports other work, such as a GXXXG packing motif within the single (TM) domain of glycophorin A [120]. Similar motifs have also been putatively identified in a number of other AMPs [121].

4.4.6
Amphiphilicity Profiling Analysis of Tilted α-Helices

Although conventional hydrophobic moment analysis can detect the potential of sequences derived from tilted α-helices for α-helix formation, the methodology has no capacity to recognize their hydrophobicity gradients. In response, a number of authors have modified hydrophobic moment analysis for the identification and characterization of tilted α-helices. One of the first attempts to identify tilted α-helices by adapting hydrophobic moment analysis led to definition of the glycine moment, $\langle\mu_G\rangle$, which characterizes these α-helices by their asymmetric molecular shape. This parameter is formally analogous to the mean hydrophobic moment but is based on the relative sizes of amino acid residues in comparison to glycine rather than residue hydrophobicity. However, although successfully identifying some tilted α-helices, it was found that analyses based on $<\mu_G>$ were limited in their ability to predict these structures [122]. Similarly, as described earlier, attempts to detect hydrophobicity gradients based on hydropathy plot analysis met with limited success and amphiphilic profile analysis was developed in response [58]. This methodology is essentially hydropathy plot analysis with $<H>$ replaced by $-<\mu_H>$ and was successfully used to identify a number of tilted α-helices in (TM) proteins [58]. Use of the technique is illustrated in Figure 4.8, which shows that a strong N → C hydrophobicity gradient is possessed by maximin H5, an AMP found in the skin and brain of the Chinese red belly toad, *Bombina maxima* [123, 124].

4.4.7
Extended Hydrophobic Moment Plot Analysis of Tilted α-Helices

It is generally accepted that conventional hydrophobic moment plot methodology has no capacity to classify tilted α-helices. In response, an extended form of hydrophobic moment plot methodology was developed based on a statistical analysis of a series of sequences known to form tilted α-helices [122]. These sequences were characterized by $<\mu_H>$ and $<H>$, and regression analysis showed that there was a strong linear correlation between these parameters, which could be described by the regression line: $<\mu_H> = 0.508 - 0.422<H>$. The ordinates of a 99% prediction interval around this regression line were determined and used to define an area on the hydrophobic moment plot diagram (Figure 4.9), which although unable to definitively identify tilted α-helices, was able to exclude candidate sequences unlikely to form such α-helices. Use of the technique is illustrated in Figure 4.9, which shows the results of a theoretical study in which Dennison *et al.* [34] analyzed a database of around 160 AMPs and predicted that the use of tilted peptide structure by these peptides may be a commonly used mechanism in their membrane interactions [64]. As a specific example, maximin H5 has been characterized by its hydrophobic moment plot parameters ($<\mu_H> = 0.29$; $<H> = 0.44$) and plotted on the extended hydrophobic moment plot diagram. It can be seen that the data point representing this peptide is within the blue area of the diagram, predicting that maximin H5 is a candidate to form a tilted structure. This prediction is consistent with the peptide's possession of a hydrophobicity gradient and suggests that it may interact with membranes via tilted insertion.

4.4.8
Amphiphilicity Quantified Using the Approach of Keller

HeliQuest, which is an online program based on hydrophobic moment analysis, detects amphiphilic α-helices with the potential for lipid binding [125] and has

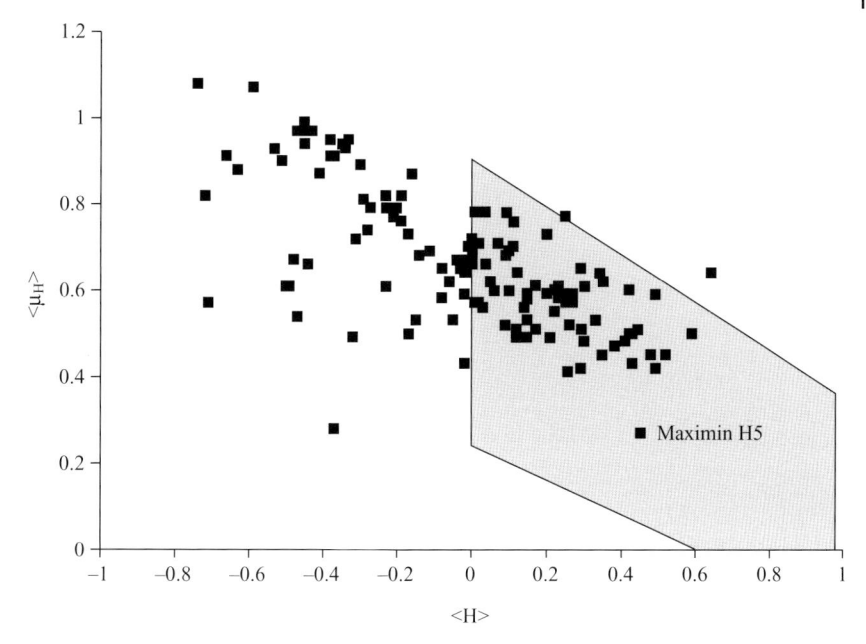

Figure 4.9 Extended hydrophobic moment plot analyses of AMPs. Maximin H5 lies in the shaded area indicating potential for oblique oriented α-helix formation.

recently been successfully used to identify such helical regions in a variety of translocation motor proteins [126]. Essentially, this program uses the hydrophobicity scale of Fauchere and Pliska [127] and a sliding 18-residue window, and, for each window in a protein sequence, computes the mean $<H>$, the mean $<\mu_H>$, and the net charge (z). For each window, the discrimination factor (D) is then calculated using Eq. (4.4) and windows with $D \geq 0.68$ are taken to be to be potential lipid-binding α-helices [125, 126]:

$$D = 0.944(<\mu_H>) + 0.33(z). \qquad (4.4)$$

In an effort to improve the efficacy of conventional hydrophobic moment plot methodology in the identification and classification of membrane-interactive α-helices, Keller [101] reanalyzed the dataset used by Eisenberg et al. [96] to construct the plot diagram using values of $<H>$ and $<\mu_H>$ generated by HeliQuest. The resulting HeliQuest-derived plot diagram was found to show an efficacy in identifying and classifying lipid interactive α-helices that was comparable to that of the conventional plot diagram. However, when the HeliQuest-derived plot diagram was used in conjunction with the discrimination factor defined in Eq. (4.4), it was found that potential lipid-binding α-helices could be identified, which were missed by the conventional plot diagram. It was concluded that this combined approach provides a powerful, user-friendly tool for identifying putative lipid-binding regions in proteins [101].

4.4.9
Amphiphilicity Quantified Using the Approach of Brasseur

In an effort to refine conventional hydrophobic moment analysis, Brasseur [128] introduced a new measure of amphiphilicity, which extended the work of previous authors [129, 130], and assumed that the hydrophobic interaction between a given point M and an atom i decreases exponentially with distance [128]. This measure of amphiphilicity was defined as the molecular hydrophobic potential (MHP):

$$MHP = \Sigma E^{tr}_i \exp(r_i - d_i), \tag{4.5}$$

where E^{tr}_i represents the energy of transfer from a hydrophobic to a hydrophilic phase for an atom i, r_i is the radius of the atom i, and d_i is the distance between atom i and the point M, measured in Å.

Based on the molecular hydrophobic potential, Brasseur [128] developed a methodology to generate a three-dimensional representation of the hydrophobicity profile of a molecule or protein segment. Using this methodology, the hydrophobicity profile of an α-helical segment can be represented graphically as lines of hydrophobic isopotential, which envelope the α-helix and characterize its amphiphilicity by defining the α-helical hydrophobic and hydrophilic faces. A quantitative measure of this amphiphilicity can then be obtained by projecting the limits of the contours forming the hydrophobic helical face perpendicular to the long axis of the α-helix and calculating the angle, ω, subtended by these projections [128]. Using ω, Brasseur [128] attempted to classify amphiphilic α-helices and according to this classification scheme, if $\omega > 180°$, the α-helix is considered to possess the potential for self-association in the membrane and the formation of a transmembrane pore. However, if $\omega = 180°$, then the α-helix is predicted likely to orient with the helical long axis parallel to the membrane surface as observed with many surface active α-helices and AMPs (Chapters 2, 3 and 5). Finally, if $\omega < 180°$, then the α-helix is predicted to be able to stabilize the edge of discoidal bilayer structures, as described above for class A α-helices found in apolipoproteins [104–107].

This analytical approach has the advantage that it can provide a three-dimensional representation of amphiphilicity but nonetheless, to quantify this amphiphilicity, a three-dimensional representation is reduced to a two-dimensional representation with the resulting loss of spatial information. Moreover, the methodology is, again, based on conventional hydrophobic moment analysis and difficulties can arise from problems associated with the latter analysis, such as the fact that the segment of interest is identified using a fixed window size [12]. Despite these problems, the molecular hydrophobic potential, has been used to investigate the structure–function relationships of proteins in a variety of contexts [131]. In earlier studies, the molecular hydrophobic potential was used to predict antigenic sites in the cystic fibrosis TM conductance regulator protein [132] and elucidate the molecular mechanism of lysosomal phospholipase inhibition induced by aminoglycosides, thereby providing insight into the formation of drug–lipid complexes [133]. Also in the 1990s, the molecular hydrophobic potential was used as a basis for the prediction of tilted α-helices (Chapter 5), which were identified in a variety of molecules, ranging from viral peptides to TM proteins [134]. More recently, it was shown that molecular hydrophobic potential analysis of AMPs with a helix–hinge–helix structure could help discriminate between those likely to interact with membranes via carpet-type mechanisms and those likely to utilize pore-forming mechanisms. Essentially, AMPs likely to use the former mechanisms possessed a strongly amphiphilic α-helix at the N-terminus only, whereas those likely to use the latter mechanism possessed strongly amphiphilic α-helices at both their N- and C-termini [135]. In an extension of these latter studies, the molecular hydrophobic potential was used to investigate the hemolytic potential of AMPs with helix–hinge–helix structure and this potential appeared to correlate with the hydrophobic properties of their N-terminal α-helices [136]. To quantify this relationship, a scoring value, F, was defined according to according to:

$$F = (<MHP> \times A_{phob})^{1/2}, \quad (4.6)$$

where A_{phob} is the estimated value of the N-terminal hydrophobic areas ($MHP > 0.09$).

These studies found that AMPs with strong hemolytic activity had high F values (greater than 300), whereas those that were non-hemolytic had lower F values. It was suggested that by optimizing their F values, therapeutically useful AMPs could be designed, which featured high antimicrobial activity but with minimal hemolytic and cytotoxic action [136]. Most recently, the molecular hydrophobic potential has been used to describe hydrophobic regions in protein pockets and guide ligand hydrophobic moieties toward favorable areas when it was found that this methodology gave more reliable results than other molecular docking methodologies that are widely used in drug design [137].

4.5
Discussion

This chapter has shown that over the last 40 years a variety of graphical, sequence-based techniques have been developed for the visualization, prediction, and classification of amphiphilic α-helices. Earlier techniques such as the helical wheel of Schiffer and Edmundson [60], while useful, were qualitative and of limited use. In response, quantitative techniques were introduced of which hydrophobic moment analysis [93, 94] and its variants, as described here, are by far the most commonly used. Essentially, this methodology uses an approach based on Fourier transforms to detect periodic hydrophobicity in the sequences of these peptides and then quantifies amphiphilicity as a function of this hydrophobicity. The underlying logic of this approach seems robust; nonetheless, problems can clearly arise from the fact that the use of different hydrophobicity scales can lead to significant variations in the determined magnitude of the $<\mu_H>$. A major assumption made by the methodology is that all amphiphilic α-helices can be represented by a fixed 100° angular frequency and it is known that the angular frequencies of naturally occurring α-helices show significant deviations from this value. Based on this fixed angle, hydrophobic moment analysis would appear only able to provide a measure of the amphiphilicity of an averaged representation of α-helical structure. Furthermore, the analysis assumes a fixed 11-residue window to identify all amphiphilic α-helix-forming segments, yet it has been shown that the optimal window size for the identification of different classes of amphiphilic α-helices varies from this value. Given the fact that $<\mu_H>$ depends upon on window size, this could clearly limit the ability of the parameter to characterize amphiphilic α-helices [12]. However, despite these difficulties, it is beyond dispute that methodologies based on hydrophobic moment analysis have made many successful predictions with regard to the membrane-interactive potential of AMPs, ranging from the identification of tilted peptides to investigating ligand–receptor interactions. Indeed, the utility of these methodologies is reinforced by the fact that the $<\mu_H>$ is now taken as a structural property of AMPs and is included within databases of these peptides [138].

It is generally accepted that knowledge-based methods are the most successful for predicting the structures of AMPs and other proteins, and analytical techniques based on the $<\mu_H>$ may be regarded as first-generation examples of these methods. However, over the last few decades, the use of methodologies based on the $<\mu_H>$ have been increasingly superseded by the introduction of more varied and sophisticated knowledge-based techniques for the prediction of protein structures [139–143]. A number of these newer techniques have been applied to the prediction and design of AMPs [144] and a major example is provided by quantitative structure–activity relationship (QSAR) models, which were introduced in the late 1980s and relate measurements on a set of predictor variables, or descriptors, to the behavior of a response variable [145]. A number of novel AMPs have recently been designed based on predictions from QSAR models, which established a relationship between descriptors, such as the contact energy between neighboring residues, along with response variables, such as the antimicrobial activity of these peptides [146, 147].

Since the 1990s, QSAR modeling has been used in conjunction with machine learning methods that can automatically extract knowledge from databases, forming major tools that are widely used in protein structure prediction, with the most commonly used examples including hidden Markov models (HMMs) and artificial neural networks (ANNs) [140, 144]. HMMs are statistical Markov models in which the system being modeled is assumed to be a Markov process with unobserved states, and provides a probabilistic method to learn temporal and sequential patterns based on observations, which can be used for amino acid sequence analysis and prediction [144, 148]. The use of HMMs as a basis for the prediction and characterization of AMPs has been well established over the last three decades [149–151] and were recently used to identify individual classes of these peptides with high accuracy, which led to the development of AMPer, a database and automated tool for the discovery of AMPs [152]. In another recent application, HMMs were used to scan the bovine genome and in addition to identifying a large number of potential AMPs, showed that cattle lack α-defensins, an important class of AMPs found in many other mammals [153]. ANNs are non-linear statistical data-modeling tools that are loosely based on nervous systems, and are usually used to find patterns in sequence data and for modeling structure–activity relationships [144, 154]. In contrast to HMMs, ANNs were first used for modeling AMPs just over a decade ago and since then have been successfully used to design a number of these peptides with antimicrobial activity more potent than that of some conventional antibiotics [155, 156]. In another study, ANNs were used to identify novel AMPs that were highly effective against a broad array of multidrug-resistant pathogens such as methicillin-resistant *Staphylococcus aureus* (MRSA). Moreover, these peptides were more effective against these organisms than MX-226, which is a derivative of indolicidin and the most advanced clinical candidate from AMPs yet reported [157]. More recently, it was shown that the use of ANNs in conjunction with descriptors derived from sequence information alone were not only able to discriminate between AMPs and peptides devoid of antimicrobial activity with high accuracy, but also to assess the antimicrobial potency of AMPs. A pioneering feature of this predictive technique was that by including concepts such as peptide aggregation within the descriptors used, a likely mechanism of membrane interaction could be predicted for AMPs [158]. It would therefore appear that the future development of such models will continue to have a significant impact on understanding AMP structure–function relationships as well as having further embedded in the design of peptide therapeutics.

References

1 Mason, J.M. (2010) Design and development of peptides and peptide mimetics as antagonists for therapeutic intervention. *Future Medicinal Chemistry*, **2**, 1813–1822.

2 Vlieghe, P., Lisowski, V., Martinez, J., and Khrestchatisky, M. (2010) Synthetic therapeutic peptides: science and market. *Drug Discovery Today*, **15**, 40–56.

3 Croft, N.P. and Purcell, A.W. (2011) Peptidomimetics: modifying peptides in the pursuit of better vaccines. *Expert Review of Vaccines*, **10**, 211–226.

4 Giuliani, A. and Rinaldi, A.C. (2011) Beyond natural antimicrobial peptides: multimeric peptides and other peptidomimetic approaches. *Cellular and Molecular Life Sciences*, **68**, 2255–2266.

5 Lico, C., Santi, L., Twyman, R.M., Pezzotti, M., and Avesani, L. (2012) The use of plants for the production of therapeutic human peptides. *Plant Cell Reports*, **31**, 439–451.

6 Huang, Y., Huang, J., and Chen, Y. (2010) Alpha-helical cationic antimicrobial peptides: relationships of structure and function. *Protein & Cell*, **1**, 143–152.

7 Strömstedt, A.A., Ringstad, L., Schmidtchen, A., and Malmsten, M. (2010) Interaction between amphiphilic peptides and phospholipid membranes. *Current Opinion in Colloid & Interface Science*, **15**, 467–478.

8 Nguyen, L.T., Haney, E.F., and Vogel, H.J. (2011) The expanding scope of antimicrobial peptide structures and their modes of action. *Trends in Biotechnology*, **29**, 464–472.

9 Teixeira, V., Feio, M.J., and Bastos, M. (2012) Role of lipids in the interaction of antimicrobial peptides with membranes. *Progress in Lipid Research*, **51**, 149–177.

10 Wang, G.S., Li, X., and Wang, Z. (2009) APD2: the updated antimicrobial peptide database and its application in peptide design. *Nucleic Acids Research*, **37**, D933–D937.

11 Phoenix, D.A. and Harris, F. (2002) The hydrophobic moment and its use in the classification of amphiphilic structures [review]. *Molecular Membrane Biology*, **19**, 1–10.

12 Phoenix, D.A., Harris, F., Daman, O.A., and Wallace, J. (2002) The prediction of amphiphilic alpha-helices. *Current Protein & Peptide Science*, **3**, 201–221.

13 Ulmschneider, M.B., Sansom, M.S.P., and Di Nola, A. (2005) Properties of integral membrane protein structures: derivation of an implicit membrane potential. *Proteins*, **59**, 252–265.

14 von Heijne, G. (2007) Formation of transmembrane helices *in vivo* – is hydrophobicity all that matters? *The Journal of General Physiology*, **129**, 353–356.

15 Paulet, D., Claustres, M., and Beroud, C. (2011) Hydrophobic pulses predict transmembrane helix irregularities and channel transmembrane units. *BMC Bioinformatics*, **12**, 135.

16 Staubitz, P., Peschel, A., Nieuwenhuizen, W.F., Otto, M., Gotz, F., Jung, G., and Jack, R.W. (2001) Structure–function relationships in the tryptophan-rich, antimicrobial peptide indolicidin. *Journal of Peptide Science*, **7**, 552–564.

17 Yang, S.-T., Yub Shin, S., Kim, Y.-C., Kim, Y., Hahm, K.-S., and Kim, J.I. (2002) Conformation-dependent antibiotic activity of tritrpticin, a cathelicidin-derived antimicrobial peptide. *Biochemical and Biophysical Research Communications*, **296**, 1044–1050.

18 Melo, M.N., Ferre, R., and Castanho, M.A.R.B. (2009) Antimicrobial peptides: linking partition, activity and high membrane-bound concentrations. *Nature Reviews Microbiology*, **7**, 245–250.

19 Tamm, L.K., Hong, H., and Liang, B. (2004) Folding and assembly of beta-barrel membrane proteins. *Biochimica et Biophysica Acta*, **1666**, 250–263.

20 Zeth, K. and Thein, M. (2010) Porins in prokaryotes and eukaryotes: common themes and variations. *Biochemical Journal*, **431**, 13–22.

21 Briolat, J., Wu, S.D., Mahata, S.K., Gonthier, B., Bagnard, D., Chasserot-Golaz, S., Helle, K.B., Aunis, D., and Metz-Boutigue, M.H. (2005) New antimicrobial activity for the catecholamine release-inhibitory peptide from chromogranin A. *Cellular and Molecular Life Sciences*, **62**, 377–385.

22 Jean-Francois, F., Desbat, B., and Dufourc, E.J. (2009) Selectivity of cateslytin for fungi: the role of acidic lipid-ergosterol membrane fluidity in antimicrobial action. *FASEB Journal*, **23**, 3692–3701.

23 Varkey, J., Singh, S., and Nagaraj, R. (2006) Antibacterial activity of linear peptides spanning the carboxy-terminal beta-sheet domain of arthropod defensins. *Peptides*, **27**, 2614–2623.

24 Hong, J., Oren, Z., and Shai, Y. (1999) Structure and organization of hemolytic and nonhemolytic diastereomers of antimicrobial peptides in membranes. *Biochemistry*, **38**, 16963–16973.

25 Oren, Z., Hong, J., and Shai, Y. (1999) A comparative study on the structure and function of a cytolytic alpha-helical peptide and its antimicrobial beta-sheet diastereomer. *European Journal of Biochemistry*, **259**, 360–369.

26 Castano, S., Desbat, B., and Dufourcq, J. (2000) Ideally amphipathic beta-sheeted peptides at interfaces: structure, orientation, affinities for lipids and hemolytic activity of (KL)(m)K peptides. *Biochimica et Biophysica Acta – Biomembranes*, **1463**, 65–80.

27 Blazyk, J., Wiegand, R., Klein, J., Hammer, J., Epand, R.M., Epand, R.F.,

Maloy, W.L., and Kari, U.P. (2001) A novel linear amphipathic beta-sheet cationic antimicrobial peptide with enhanced selectivity for bacterial lipids. *The Journal of Biological Chemistry*, **276**, 27899–27906.

28 Jin, Y., Hammer, J., Pate, M., Zhang, Y., Zhu, F., Zmuda, E., and Blazyk, J. (2005) Antimicrobial activities and structures of two linear cationic peptide families with various amphipathic beta-sheet and alpha-helical potentials. *Antimicrobial Agents and Chemotherapy*, **49**, 4957–4964.

29 Perutz, M.F., Kendrew, J.C., and Watson, H.C. (1965) Structure and function of haemoglobin: II. Some relations between polypeptide chain configuration and amino acid sequence. *Journal of Molecular Biology*, **13**, 669–678.

30 Gangabadage, C.S., Zdunek, J., Tessari, M., Nilsson, S., Olivecrona, G., and Wijmenga, S.S. (2008) Structure and dynamics of human apolipoprotein CIII. *The Journal of Biological Chemistry*, **283**, 17416–17427.

31 Gursky, O. (2005) Apolipoprotein structure and dynamics. *Current Opinion in Lipidology*, **16**, 287–294.

32 Bolanos-Garcia, V.M. and Miguel, R.N. (2003) On the structure and function of apolipoproteins: more than a family of lipid-binding proteins. *Progress in Biophysics and Molecular Biology*, **83**, 47–68.

33 Bechinger, B. (2004) Structure and function of membrane-lytic peptides. *Critical Reviews in Plant Sciences*, **23**, 271–292.

34 Dennison, S.R., Wallace, J., Harris, F., and Phoenix, D.A. (2005) Amphiphilic alpha-helical antimicrobial peptides and their structure–function relationships. *Protein and Peptide Letters*, **12**, 31–39.

35 Dennison, S.R., Wallace, J., Harris, F., and Phoenix, D.A. (2005) Relationships between the physiochemical properties, microbial specificity and antimicrobial activity of alpha-helical antimicrobial peptides: a statistical investigation. *Current Topics in Peptide and Protein Research*, **7**, 53–62.

36 Davidson, W.S. and Silva, R.A. (2005) Apolipoprotein structural organization in high density lipoproteins: belts, bundles, hinges and hairpins. *Current Opinion in Lipidology*, **16**, 295–300.

37 Saito, H., Lund-Katz, S., and Phillips, M.C. (2004) Contributions of domain structure and lipid interaction to the functionality of exchangeable human apolipoproteins. *Progress in Lipid Research*, **43**, 350–380.

38 Phoenix, D.A. and Harris, F. (1998) Amphiphlic α-helices and lipid interactions, in *Protein Targeting and Translocation* (ed. D.A. Phoenix), Portland Press, London, pp. 19–36.

39 Orgel, J.P. (2006) Surface-active helices in transmembrane proteins. *Current Protein & Peptide Science*, **7**, 553–560.

40 Thomas, A. and Brasseur, R. (2006) Tilted peptides: the history. *Current Protein & Peptide Science*, **7**, 523–527.

41 Pranting, M., Loov, C., Burman, R., Goransson, U., and Andersson, D.I. (2010) The cyclotide cycloviolacin O2 from *Viola odorata* has potent bactericidal activity against Gram-negative bacteria. *The Journal of Antimicrobial Chemotherapy*, **65**, 1964–1971.

42 Gran, L., Sletten, K., and Skjeldal, L. (2008) Cyclic peptides from *Oldenlandia affinis* DC. Molecular and biological properties. *Chemistry & Biodiversity*, **5**, 2014–2022.

43 Pinto, M.F.S., Almeida, R.G., Porto, W.F., Fensterseifer, I.C.M., Lima, L.A., Dias, S.C., and Franco, O.L. (2011) Cyclotides: from gene structure to promiscuous multifunctionality. *Journal of Evidence-Based Complementary & Alternative Medicine*, **17**, 40–53.

44 Nair, S.S., Romanuka, J., Billeter, M., Skjeldal, L., Emmett, M.R., Nilsson, C.L., and Marshall, A.G. (2006) Structural characterization of an unusually stable cyclic peptide, kalata B2 from *Oldenlandia affinis*. *Biochimica et Biophysica Acta*, **1764**, 1568–1576.

45 Jennings, C.V., Rosengren, K.J., Daly, N.L., Plan, M., Stevens, J., Scanlon, M.J., Waine, C., Norman, D.G., Anderson,

M.A., and Craik, D.J. (2005) Isolation, solution structure, and insecticidal activity of kalata B2, a circular protein with a twist: do Mobius strips exist in nature? *Biochemistry*, **44**, 851–860.

46 Colgrave, M.L., Kotze, A.C., Huang, Y.H., O'Grady, J., Simonsen, S.M., and Craik, D.J. (2008) Cyclotides: natural, circular plant peptides that possess significant activity against gastrointestinal nematode parasites of sheep. *Biochemistry*, **47**, 5581–5589.

47 Plan, M.R., Saska, I., Cagauan, A.G., and Craik, D.J. (2008) Backbone cyclised peptides from plants show molluscicidal activity against the rice pest *Pomacea canaliculata* (golden apple snail). *Journal of Agricultural and Food Chemistry*, **56**, 5237–5241.

48 Raj, P.A. and Dentino, A.R. (2002) Current status of defensins and their role in innate and adaptive immunity. *FEMS Microbiology Letters*, **206**, 9–18.

49 Arnett, E. and Seveau, S. (2011) The multifaceted activities of mammalian defensins. *Current Pharmaceutical Design*, **17**, 4254–4269.

50 Pazgier, M., Hoover, D.M., Yang, D., Lu, W., and Lubkowski, J. (2006) Human beta-defensins. *Cellular and Molecular Life Sciences*, **63**, 1294–1313.

51 Hazlett, L. and Wu, M. (2011) Defensins in innate immunity. *Cell and Tissue Research*, **343**, 175–188.

52 Ganz, T. (2003) The role of antimicrobial peptides in innate immunity. *Integrative and Comparative Biology*, **43**, 300–304.

53 Yang, D., Biragyn, A., Hoover, D.M., Lubkowski, J., and Oppenheim, J.J. (2004) Multiple roles of antimicrobial defensins, cathelicidins, and eosinophil-derived neurotoxin in host defense. *Annual Review of Immunology*, **22**, 181–215.

54 Diamond, G. and Ryan, L.K. (2011) Beta-defensins: what are they REALLY doing in the oral cavity? *Oral Diseases*, **17**, 628–635.

55 Lai, Y. and Gallo, R.L. (2009) AMPed up immunity: how antimicrobial peptides have multiple roles in immune defense. *Trends in Immunology*, **30**, 131–141.

56 Dennison, S.R., Whittaker, M., Harris, F., and Phoenix, D.A. (2006) Anticancer alpha-helical peptides and structure–function relationships underpinning their interactions with tumour cell membranes. *Current Protein & Peptide Science*, **7**, 487–499.

57 Harris, F., Daman, A., Wallace, J., Dennison, S.R., and Phoenix, D.A. (2006) Oblique orientated alpha-helices and their prediction. *Current Protein & Peptide Science*, **7**, 529–537.

58 Harris, F., Dennison, S., and Phoenix, D.A. (2006) The prediction of hydrophobicity gradients within membrane interactive protein alpha-helices using a novel graphical technique. *Protein and Peptide Letters*, **13**, 595–600.

59 Phoenix, D.A. and Harris, F. (2003) Is use of the hydrophobic moment a sound basis for predicting the structure–function relationships of membrane interactive alpha-helices? *Current Protein & Peptide Science*, **4**, 357–366.

60 Schiffer, M. and Edmundson, A.B. (1967) Use of helical wheels to represent the structures of proteins and to identify segments with helical potential. *Biophysical Journal*, **7**, 121–135.

61 Jack, R.W., Tagg, J.R., and Ray, B. (1995) Bacteriocins of Gram-positive bacteria. *Microbiological Reviews*, **59**, 171–200.

62 Dennison, S.R., Harris, F., and Phoenix, D.A. (2003) Factors determining the efficacy of alpha-helical antimicrobial peptides. *Protein and Peptide Letters*, **10**, 497–502.

63 Giangaspero, A., Sandri, L., and Tossi, A. (2001) Amphipathic alpha helical antimicrobial peptides. *European Journal of Biochemistry*, **268**, 5589–5600.

64 Dennison, S.R., Harris, F., and Phoenix, D.A. (2005) Are oblique orientated alpha-helices used by antimicrobial peptides for membrane invasion? *Protein and Peptide Letters*, **12**, 27–29.

65 Dunhill, P. (1968) The use of helical net-diagrams to represent protein structures. *Biophysical Journal*, **8**, 865–875.

66 Jiang, Z., Vasil, A.I., Gera, L., Vasil, M.L., and Hodges, R.S. (2011) Rational design of α-helical antimicrobial peptides to target Gram-negative pathogens, *Acinetobacter baumannii* and *Pseudomonas aeruginosa*: utilization of charge, "specificity determinants," total hydrophobicity, hydrophobe type and location as design parameters to improve the therapeutic ratio. *Chemical Biology & Drug Design*, **77**, 225–240.

67 Chou, K.C., Zhang, C.T., and Maggiora, G.M. (1997) Disposition of amphiphilic helices in heteropolar environments. *Proteins*, **28**, 99–108.

68 Chou, K.-C., Lin, W.-Z., and Xiao, X. (2011) Wenxiang: a web-server for drawing wenxiang diagrams. *Natural Science*, **3**, 862–865.

69 Zhou, G.P. (2011) The disposition of the LZCC protein residues in wenxiang diagram provides new insights into the protein–protein interaction mechanism. *Journal of Theoretical Biology*, **284**, 142–148.

70 Kyte, J. and Doolittle, R.F. (1982) A simple method for displaying the hydropathic character of a protein. *Journal of Molecular Biology*, **157**, 105–132.

71 Vandergeer, P., Hunter, T., and Lindberg, R.A. (1994) Receptor protein-tyrosine kinases and their signal transduction pathways. *Annual Review of Cell Biology*, **10**, 251–337.

72 Alberts, B., Johnson, A., Lewis, J., Raff, M., Roberts, K., and Walter, P. (2010) Membrane proteins, in *Molecular Biology of the Cell* (eds B. Alberts, A. Johnson, J. Lewis, M. Raff, K. Roberts, and P. Walter), Taylor & Francis, New York, pp. 629–650.

73 Malbon, C.C. (2004) Frizzleds: new members of the superfamily of G-protein-coupled receptors. *Frontiers in Bioscience*, **9**, 1048–1058.

74 Strader, C.D., Candelor, M.R., Tota, M.R., Fong, T.M., and Underwood, D. (1995) Molecular biological approaches to drug design for G protein coupled receptors. *Journal of Cellular Biochemistry*, **19B**, 348.

75 Strader, C.D., Fong, T.M., Tota, M.R., Underwood, D., and Dixon, R.A.F. (1994) Structure and function of G-protein-coupled receptors. *Annual Review of Biochemistry*, **63**, 101–132.

76 White, S.H. (2003) Translocons, thermodynamics, and the folding of membrane proteins. *FEBS Letters*, **555**, 116–121.

77 Dutzler, R., Campbell, E.B., Cadene, M., Chait, B.T., and MacKinnon, R. (2002) X-ray structure of a ClC chloride channel at 3.0 angstrom reveals the molecular basis of anion selectivity. *Nature*, **415**, 287–294.

78 Jiang, Y.X., Lee, A., Chen, J.Y., Ruta, V., Cadene, M., Chait, B.T., and MacKinnon, R. (2003) X-ray structure of a voltage-dependent K^+ channel. *Nature*, **423**, 33–41.

79 Li, Z., Zhou, X., Dai, Z., and Zou, X. (2012) Classification of G proteins and prediction of GPCRs–G proteins coupling specificity using continuous wavelet transform and information theory. *Amino Acids*, **43**, 793–804.

80 Marks, D.S., Colwell, L.J., Sheridan, R., Hopf, T.A., Pagnani, A., Zecchina, R., and Sander, C. (2011) Protein 3D structure computed from evolutionary sequence variation. *PLoS ONE*, **6**, e28766.

81 Selent, J. and Kaczor, A.A. (2011) Oligomerization of G protein-coupled receptors: computational methods. *Current Medicinal Chemistry*, **18**, 4588–4605.

82 Ziaur, R. and Khan, A. (2011) Prediction of GPCRs with pseudo amino acid composition: employing composite features and grey incidence degree based classification. *Protein and Peptide Letters*, **18**, 872–878.

83 Abrol, R., Bray, J.K., and Goddard, W.A., III (2012) Bihelix: towards *de novo* structure prediction of an ensemble of G-protein coupled receptor conformations. *Proteins*, **80**, 505–518.

84 Rodriguez, D., Bello, X., and Gutierrez-de-Teran, H. (2012) Molecular modelling of G protein-coupled receptors through the web. *Molecular Informatics*, **31**, 334–341.

85 Jähnig, F. (1990) Structure predictions of membrane proteins are not that bad. *Trends in Biochemical Sciences*, **15**, 93–95.

86 Lim, V.I. (1974) Structural principles of the globular organization of protein chains. A stereochemical theory of globular protein secondary structure. *Journal of Molecular Biology*, **88**, 857–872.

87 Harris, F., Brandenburg, K., Seydel, U., and Phoenix, D. (2002) Investigations into the mechanisms used by the C-terminal anchors of *Escherichia coli* penicillin-binding proteins 4, 5, 6 and 6b for membrane interaction. *European Journal of Biochemistry*, **269**, 5821–5829.

88 Brandenburg, K., Harris, F., Dennison, S., Seydel, U., and Phoenix, D. (2002) Domain V of m-calpain shows the potential to form an oblique-orientated alpha-helix, which may modulate the enzyme's activity via interactions with anionic lipid. *European Journal of Biochemistry*, **269**, 5414–5422.

89 Marcotte, I., Wegener, K.L., Lam, Y.H., Chia, B.C.S., de Planque, M.R.R., Bowie, J.H., Auger, M., and Separovic, F. (2003) Interaction of antimicrobial peptides from Australian amphibians with lipid membranes. *Chemistry and Physics of Lipids*, **122**, 107–120.

90 Auger, I.E. (1993) Design of amphipathic helices: computational techniques to predict amphipathic helical segments, in *The Amphipathic Helix* (ed. R.M. Epand), CRC Press, Boca Raton, FL, pp. 7–20.

91 Cornette, J.L., Cease, K.B., Margalit, H., Spouge, J.L., Berzofsky, J.A., and DeLisi, C. (1987) Hydrophobicity scales and computational techniques for detecting amphipathic structures in proteins. *Journal of Molecular Biology*, **195**, 659–685.

92 McLachlan, A.D. and Stewart, M. (1976) The 14-fold periodicity in alpha-tropomyosin and the interaction with actin. *Journal of Molecular Biology*, **103**, 271–298.

93 Eisenberg, D., Weiss, R.M., and Terwilliger, T.C. (1982) The helical hydrophobic moment: a measure of the amphiphilicity of a helix. *Nature*, **299**, 371–374.

94 Eisenberg, D., Weiss, R.M., Terwilliger, T.C., and Wilcox, W. (1982) Hydrophobic moment and protein structure. *Faraday Symposia of the Chemical Society*, **17**, 109–120.

95 Eisenberg, D., Wilcox, W., and McLachlan, A.D. (1986) Hydrophobicity and amphiphilicity in protein structure. *Journal of Cellular Biochemistry*, **31**, 11–17.

96 Eisenberg, D., Schwarz, E., Komaromy, M., and Wall, R. (1984) Analysis of membrane and surface protein sequences with the hydrophobic moment plot. *Journal of Molecular Biology*, **179**, 125–142.

97 Daman, O., Wallace, J., Harris, F., and Phoenix, D.A. (2005) An investigation into the ability to define transmembrane protein spans using the biophysical properties of amino acid residues. *Molecular and Cellular Biochemistry*, **275**, 189–197.

98 Palliser, C.C. and Parry, D.A.D. (2001) Quantitative comparison of the ability of hydropathy scales to recognize surface beta-strands in proteins. *Proteins*, **42**, 243–255.

99 Mant, C.T., Kovacs, J.M., Kim, H.-M., Pollock, D.D., and Hodges, R.S. (2009) Intrinsic amino acid side-chain hydrophilicity/hydrophobicity coefficients determined by reversed-phase high-performance liquid chromatography of model peptides: comparison with other hydrophilicity/hydrophobicity scales. *Biopolymers*, **92**, 573–595.

100 Shamshurin, D., Spicer, V., and Krokhin, O.V. (2011) Defining intrinsic hydrophobicity of amino acids' side chains in random coil conformation. Reversed-phase liquid chromatography of designed synthetic peptides vs. random peptide data sets. *Journal of Chromatography A*, **1218**, 6348–6355.

101 Keller, R.C. (2011) New user-friendly approach to obtain an Eisenberg plot and its use as a practical tool in protein sequence analysis. *International Journal of Molecular Sciences*, **12**, 5577–5591.

102 Segrest, J.P., De Loof, H., Dohlman, J.G., Brouillette, C.G., and

Anantharamaiah, G.M. (1990) Amphipathic helix motif: classes and properties. *Proteins*, **8**, 103–117.

103 Datta, G., Epand, R.F., Epand, R.M., Chaddha, M., Kirksey, M.A., Garber, D.W., Lund-Katz, S., Phillips, M.C., Hama, S., Navab, M., Fogelman, A.M., Palgunachari, M.N., Segrest, J.P., and Anantharamaiah, G.M. (2004) Aromatic residue position on the nonpolar face of class A amphipathic helical peptides determines biological activity. *The Journal of Biological Chemistry*, **279**, 26509–26517.

104 Epand, R.M., Datta, G., Epand, R.F., Navab, M., Fogelman, M., Segrest, J.P., and Anantharamaiah, G.M. (2004) Towards the molecular mechanism for the inhibition of atherosclerosis by class A amphipathic helical peptides. *Biophysical Journal*, **86**, 42A–42A.

105 Mishra, V.K., Anantharamaiah, G.M., Segrest, J.P., Palgunachari, M.N., Chaddha, M., Sham, S.W.S., and Krishna, N.R. (2006) Association of a model class A (apolipoprotein) amphipathic alpha helical peptide with lipid – high resolution NMR studies of peptide–lipid discoidal complexes. *The Journal of Biological Chemistry*, **281**, 6511–6519.

106 Mishra, V.K., Palgunachari, M.N., Krishna, R., Glushka, J., Segrest, J.P., and Anantharamaiah, G.M. (2008) Association of a model class A (apolipoprotein) amphipathic alpha helical peptide with lipid – high resolution NMR studies of peptide–lipid discoidal complexes. *The Journal of Biological Chemistry*, **283**, 34393–34402.

107 Anantharamaiah, G.M., Mishra, V.K., Garber, D.W., Datta, G., Handattu, S.P., Palgunachari, M.N., Chaddha, M., Navab, M., Reddy, S.T., Segrest, J.P., and Fogelman, A.M. (2007) Structural requirements for antioxidative and anti-inflammatory properties of apolipoprotein A-I mimetic peptides. *Journal of Lipid Research*, **48**, 1915–1923.

108 Epand, R.M., Shai, Y., Segrest, J.P., and Anantharamaiah, G.M. (1995) Mechanisms for the modulation of membrane bilayer properties by amphipathic helical peptides. *Biopolymers*, **37**, 319–338.

109 Polozov, I.V., Polozova, A.I., Mishra, V.K., Anantharamaiah, G.M., Segrest, J.P., and Epand, R.M. (1998) Studies of kinetics and equilibrium membrane binding of class A and class L model amphipathic peptides. *Biochimica et Biophysica Acta*, **1368**, 343–354.

110 Polozov, I.V., Anantharamaiah, G.M., Segrest, J.P., and Epand, R.M. (2001) Osmotically induced membrane tension modulates membrane permeabilization by class L amphipathic helical peptides: nucleation model of defect formation. *Biophysical Journal*, **81**, 949–959.

111 Polverini, E., Casadio, R., Neyroz, P., and Masotti, L. (1998) Conformational changes of neuromedin B and delta sleep-inducing peptide induced by their interaction with lipid membranes as revealed by spectroscopic techniques and molecular dynamics simulation. *Archives of Biochemistry and Biophysics*, **349**, 225–235.

112 Polverini, E., Neyroz, P., Fariselli, P., Casadio, R., and Masotti, L. (1995) The effect of membranes on the conformation of neuromedin C. *Biochemical and Biophysical Research Communications*, **214**, 663–668.

113 Taylor, J.W. (1993) Amphiphilic helices in neuropeptides, in *The Amphipathic Helix* (ed. R.M. Epand), CRC Press, Boca Raton, FL, pp. 285–312.

114 Segrest, J.P., Jones, M.K., Deloof, H., Brouillette, C.G., Venkatachalapathi, Y.V., and Anantharamaiah, G.M. (1992) The amphipathic helix in the exchangeable apolipoproteins – a review of secondary structure and function. *Journal of Lipid Research*, **33**, 141–166.

115 Harris, F. and Phoenix, D.A. (1997) An investigation into the ability of C-terminal homologues of *Escherichia coli* low molecular mass penicillin-binding proteins 4, 5 and 6 to undergo membrane interaction. *Biochimie*, **79**, 171–174.

116 Dempsey, C.E. (1990) The actions of melittin on membranes. *Biochimica et Biophysica Acta*, **1031**, 143–161.

117 Tytler, E.M., Anantharamaiah, G.M., Walker, D.E., Mishra, V.K.,

Palgunachari, M.N., and Segrest, J.P. (1995) Molecular basis for prokaryotic specificity of magainin-induced lysis. *Biochemistry*, **34**, 4393–4401.

118 Tomich, J. (1993) Amphipathic helicies in channel-forming structures, in *The Amphipathic Helix* (ed. R.M. Epand), CRC Press, Boca Raton, FL, pp. 221–254.

119 Lazarovici, P. (2002) The structure and function of pardaxin. *Toxin Reviews*, **21**, 391–421.

120 MacKenzie, K.R. and Engelman, D.M. (1998) Structure-based prediction of the stability of transmembrane helix–helix interactions: the sequence dependence of glycophorin A dimerization. *Proceedings of the National Academy of Sciences of the United States of America*, **95**, 3583–3590.

121 Tang, M., Waring, A.J., and Hong, M. (2005) Intermolecular packing and alignment in an ordered beta-hairpin antimicrobial peptide aggregate from 2D solid-state NMR. *Journal of the American Chemical Society*, **127**, 13919–13927.

122 Harris, F., Wallace, J., and Phoenix, D.A. (2000) Use of hydrophobic moment plot methodology to aid the identification of oblique orientated alpha-helices. *Molecular Membrane Biology*, **17**, 201–207.

123 Lai, R., Liu, H., Lee, W.H., and Zhang, Y. (2002) An anionic antimicrobial peptide from toad *Bombina maxima*. *Biochemical and Biophysical Research Communications*, **295**, 796–799.

124 Liu, R., Liu, H., Ma, Y., Wu, J., Yang, H., Ye, H., and Lai, R. (2011) There are abundant antimicrobial peptides in brains of two kinds of *Bombina* toads. *Journal of Proteome Research*, **10**, 1806–1815.

125 Gautier, R., Douguet, D., Antonny, B., and Drin, G. (2008) HELIQUEST: a web server to screen sequences with specific α-helical properties. *Bioinformatics*, **24**, 2101–2102.

126 Keller, R.C.A. (2011) The prediction of novel multiple lipid-binding regions in protein translocation motor proteins: a possible general feature. *Cellular & Molecular Biology Letters*, **16**, 40–54.

127 Fauchere, J.C. and Pliska, V. (1983) Hydrophobic parameters pi of amino acid side chains from the partitioning of N-acetyl-amino-acid amides. *European Journal of Medicinal Chemistry*, **8**, 369–375.

128 Brasseur, R. (1991) Differentiation of lipid-associating helices by use of three-dimensional molecular hydrophobicity potential calculations. *The Journal of biological chemistry*, **266**, 16120–16127.

129 Cohen, A., Roth, Y., and Shapiro, B. (1988) Universal distributions and scaling in disordered systems. *Physical Review B*, **38**, 12125–12132.

130 Fauchere, J.L., Charton, M., Kier, L.B., Verloop, A., and Pliska, V. (1988) Amino acid side chain parameters for correlation studies in biology and pharmacology. *International Journal of Peptide and Protein Research*, **32**, 269–278.

131 Efremov, R.G., Chugunov, A.O., Pyrkov, T.V., Priestle, J.P., Arseniev, A.S., and Jacoby, E. (2007) Molecular lipophilicity in protein modeling and drug design. *Current Medicinal Chemistry*, **14**, 393–415.

132 Gallet, X., Benhabiles, N., Lewin, M., Brasseur, R., and Thomas-Soumarmon, A. (1995) Prediction of the antigenic sites of the cystic-fibrosis transmembrane conductance regulator protein by molecular modeling. *Protein Engineering*, **8**, 829–834.

133 Mingeot-Leclercq, M.P., Tulkens, P.M., and Brasseur, R. (1992) Accessibility of aminoglycosides, isolated and in interaction with phosphatidylinositol, to water. A conformational analysis using the concept of molecular hydrophobicity potential. *Biochemical Pharmacology*, **44**, 1967–1975.

134 Rahman, M., Lins, L., Thomas-Soumarmon, A., and Brasseur, R. (1997) Are amphipathic asymmetric peptides ubiquitous structures for membrane destabilization? *Journal of Molecular Modeling*, **3**, 203–215.

135 Dubovskii, P.V., Volynsky, P.E., Polyansky, A.A., Chupin, V.V., Efremov, R.G., and Arseniev, A.S. (2006) Spatial

structure and activity mechanism of a novel spider antimicrobial peptide. *Biochemistry*, **45**, 10759–10767.
136 Polyansky, A.A., Vassilevski, A.A., Volynsky, P.E., Vorontsova, O.V., Samsonova, O.V., Egorova, N.S., Krylov, N.A., Feofanov, A.V., Arseniev, A.S., Grishin, E.V., and Efremov, R.G. (2009) N-terminal amphipathic helix as a trigger of hemolytic activity in antimicrobial peptides: a case study in latarcins. *FEBS Letters*, **583**, 2425–2428.
137 Nurisso, A., Bravo, J., Carrupt, P.-A., and Daina, A. (2012) Molecular docking using the molecular lipophilicity potential as hydrophobic descriptor: impact on GOLD docking performance. *Journal of Chemical Information and Modeling*, **52**, 1319–1327.
138 Piotto, S.P., Sessa, L., Concilio, S., and Iannelli, P. (2012) YADAMP: yet another database of antimicrobial peptides. *International Journal of Antimicrobial Agents*, **39**, 346–351.
139 Pavlopoulou, A. and Michalopoulos, I. (2011) State-of-the-art bioinformatics protein structure prediction tools [review]. *International Journal of Molecular Medicine*, **28**, 295–310.
140 Cheng, J., Tegge, A.N., and Baldi, P. (2008) Machine learning methods for protein structure prediction. *IEEE Reviews in Biomedical Engineering*, **1**, 41–49.
141 Lee, D., Redfern, O., and Orengo, C. (2007) Predicting protein function from sequence and structure. *Nature Reviews Molecular Cell Biology*, **8**, 995–1005.
142 Mihasan, M. (2010) Basic protein structure prediction for the biologist: a review. *Archives of Biological Sciences*, **62**, 857–871.
143 Nam, H.-J., Jeon, J., and Kim, S. (2009) Bioinformatic approaches for the structure and function of membrane proteins. *BMB Reports*, **42**, 697–704.
144 Fjell, C.D., Hiss, J.A., Hancock, R.E.W., and Schneider, G. (2012) Designing antimicrobial peptides: form follows function. *Nature Reviews Drug Discovery*, **11**, 37–51.
145 Taboureau, O. (2010) Methods for building quantitative structure–activity relationship (QSAR) descriptors and predictive models for computer-aided design of antimicrobial peptides, in *Antimicrobial Peptides: Methods and Protocols* (eds A. Giuliani and A.C. Rinaldi), Humana Press, Totowa, NJ, pp. 77–86.
146 Wang, Y., Ding, Y., Wen, H., Lin, Y., Hu, Y., Zhang, Y., Xia, Q., and Lin, Z. (2012) QSAR modeling and design of cationic antimicrobial peptides based on structural properties of amino acids. *Combinatorial Chemistry & High Throughput Screening*, **15**, 347–353.
147 Jenssen, H., Fjell, C.D., Cherkasov, A., and Hancock, R.E.W. (2008) QSAR modeling and computer-aided design of antimicrobial peptides. *Journal of Peptide Science*, **14**, 110–114.
148 Ghahramani, Z. (2001) An introduction to hidden Markov models and Bayesian networks. *International Journal of Pattern Recognition and Artificial Intelligence*, **15**, 9–42.
149 Andres, E. and Dimareq, J.L. (2007) Cationic antimicrobial peptides: from innate immunity study to drug development. Up date. *Medecine et Maladies Infectieuses*, **37**, 194–199.
150 Hilpert, K., Fjell, C.D., and Cherkasov, A. (2008) Short linear cationic antimicrobial peptides: screening, optimizing, and prediction, in *Methods in Molecular Biology* (ed. L. Otvos), Springer, Berlin, pp. 127–159.
151 Polanco, C. and Samaniego, J.L. (2009) Detection of selective cationic amphipathic antibacterial peptides by hidden Markov models. *Acta Biochimica Polonica*, **56**, 167–176.
152 Fjell, C.D., Hancock, R.E., and Cherkasov, A. (2007) AMPer: a database and an automated discovery tool for antimicrobial peptides. *Bioinformatics*, **23**, 1148–1155.
153 Fjell, C.D., Jenssen, H., Fries, P., Aich, P., Griebel, P., Hilpert, K., Hancock, R.E., and Cherkasov, A. (2008) Identification of novel host defense peptides and the absence of alpha-defensins in the bovine genome. *Proteins*, **73**, 420–430.

154 Krogh, A. (2008) What are artificial neural networks? *Nature Biotechnology*, **26**, 195–197.

155 Fjell, C.D., Hancock, R.E.W., and Jenssen, H. (2010) Computer-aided design of antimicrobial peptides. *Current Pharmaceutical Analysis*, **6**, 66–75.

156 Fjell, C.D., Jenssen, H., Hilpert, K., Cheung, W.A., Pante', N., Hancock, R.E.W., and Cherkasov, A. (2009) Identification of novel antibacterial peptides by chemoinformatics and machine learning. *Journal of Medicinal Chemistry*, **52**, 2006–2015.

157 Cherkasov, A., Hilpert, K., Jenssen, H., Fjell, C.D., Waldbrook, M., Mullaly, S.C., Volkmer, R., and Hancock, R.E.W. (2009) Use of artificial intelligence in the design of small peptide antibiotics effective against a broad spectrum of highly antibiotic-resistant superbugs. *ACS Chemical Biology*, **4**, 65–74.

158 Torrent, M., Andreu, D., Nogues, V.M., and Boix, E. (2011) Connecting peptide physicochemical and antimicrobial properties by a rational prediction model. *PLoS ONE*, **6**, e16968.

5
Models for the Membrane Interactions of Antimicrobial Peptides

Summary

The ability of antimicrobial peptides (AMPs) to kill microorganisms depends primarily upon their ability to interact with the cytoplasmic membrane (CM) of these organisms. To interact with the CM, these peptides must pass through the structures that envelope the microbe, including the outer membrane of Gram-negative bacteria, and the cell walls of bacteria and fungi. Upon encountering the microbial CM, the ability of AMPs to partition into the bilayer is influenced by a number of factors, including the transmembrane potential and membrane fluidity along with the presence of lipid receptors, sterols, anionic lipids, and lipid-specific domains. The interactions of AMPs with these membranes have been investigated by a variety of techniques, which has led to the presentation of a number of models to describe them. In this chapter, we present an overview of the major established examples of these models, using cationic AMPs that adopt an α-helical structure as a major example (α-CAMPs). The models described include the barrel-stave pore, the toroidal pore and its variants such as the disordered toroidal pore, and the carpet mechanism. In addition, we review several very recently described models for the membrane interactions of α-CAMPs, including the tilted peptide mechanism and the use of amyloidogenesis.

5.1
Introduction

The cytoplasmic membrane (CM) of cells is a complex network of lipids and proteins with a molecular organization that makes a major contribution to both maintaining the structural integrity of cells and the activities of proteins that are associated with the lipid bilayer. The current view is that the lipid molecules within a lipid bilayer are dynamic and actively involved in a multitude of biological functions, and are not simply a scaffold for the activities of proteins and peptides, as was long the prevailing view [1]. Increasingly, the dynamism of membrane lipid is being shown to play a fundamental role in facilitating the interactions of antimicrobial peptides (AMPs) with the CM of microbes and it is generally accepted

Antimicrobial Peptides, First Edition. David A. Phoenix, Sarah R. Dennison, and Frederick Harris.
© 2013 Wiley-VCH Verlag GmbH & Co. KGaA. Published 2013 by Wiley-VCH Verlag GmbH & Co. KGaA.

that these interactions are a crucial step in the antimicrobial activity of these peptides [2–5]. The association of AMPs with CM lipid can lead directly to the inactivation of microbes via cell lysis [6–8] or indirectly through internalization of these peptides, so facilitating attack on intracellular targets [9, 10]. However, in order to reach and interact with microbial CMs, AMPs have to negotiate and traverse a variety of enveloping structures (Figure 5.1). These structures vary across target organisms in their physiochemical properties, permeability, and thickness, and therefore influence the susceptibility of these organisms to AMPs. For example, it has been reported that although the CMs of some bacterial strains have similar phospholipid compositions, the susceptibility of these microbes to given AMPs varied considerably and appeared to be related to architectural differences in the envelopes of these prokaryotes [13].

In general, cationic AMPs (CAMPs) are by far the best characterized (Chapter 2) and the passage of these peptides through the microbial envelope appears to involve low-affinity, electrostatic interactions between these peptides and a range of anionic moieties that promote the migration of these peptides to the CM of the host microbe [14]. The affinity level of CAMPs for these anionic moieties appears to be crucial in that low affinity can reduce the targeting efficacy, but higher affinities can lead these peptides to bind anionic envelope components sufficiently strongly to prevent their passage to the microbial CM and thereby inhibit their antimicrobial action. Similar inhibitory actions have been reported for the interaction of CAMPs with a variety of anionic cell surface moieties, which bind and thereby reduce the toxicity of these peptides to erythrocytes [15], non-cancerous cells [16, 17], and transformed cells [18] (Chapter 6).

In the case of fungi, migration through the cell envelope by CAMPs involves interaction with a number of anionic components including chitin chains, (1 → 3)-β-glucan, and phosphomannans or related compounds (Figure 5.1) [10, 19]. In

Figure 5.1 Schematic representations of the cell envelopes of Gram-negative bacteria (a), Gram-positive bacteria (b), and fungi (c) (a and b adapted from [11]; c adapted from [12]). In the case of Gram-negative bacteria this envelope is primarily formed from an inner plasma membrane, a thin intermediate peptidoglycan layer, and outer membrane, which is covalently linked to the peptidoglycan layer by lipoprotein and has molecules of LPS on its external surface (a). In contrast, Gram-positive bacteria possess an envelope that comprises a plasma membrane that is closely opposed by a thick peptidoglycan layer, which has molecules of teichoic and teichuronic acids covalently bound to its external surface (b). In both cases, membranes are populated by a variety of proteins such as porins, which facilitate the passive diffusion of molecules across the membrane. Fungi possess a cell envelope, which consists of the plasma membrane and an extracellular matrix, or cell wall, that, for most fungi, comprises five major components organized in a layered structure: (1 → 3)-β-glucan, (1 → 6)-β-glucan, (1 → 3)-α-glucan, and chitin, which are polysaccharides, along with glycoproteins (c). However, the composition of the cell wall can vary between different fungal species and in the example shown above, which represents the cell envelope of the model yeast, *Saccharomyces cerevisiae*, chitin is present but α-glucan is absent.

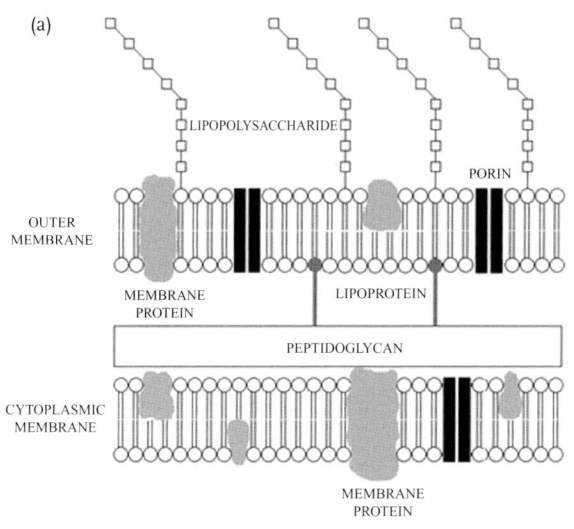

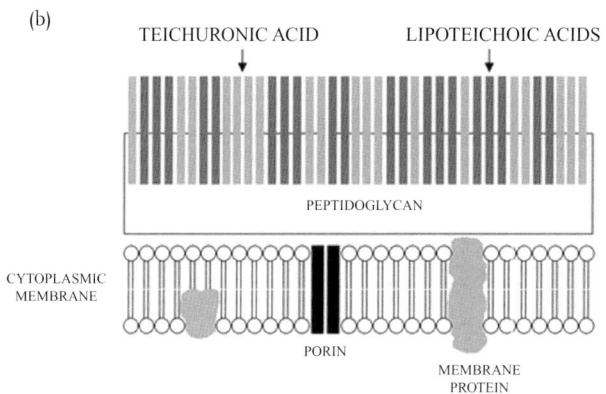

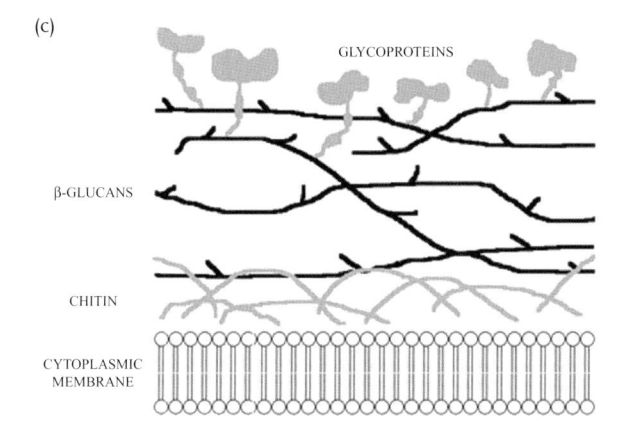

(a)

LIPOPOLYSACCHARIDE

PORIN

OUTER MEMBRANE

MEMBRANE PROTEIN

LIPOPROTEIN

PEPTIDOGLYCAN

CYTOPLASMIC MEMBRANE

MEMBRANE PROTEIN

(b)

TEICHURONIC ACID   LIPOTEICHOIC ACIDS

PEPTIDOGLYCAN

CYTOPLASMIC MEMBRANE

PORIN

MEMBRANE PROTEIN

(c)

GLYCOPROTEINS

β-GLUCANS

CHITIN

CYTOPLASMIC MEMBRANE

order to reach the CM of fungi and bacteria that participate in biofilms, CAMPs pass through the structural frameworks associated with these sessile organisms and interact with anionic exopolysaccharides such as alginate or poly-γ-D,L-glutamic acid [20, 21]. In the case of Gram-positive bacteria, such as *Streptococcus* and *Bacillus* spp., CAMPs interact with teichoic and lipoteichoic acids, which are anionic components of the bacterial cell wall, resulting in migration of these peptides across the peptidoglycan layer to interact directly with the CM (Figure 5.1) [22]. It has been shown that interaction with these anionic polymers is important for the antimicrobial activity some CAMPs, such as defensins, which are β-sheet peptides produced across eukarya (Chapter 1), by the demonstration that reducing their negative charge decreases the susceptibility of target bacteria [23]. With Gram-negative bacteria, CAMPs first encounter the outer membrane, which possesses anionic lipopolysaccharide (LPS) as the major component of its external leaflet (Figure 5.1) [14]. The majority of CAMPs have a high affinity for LPS, which allows these peptides to traverse the outer membrane via a self-promoted uptake pathway. Essentially, these CAMPs interact with lipid A moieties of LPS, which serve as anionic binding sites, thereby competitively replacing Mg^{2+} and Ca^{2+} ions that help maintain cell surface stability via the cross-linking of carboxylated and phosphorylated head-groups of lipids. Removal of these ions leads to disruption of the outer membrane, and allows the peptides to migrate across the peptidoglycan layer and accumulate in the periplasmic space before interacting with components of the CM [14, 24]. Use of this pathway by CAMPs can also facilitate the co-uptake of other small molecules, which appears to explain the fact that when directed against Gram-negative bacteria, many of these peptides are able to act in synergy with other antimicrobial agents [25–28]. However, synergistic action between CAMPs and other antimicrobial agents is not limited to these latter bacteria, and has also been demonstrated for Gram-positive bacteria [27–30] and fungi [30–32]. A further consequence of CAMPs binding to LPS and lipoteichoic acid is the neutralization of the endotoxic activity associated with these molecules in bacterial sepsis, thereby avoiding the exalted inflammatory response induced by the antibacterial activity of other antibiotics [33, 34].

The generally held view is that the antimicrobial action of CAMPs does not involve the use of protein and lipid membrane receptors [35], although such use is well established for some prokaryotic AMPs [36–38]. Recent studies have suggested that some CAMPs with non-lytic antibacterial mechanisms utilize protein receptors for membrane translocation and interaction with cytoplasmic targets [10, 39, 40]. Most recently, evidence has been presented to show that lipoprotein in the outer membrane of Gram-negative bacteria serves as a cell surface receptor for a number of CAMPs [41], including SMAP-29 from sheep [42], and CAP-18 and LL-37 from humans [43, 44]. There is also evidence to suggest that phosphatidylethanolamine (PE) in the membranes of Gram-negative bacteria may serve as a receptor in the antimicrobial action of kalata B1 and cycloviolacin O2 [45, 46], which are CAMPs from plants with potent membranolytic activity towards this bacterial class [47, 48]. It has also recently been shown that interaction with microbe-specific lipid receptors is a key step in the antimicrobial action of a number of defensins [49], which are CAMPs found across the eukaryotic kingdom [50–54]. For example, lipid II has been shown to facilitate the antibacterial activity of fungal defensins such as plectasin from *Pseudoplectania nigrella* via the inhibition of cell wall biosynthesis [49, 55]. In the case of fungi, it has been demonstrated that fungal-specific sphingolipids promote the antifungal activity of plant defensins such as FsAFP2 from radish by a variety of mechanisms, including membranolysis and apoptosis induction [19, 49, 51]. These observations clearly indicate that lipid receptors mediate the membrane interactions and antimicrobial mechanisms of some CAMPs. Nonetheless, based on current evidence, the antimicrobial mechanisms of the vast majority of these peptides are mediated by relatively non-specific interactions with membrane lipids. In Chapter 2, the dependence of these interactions on the structural and physiochemical properties of CAMPs was described; in this chapter, the role of membrane-based factors in these interactions is discussed using peptides that adopt α-helical secondary structure (α-CAMPs) as a major example. Moreover, based on data provided by theoretical analyses, experimental techniques, and molecular dynamics (MD) simulations, a number of models to describe these interactions have been proposed, which are described here, including major established examples of these models, such as the barrel-stave pore, the toroidal pore and its variants, and the carpet mechanism. In addition, we review several very recently described models for the membrane interactions of α-AMPs, including the tilted peptide mechanism and the use of amyloidogenesis.

5.2
CM-Associated Factors That Affect the Antimicrobial Action of α-CAMPs

Originally suggested by Shai [56], it is now generally accepted that the initial association of CAMPs with the microbial CM is driven by electrostatic interactions between the net positive charge possessed by these peptides and the net negative charge carried by these membranes [57, 58]. The CMs of bacteria are generally rich in anionic phospholipids, primarily phosphatidylglycerol (PG) and cardiolipin (CL), which are either present at low levels or absent from mammalian cells (Figure 5.2) [57, 59]. Indeed, it is the possession of this net negative charge that is believed to underpin the selectivity of CAMPs for bacterial cells over eukaryotic cells, since the latter carry a net neutral charge due to the presence of zwitterionic lipids such as PE, sphingomyelin (SM), and phosphatidylcholine (PC) [61]. Nonetheless, there are considerable differences between the levels of anionic lipids found in the membranes of different bacterial species and classes as can be seen from Figure 5.2. The CMs of Gram-positive bacteria are primarily composed of lipids that are derived from PG, exemplified by that of *Staphylococcus aureus*, which contains around 57% PG, 38% lysylated PG (LysylPG), and 5% CL, a dimer of PG (Figure 5.2). In contrast, the CMs of Gram-negative bacteria are mainly

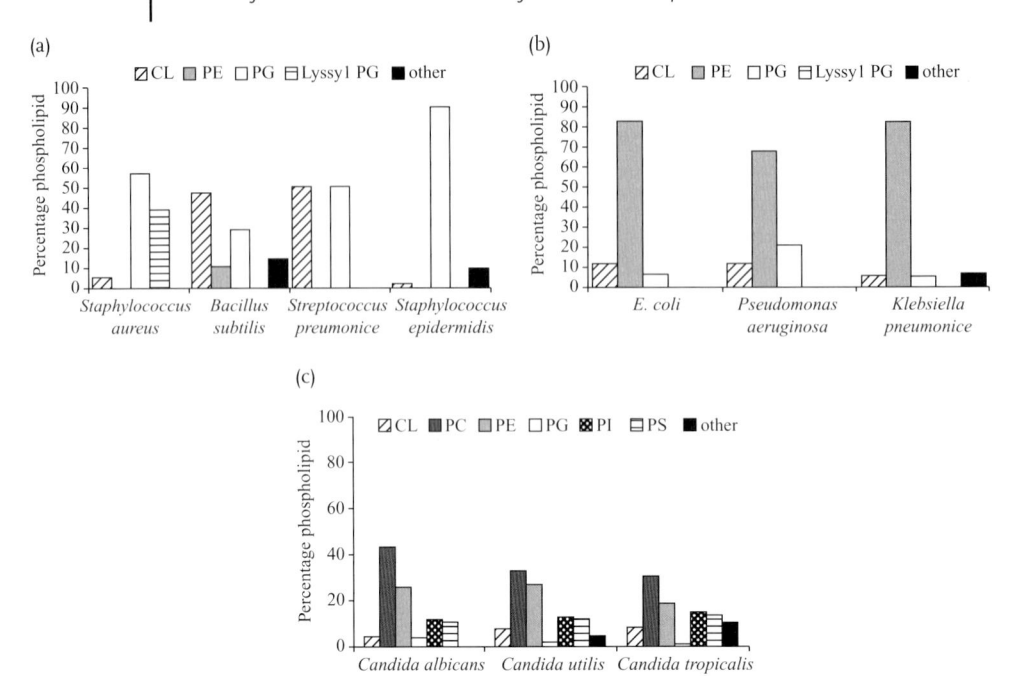

Figure 5.2 Lipid composition of phospholipids present in the CMs of Gram-positive bacteria (a), Gram-negative bacteria (b), and fungi (c). These lipids include cardiolipin (CL), phosphatidylethanolamine (PE), phosphatidylglycerol (PG), lysylated phosphatidylglycerol (Lysl PG), phosphatidylinositol (PI), and phosphatidylserine (PS). The data were taken from Ratledge and Wilkinson [60].

composed of PE, typically that of *Escherichia coli*, which contains approximately 80% PE, 6% PG, and 12% CL (Figure 5.2). In general, anionic lipids constitute over 80% of the total lipid found in the membranes of Gram-positive bacteria, but less than 30% of the total lipid present in those of Gram-negative bacteria [57]. It would therefore seem that varying levels of anionic lipid contribute to the differing bacterial specificities shown across CAMPs given that many of these peptides show a preference for either Gram-positive or Gram-negative bacteria [62, 63]. An example of this differential antibacterial activity is found in defensins from plants where AhAMP1 from the horse chestnut, *Aesculus hippocastanum*, specifically inhibits Gram-positive bacteria, but VrD1 from the mung bean, *Vigna radiate*, only has activity towards Gram-negative bacteria. In contrast, VrD2, also from *V. radiate*, displays inhibitory activity against both Gram-positive and Gram-negative bacteria [64].

Upon interaction with the microbial CM, a number of factors are able to modulate the activity of CAMPs and their ability to partition into these membranes. A major factor is the transmembrane potential ($\Delta \psi$), which results from differing extents and rates of proton flux across the membrane [65]. Mammalian membranes maintain a weak $\Delta \psi$, but due in part to their high levels of anionic lipids, the CM of fungi [66] and bacteria [7, 67] maintain a relatively strong $\Delta\psi$, which is of the order of $-120\,mV$ in the latter case. This high $\Delta\psi$ strongly attracts CAMPs and enhances both their selectivity and toxicity for microbial cells [65]. A similar mechanism appears to underpin the ability of CAMPs to target cancer cell membranes, which exhibit loss of lipid asymmetry, resulting in the presence of relatively high levels of anionic lipid on their outer surfaces and an accompanying high $\Delta\psi$ [68, 69].

Another factor that is important to the antimicrobial activity of CAMPs is the propensity of some lipids within the CM to adopt non-lamellar phases and induce membrane curvature. This propensity is related to the polymorphism of membrane lipids or their differing "molecular shapes," which can lead to differences in their packing arrangements [70]. For example, lipids such as PC that possess head-groups and acyl chains that generate the same cross-sectional area can be represented as cylinders and pack to form a planar bilayer (Figure 5.3, Chapter 6).

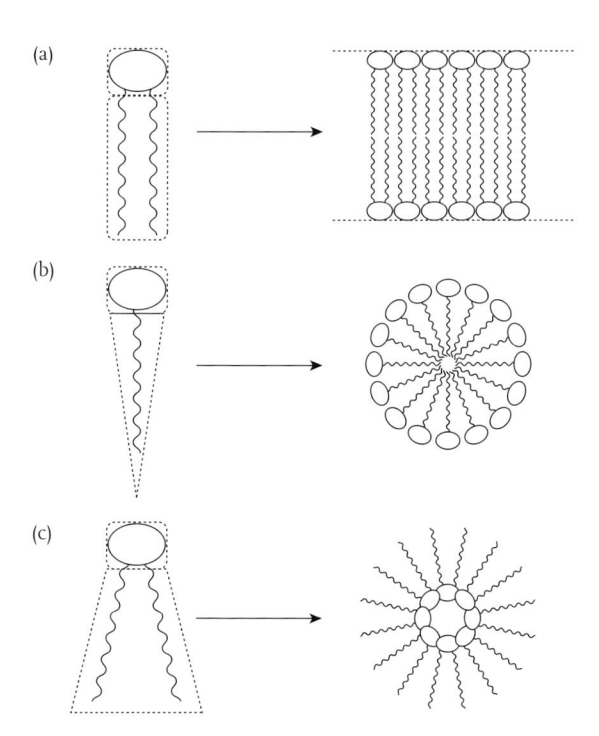

Figure 5.3 Polymorphism of membrane lipids or their differing "molecular shapes." Lipids such as PC that possess head-groups and acyl chains that generate the same cross-sectional area can be represented as cylinders and pack to form a planar bilayer (a). For lipids such as LPC, the cross-sectional area of their lipid head-group is larger than that of their acyl chains and these lipids form a cone or a wedge. These lipids are referred to as H_I phase-preferring lipids and favor positive membrane curvature (b). Lipids such as PE possess head-groups with a cross-sectional area that is larger than that of their acyl chains and form an inverted cone shape. These lipids are referred to as H_{II} phase-preferring lipids and favor negative membrane curvature (c).

However, in the case of lipids such as lysophosphatidylcholine (LPC), the cross-sectional area of their lipid head-group is larger than that of their acyl chains and these lipids form a cone or a wedge (Figure 5.3). These lipids are referred to as H_I phase-preferring lipids and favor positive membrane curvature. In contrast, lipids such as PE possess head-groups with a cross-sectional area that is larger than that of their acyl chains and form an inverted cone shape (Figure 5.3). These lipids are referred to as H_{II} phase-preferring lipids and favor negative membrane curvature. Evidence has been presented that suggests that CAMPs are able to preferentially bind various CM lipids, thereby promoting non-lamellar structures and membrane curvature that can be associated with particular antimicrobial mechanisms [2]. For example, the induction of positive membrane curvature by CAMPs appears to be involved in the formation of toroidal pores and carpet-type mechanisms as recently demonstrated for aurein 1.2 and citropin 1.1, which are amphibian α-CAMPs [71]. In contrast, the induction of negative membrane curvature by the interaction of CAMPs with CM lipids appears to feature in the aggregate model of membrane interaction where CAMPs are internalized without causing significant damage to the bilayer structure [2], as suggested for indolicidins, which are extended bovine AMPs [8, 72].

The propensity for some CM lipids to form non-lamellar structures also appears to contribute to a recently described mechanism of antimicrobial action in which CAMPs induce the clustering of anionic lipids within the membrane and the resulting lipid segregation leads to the slow leakage of intracellular contents and/or membrane depolarization [22]. This mechanism of membrane perturbation appears to be favored in the case of Gram-negative bacteria, which generally possess a CM containing high levels of PG and PE (Figure 5.1) where the latter lipid has no great predilection to form stable bilayers after lipid demixing by CAMPs [6]. In contrast, Gram-positive bacteria possess a CM that is predominantly formed from PG and CL, and in these cases lipid segregation by CAMPs would not be generally expected. The preferential use of this mechanism of membrane perturbation by CAMPs would seem to be reflected by the fact that many of these peptides have higher levels of toxicity to Gram-negative bacteria as compared to Gram-positive bacteria [3, 57].

In contrast to bacteria, the CM of fungal cells tend to share similar properties to mammalian membranes, having a net neutral charge due to the presence of zwitterionic lipids including PC, PE, and SM (Figure 5.1) although possessing a number of fungal-specific lipids [73, 74]. In comparison to bacteria, there has been far less research into the antifungal mechanisms of CAMPs, but there is evidence to show that interaction of these peptides with anionic lipid and the fungal CM leads to membrane lysis and plays a role in their antifungal mechanisms [75–81] as for ABP-CM4 from the silkworm, *Bombyx mori* [78, 79]. In addition, it has been observed that CAMPs with primarily antifungal activity tend to possess a high proportion of neutral polar residues, which led to the suggestion that the interactions of these peptides with the fungal CM membrane may show different structure–activity relationships to those used by CAMPs in their other antimicrobial activities [7, 82–84], possibly involving an ability to form complexes with fungal lipids [85]. However, other CAMPs show only a weak ability to permeabilize the fungal CM and appear to inactivate fungi by other mechanisms involving multiple activities such as facilitating the efflux of ATP from the cell [32]. Taken with the fact that, as described above, some CAMPs utilize lipid receptors to promote their antifungal action, these peptides appear to use a diverse range of antifungal mechanisms as compared to their antibacterial mechanisms.

A major difference between the lipid composition of bacterial membranes and those of eukaryotes is the presence of sterols, with the major sterol in fungal membranes being ergosterol while in the case of mammalian membranes it is cholesterol [86]. An important property of both cholesterol and ergosterol is that they impart significant changes to the biophysical properties of lipid bilayers, including the regulation of membrane fluidity. These sterols are able to intercalate with membrane lipids and their rigid ring structures appear to limit the *trans* to *gauche* isomerization of vicinal phospholipid acyl chains, thereby favoring the liquid ordered membrane phase. These effects are believed to lead to increased membrane cohesion and mechanical stiffness, thereby reducing the ability of CAMPs to partition into the bilayer and exert their lytic action [87–91]. This inhibitory effect has been demonstrated for cholesterol and a variety of CAMPs [91], including DDK from amphibians [92] and S-thanatin derived from insects [93]. Ergosterol has been shown to have a similar inhibitory effect on the action of CAMPs, although the level of this effect was decreased as compared to that of cholesterol. Taken with other work, these results led to the general view that compared to mammals the absence of cholesterol in bacterial membranes contributes to the relatively higher susceptibility of these organisms to the action of CAMPs. However, the presence of ergosterol in fungal membranes appears to endow these microbes with a susceptibility to the action of these peptides that is intermediate between those of bacteria and mammals [77, 89, 94]. The capacity of cholesterol to protect mammalian cells from the action of CAMPs [91] along with its ability to inhibit the hemolytic activity of magainins, which are amphibian CAMPs [95], led to the view that the presence of the sterol in mammalian membranes is a major factor contributing to the specificity of these peptides for microbes [90, 92]. Clearly, the presence of cholesterol in mammalian membranes also contributes to both the specificity and toxicity of CAMPs to cancer cells, which is described in Chapter 6. Pertaining to both the antimicrobial and anticancer activity of CAMPs, recent studies have suggested that factors in addition to the structure of the sterol may contribute to its ability to regulate membrane fluidity and thereby its inhibitory effects on the action of these peptides [91], such as the cholesterol-mediated dehydration of the lipid head-group region [96]. There is also evidence to suggest that direct interaction between cholesterol and CAMPs can inhibit the antimicrobial action of these peptides [97–101] while in some cases the localization of the sterol into lipid rafts of mammalian membranes can increase the susceptibility of non-raft domains to CAMPs, effectively reducing the microbial specificity of these peptides [91]. There is also evidence to suggest that a similar mechanism involving lipid rafts rich in ergosterol can increase the susceptibility of some fungi to some CAMPs [76].

5.3
Mechanisms Used by CAMPs for Microbial Membrane Interaction

The ability to target the microbial cell membrane has been investigated for CAMPs with a variety of secondary structures, including those that adopt β-sheet and extended configurations, but most studies on this ability have focused on α-CAMPs [6]. These peptides may be intrinsically α-helical or may adopt such structure upon interaction with the outer surface of either Gram-negative or Gram-positive bacteria, and upon binding to the CM of these organisms [7, 58]. Upon interaction with the bacterial membrane, α-CAMPs have been shown to induce inactivation of the host organism using a number of modes of action. Some α-CAMPs translocate the bilayer in a lipid-mediated, non-membranolytic manner to attack intracellular targets [9]. A variety of mechanisms have been proposed to describe these translocation processes and, in general, these they appear to involve the use of transient pores formed in the bilayer [102, 103]. For example, it is generally believed that buforin II, which is derived from histones of the Asian toad, *Bufo bufo gargarizans* [104], translocates the bacterial membrane via transient toroidal pore formation (discussed below) to inhibit the cellular functions of the host organism through binding to RNA and DNA [105, 106]. In contrast, other α-CAMPs appear to kill bacteria via multiple mechanisms of membrane lysis [6], such as clavanin A from the tunicate, *Styela clava* [107]. Under mildly acidic conditions, this peptide induces bacterial membrane disruption via lipid-mediated interactions but under neutral conditions, apparently through inhibitory interactions with proteins involved in maintaining the pH gradient across the bilayer [108, 109]. Yet other α-CAMPs are known to exert their antibacterial effects by multiple mechanisms of a more diverse nature: indirectly through the stimulation of immune responses (Chapter 1) [110, 111], and directly through attack on membranes, outer surface lipids and proteins, and intracellular targets such as nucleic acids [6, 49]. As an example, the human α-AMP, LL-37, appears to utilize a variety of membrane pore-forming mechanisms in its action against bacteria [112–114] and is able to mitigate the immune response to pathogens. It has been shown that the peptide is able to induce the selective movement of neutrophils, monocytes, and T lymphocyte cells, thereby promoting an increased innate and adaptive immune response [110, 115]. Based on these, and other, examples, it is becoming increasingly clear that the antibacterial mechanisms of α-CAMPs are more diverse and complex than at first thought [6, 116]. This observation is reflected by the growing number of models presented to describe the lipid-mediated permeabilization of the target bacterial membrane by α-CAMPs (Table 5.1), generally accepted as a factor common to the majority of their antibacterial mechanisms [116]. Moreover, in many cases, these models have also been proposed to describe the anticancer action of a number of CAMPs (Chapter 6).

Table 5.1 Models proposed for the membrane interactions of α-AMPs.

Antibacterial mechanism	Representative α-AMPs	Key references
Non-permeabilizing		
Translocation via a transient pore	Buforin II	[117]
Translocation via sinking raft	δ-Lysin	[118]
Permeabilizing		
Barrel-stave pore	Almethicin	[4]
Toroidal pore	Magainin 2	[4]
Huge toroidal pore	Lacticin Q	[119]
Disordered toroidal pore	Melittin	[120]
Aggregate model	Magainins	[121]
Interfacial activity model	Magainin 2	[122]
Chaotic or non-stoichiometric model	Magainin 2	[123]
Carpet mechanism	Cecropin	[4]
Detergent model		[124]
Membrane discrimination model	V13K$_L$	[125]
SHM model		[126]
Membrane thinning/thickening	LL-37	[127]
Charged lipid clustering	Magainin analogs	[57]
Sand in a gearbox model	Synthetic α-AMPs	[128]
Oxidized phospholipid targeting	Temporin L	[129]
Electroporation	NK-lysin	[130]
Tilted peptide mechanism	Aurein 1.2	[131]
Amyloid formation	Temporin B	[132]
Leaky slit model		[133]

5.4
Established Models for the Membrane Interactions of α-AMPs

Over the last four decades, three major models have become established as descriptions of the mechanisms of membrane invasion used by α-CAMPs to exert their antibacterial activity. These models are generally referred to as the barrel-stave pore model, the toroidal pore model, and the carpet mechanism, and have been extensively reviewed elsewhere [4, 84, 134]. Here, an overview of these models is presented along with a discussion of variants that have been recently described.

5.4.1
Barrel-Stave and Toroidal Pore Models

The barrel-stave pore model for the membrane interaction of α-CAMPs was first proposed in the late 1970s [135]. According to this model, α-AMPs initially bind electrostatically to the lipid head-group region of the bacterial membrane in an

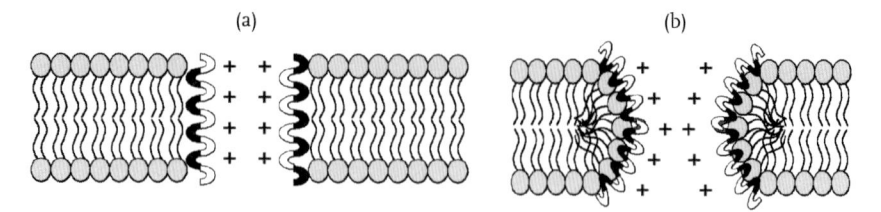

Figure 5.4 Schematic representation of pores formed by α-AMPs (adapted from [5]). (a) "Barrel-stave" pore. The hydrophobic surfaces of α-AMPs (black) are in direct contact with the membrane lipid core while their hydrophilic surfaces (white) point inwards producing an aqueous pore. (b) "Toroidal" pore. The lipid head-group regions of opposing membrane surfaces are continuous and an aqueous pore is formed, which is lined by polar lipid head-groups and the hydrophilic surfaces of closely associated α-AMPs.

orientation parallel to the bilayer surface. Aggregation and insertion of these peptides into the membrane with an orientation perpendicular to the bilayer surface leads to the creation of a transmembrane pore in which arrangement of the participating α-AMPs resemble staves or barrel-like clusters. The pore is constructed with the hydrophobic surfaces of these peptides interacting with the lipid core of the membrane while their hydrophilic surfaces line the pore interior (Figure 5.4) [4]. The ability to form barrel-stave pores appears to have only been generally accepted for alamethicin and other peptaibols from the fungal genus *Trichoderma* [136]. Even in this case, variants on the barrel-stave model have been proposed [137], such as a recent study that suggested that alamethicin forms transient pores, which result from random associations of the peptide inserted perpendicular to the membrane [138].

The toroidal pore, or wormhole, model for the membrane interaction of α-CAMPs was first proposed in the late 1990s for magainins [139, 140], which are derived from the frog, *Xenopus laevis* [141], and is the most cited pore mechanism for membrane permeabilization by these peptides [116]. The model assumes that α-AMPs utilize a mode of initial membrane binding and orientation that is similar to that proposed by the barrel-stave model (Figure 5.4). However, the toroidal pore model proposes that the aggregation of α-AMPs on the membrane surface imposes a positive curvature strain by increasing the distance between membrane lipid head-groups (Figure 5.4b) – an effect usually described as membrane thinning [127]. The toroidal pore model further proposes that when the aggregation of α-AMPs reaches a critical concentration, these peptides realign perpendicular to the bilayer, causing the membrane surface to cavitate inwards and ultimately form a pore. In contrast to the barrel-stave pore model, the formation of these toroidal pores requires that α-AMPs remain in close association with membrane lipid head-groups such that these pores are lined by polar lipid head-groups and hydrophilic surfaces of α-AMPs, as shown in Figure 5.4b. It has been proposed that toroidal pores are transient and, in some cases, dissolution of these pores can lead to the internalization of α-AMPs via mechanisms that effectively are non-membranolytic [142, 143]. For example, the membrane translocation of buforin II is believed to involve the formation of unstable toroidal pores in the bacterial bilayer, which rapidly disintegrate with minimal disruption of the membrane while facilitating translocation of the peptide to the cell interior [117, 144].

Since its inception, a number of variants of the toroidal pore model have been presented (Table 5.1). Recent studies on lacticin Q, which is an α-AMP produced by *Lactococcus lactis* [145], showed that the peptide exerted its antibacterial activity via the formation of very large toroidal pores in the membrane [119]. These pores were up to twice the size of those produced by magainin [146] and this novel process of pore formation was termed the "Huge Toroidal Pore" model [119, 145]. Most variants of the toroidal pore mechanism that have been presented differ primarily from the model described above in that the pores they depict are less ordered. The aggregate model was presented in the late 1990s for magainins and other AMPs [121], and proposes a mechanism of membrane permeabilization that is similar to that of the toroidal pore model [24]. However, in the former model, peptides do not adopt any specific orientation upon membrane insertion, and form membrane-spanning aggregates composed of lipid and peptide micelles [83]. Extensions to this model propose that channels formed by these aggregates can vary greatly in their structural characteristics and that peptide-induced negative membrane curvature, or membrane thickening [127], can promote the formation of non-bilayer intermediates that facilitate the translocation of aggregated AMPs across the membrane [2, 147]. More recent work on dye-filled vesicles led to the presentation of two chaotic or non-stoichiometric models of membrane interaction in which peptides initially bind to the membrane surface and accumulate, creating a mass imbalance across the lipid bilayer, thereby perturbing the membrane [148]. In the "all-or-none" model, membrane perturbation by α-AMPs such as magainins and cecropin A, which is derived from the moth, *Hyalophora cecropia* [149], are believed to lead to disturbance of the lipid-packing order and the formation of transient pores, which facilitate the rapid release of intravesicular contents. These pores, which are described as disordered or "chaotic" toroidal pores, are mostly lined by lipids with some associated α-AMPs and their formation does not involve a significant oligomerization of these peptides [123, 150]. In the "graded release" model, membrane perturbation by α-AMPs is believed to lead to their insertion into the membrane hydrophobic core. This insertion of peptides constitutes a process of transient pore formation and facilitates the translocation of α-AMPs across the membrane, either as monomers or small aggregates in an orientation parallel to the bilayer surface. This process of translocation can be accompanied by the release of intravesicular contents; however, when the process is completed, the mass imbalance across the membrane is restored and the transient pore effectively sealed with the result that dye efflux either slows or ceases [148]. The best documented example of an α-AMP using this latter model is δ-hemolysin from

S. aureus, which is believed to translocate the membrane as a trimer – a process that has also been described as the sinking raft model [118, 151]. Based on MD simulations, the disordered toroidal pore model was introduced in the mid-2000s [152] for membrane permeabilization by magainins and their analogs [120, 152, 153]. In this model, pore formation is more stochastic than in the toroidal pore model, with only a few peptide molecules located near the center of the pore and more peptides situated at the mouth of the pore on the external leaflet. As they translocate to the cell interior, peptides remain mostly parallel to the bilayer plane and only a minor tilt is observed [120]. In other MD simulations [154], similar tilted orientations were observed for pore formation by magainin analogs, and the α-AMPs, melittin from the bee, *Apis mellifera* [155], and piscidin 1 from hybrid striped sea bass [156]. These latter MD studies also introduced the view that the imperfect amphiphilicity of α-AMPs, which is the presence of charged or polar residues in the hydrophobic face of these peptides, could play a role in pore construction [154]. Imperfect amphiphilicity also features in the interfacial activity model, which suggests that peptides with this structural property, such as magainin 2, are able to partition into the membrane and drive the vertical rearrangement of lipid polar and non-polar groups, thereby facilitating pore construction. The interfacial activity model closely resembles the disordered toroidal pore model with the exception that it describes membrane disruption by α-AMPs without a need for oligomerization by these peptides [122]. Most recently, MD studies on magainin 2 have suggested that disordered toroidal pores formed by the peptide in the bacterial membrane represent equilibrium pore structures. These studies showed that the initial binding of the peptide to the bacterial membrane leads to an imbalance of surface concentrations across the bilayer. This imbalance provides the driving force for bilayer buckling and vesicle budding, accompanied by the nucleation of giant transient pores, which gradually decrease in diameter to equilibrium values [157].

5.4.2
Carpet Mechanism and the Shai–Huang–Matsazuki Model

The "carpet" model was first proposed to describe the membrane interactions of dermaseptin [158], which was isolated from the tree frog, *Phyllomedusa sauvagii* [159], and then from many other α-AMPs with the result that today it is the most commonly cited mechanism of membrane perturbation for these peptides [116]. According to the carpet mechanism, α-AMPs "carpet" the surface a bacterial membrane in an orientation parallel to this surface (Figure 5.5). When the accumulation of these peptides has reached a critical level, they reorientate to form toroidal pores and, eventually, these pores coalesce forming "islands" in the membrane, which ultimately leads to membrane fragmentation via global destablization and micellization of the bilayer [4]. The collapse of membrane integrity described in this latter stage of the carpet mechanism has also been presented as the detergent model [124] while the carpet mechanism itself is generally considered to be an extension or extreme case of the toroidal pore model [5]. In the

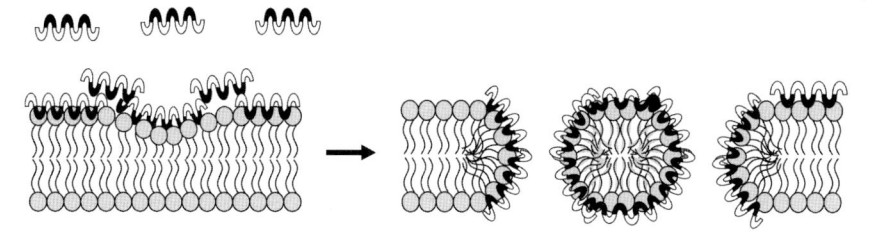

Figure 5.5 "Carpet" mechanism of AMPs antimicrobial action (adapted from [5]) showing α-AMPs "carpeting" the surface of a target microbial membrane. At a critical concentration the membrane is saturated with peptides, which leads to the formation of multiple toroidal pores, micellization of the membrane, and ultimately cell lysis.

mid-2000s, the "membrane discrimination model" was proposed for α-AMPs [125, 160]. Essentially, this model proposes that these peptides permeabilize bacterial membranes via the carpet mechanism and eukaryotic membranes via the barrel-stave pore model [134], as reported for temporins A and L [161], which are α-AMPs from the frog, *Rana temporaria* [162]. Also towards the mid-2000s, a novel model was attributed to Shai, Huang, and Matsazuki (the SHM model), which appears to be one of the first attempts to incorporate models described above into a multicomponent paradigm of membrane interaction for α-AMPs (Figure 5.6) [126]. Essentially, the SHM model proposes that "carpeting" of the microbial membrane with α-AMPs leads to the displacement of membrane lipid, alterations to membrane structure, and either microbial membrane destruction or the internalization of these peptides. The SHM model would appear able to describe the membrane interactions of α-AMPs that form toroidal pores, such as various dermaseptins from the *Phyllomedusa* genus of frogs [163]; translocate the membrane via transient toroidal pores, such as buforin II [104] ; and permeabilize membranes via the carpet mechanism and detergent models, such as cecropin and melittin, respectively [4, 97].

5.5
Recent Novel Models for the Membrane Interactions of α-AMPs

Several authors have recently discussed "molecular shape models," which take into account the shapes of the lipids and peptides involved in the membrane interactions of α-AMPs. Essentially, the model proposes that the interactions of these differently shaped peptide and lipid molecules can lead to the adoption of a wide variety of bilayer morphologies; to describe the supramolecular complexes formed, the interactions of α-CAMPs with membranes are best described by phase diagrams. These diagrams include distinct areas that correspond to various models described above, such as the barrel-stave pore, toroidal pore, and carpet mechanism, and can therefore indicate the propensity of α-CAMPs to use a given model

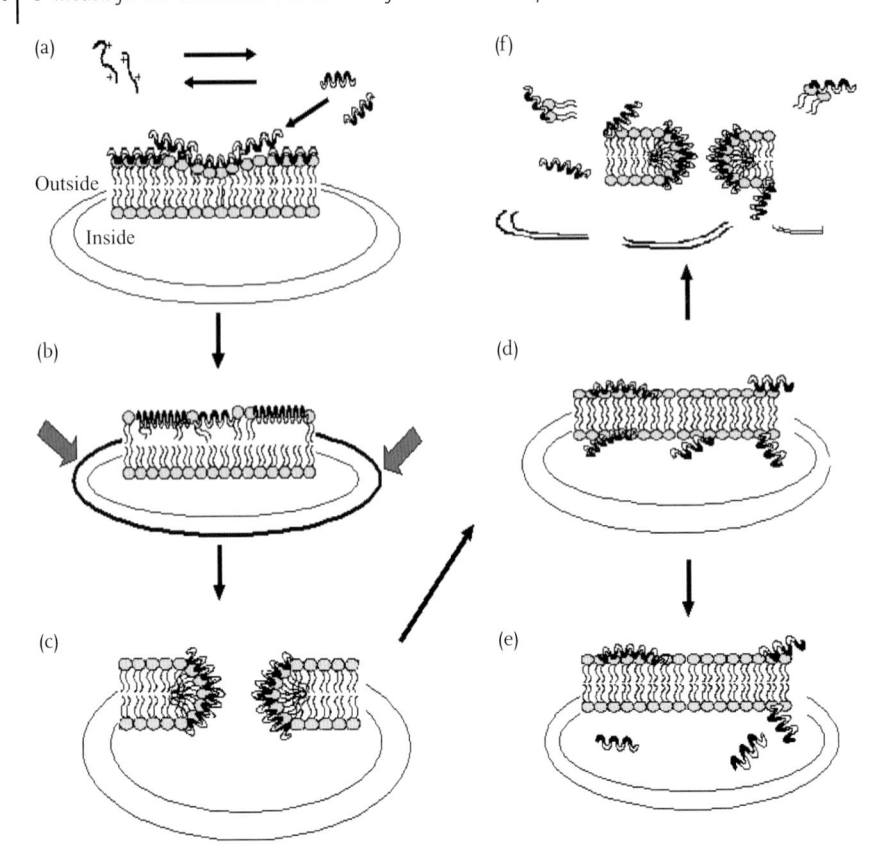

Figure 5.6 SHM model for the antimicrobial action of α-AMPs (adapted from [126]).
(a) α-CAMPs target and "carpet" the outer surface of a microbial membrane.
(b) Incorporation of these α-CAMPs into the membrane, accompanied by thinning of the outer leaflet and an increase in its surface area relative to the inner leaflet, resulting in strain within the bilayer (indicated by broad arrows). (c) Membrane phase transition, the appearance of membrane "wormhole" lesions, and transient pores, which facilitates the translocation of lipids and α-CAMPs to the inner leaflet (d). In some cases, the diffusion of α-CAMPs to intracellular targets may now occur (e) while in other cases, fragmentation and physical destruction of the microbial cell membrane can occur (f).

to permeabilize membranes or perturb the lipid packing [124, 164, 165]. In what appears to be a unique form of membrane association, temporin L and temporin B, also from the frog, *R. temporaria* [162], have been shown to intercalate more efficiently into membranes containing oxidized PC, which appeared to involve Schiff base formation between the amino groups of these peptides and aldehyde groups of the lipid. It was suggested that with the release of reactive oxygen species during phagocytosis, it is possible that lipid oxidation could increase bacterial membrane susceptibility to these α-CAMPs [129]. The "molecular electroporation model" proposes that binding of some α-CAMPs to the bacterial membrane generates an electrostatic potential across the bilayer sufficient to induce the formation of pores via electroporation [130]. Permeabilization of the membrane by electroporation requires a high charge density and it has been suggested that this mechanism is therefore likely to be used by strongly cationic α-CAMPs [166]. Currently, the best documented use of this model appears to be in the antibacterial action of NK-lysin [167], which is a highly basic porcine α-AMP [168]. As described earlier, another recently proposed model suggests that the high cationic charge of some α-AMPs, such as magainin analogs, induces the clustering of anionic lipids within the bacterial membrane, which then leads to the leakage of intracellular contents and/or depolarization of the bilayer. Alternatively, phase boundary effects between domains of differently charged lipids may compromise the overall membrane stability and lead to leakage [57]. In contrast to many of the above models, the "sand in a gearbox" model proposed that the antibacterial mechanisms of some α-CAMPs may involve membrane attack via non-pore-forming mechanisms. A series of synthetic α-CAMPs were found to induce perturbation of the lateral bacterial membrane structure, which disrupted processes vital to the cell, such as the function of membrane-bound multienzyme complexes, electron transport chains, and cell wall or lipid biosynthesis machineries [128]. In general, most of these novel models have been reported for the antibacterial action of only a few α-CAMPs, but two models, which appear to have a more generic utilization by these peptides, are the tilted peptide mechanism and the formation of amyloid structure.

5.6
Tilted Peptide Mechanism

As a part of the viral infection process, fusion proteins in the viral envelope mediate fusion between this envelope and the host cell membrane, which leads to internalization of the viral genome [169]. In the late 1980s, studies on the Newcastle Disease Virus showed that its fusion protein included an α-helical segment with a strong gradient in hydrophobicity along the helical long axis [170]. Similar segments have since been identified in the fusion proteins of a variety of other viruses [171] and it is now known that their hydrophobicity gradients allow them to penetrate the membrane at a shallow angle, thereby destabilizing membrane lipid organization and promoting the fusion of viral and host cell membranes [172]. These α-helical segments, or fusion peptides, are generally regarded as a novel class of membrane interactive α-helices [173]; until two decades ago, it was generally believed that protein α-helical structures involved in membrane interactions possessed an approximately constant level of hydrophobicity along their long axis [174]. This belief, which is inherent to the models described earlier in this chapter, was based on the observation that known α-helical protein architectures adopted an orientation that was approximately either parallel or perpendicular to the bilayer surface [172, 175].

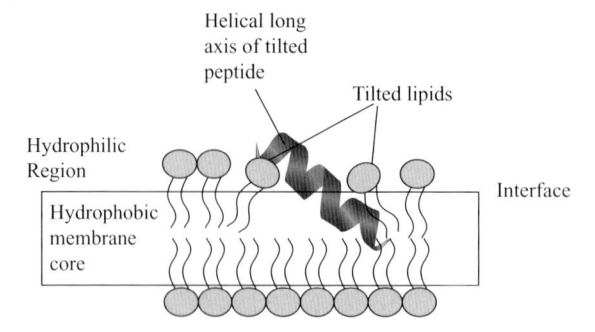

Figure 5.7 Tilted peptide mechanism: membrane insertion of a tilted α-CAMP with a hydrophobicity gradient increasing from the C-terminus to the N-terminus (adapted from [175]).

A variety of biophysical experimental techniques and molecular modeling have been used to study the process of membrane insertion by fusion peptides. These studies have shown that fusion peptides generally penetrate membranes at an angle between 20° and 80° to the bilayer normal, which creates negative curvature in the membrane, promoting the disturbance of lipid packing and hemifusion of the outer leaflet with the viral envelope [169]. Most recently, a model was presented to describe the mechanisms underpinning the process of membrane insertion by fusion peptides by considering a tilted peptide with a hydrophobicity gradient increasing from the C-terminus to the N-terminus (Figure 5.7). In this model, the depth of membrane penetration by the peptide was governed by electrostatic interactions and hydrogen bonding between its C-terminal region and moieties within the bilayer head-group region, effectively anchoring the peptide. The higher hydrophobicity of the peptide's extreme N-terminal region was found to promote deeper insertion of this region into the bilayer. However, the hydrophobic forces driving this effect were significantly diminished as the N-terminus approached the membrane core region with its low dielectric constant, indicating that this effect was not the major factor governing the tilt angle of the peptide. The tilt angle of the peptide appeared to be determined by both the length of the peptide and the distribution properties of its hydrophobicity gradient with significant hydrophobicity near the geometric midpoint of the peptide promoting deeper membrane insertion by the N-terminus and therefore a larger tilt angle (Figure 5.7) [176]. This model was generally supported by experimental data on viral fusion peptides [171, 176], and appears to generally represent tilted peptides reported in a variety of non-viral proteins and peptides [173, 177, 178], including α-CAMPs (Table 5.2).

Several authors have observed that there are sequence, structural, and functional similarities between viral fusion peptides and α-CAMPs [185], which led to the suggestion that convergent evolution could have occurred between these two peptide classes [186]. Indeed, C-terminal amidation is a common structural modification of α-CAMPs [187] and it has been shown that the addition of a C-terminal

Table 5.2 AMPs that use tilted peptide structure in their membrane interactions.

AMP	Key references
α-CAMPs	
Aurein 1.2	[131]
Citropin 1.1	[131]
VP1	[179]
AP1	[180]
Other AMPs	
Colicin E1	[181]
Human prion protein	[182]
α-Synuclein	[183]
Aβ40	[184]
Aβ42	[184]

amide to viral fusion peptides endows many of these peptides with antimicrobial activity [188]. This observation led Dennison *et al.* [131] to analyze aurein 1.2 and citropin 1.1, which are C-terminally amidated α-AMPs found in a variety of Australian tree frogs [189], and VP1 [190], a synthetic AMP derived from m-calpain for the possession of hydrophobicity gradients [131]. These analyses predicted that these three AMPs would possess tilted structure with similar characteristics to that of the HA2 peptide, which has been extensively characterized and is the fusion peptide of the influenza virus [191]. In all three cases, this prediction was strongly supported by experimental data, which showed that the antimicrobial activity of these AMPs involved the perturbation of microbial membranes via mechanisms consistent with tilted insertion [131, 192–194]. Dennison *et al.* [131] further analyzed a database of around 160 AMPs and predicted that the use of tilted peptide structure by these peptides in their membrane interactions may be a commonly used mechanism. Consistent with this suggestion, both theoretical analyses and experimental data have shown a variety of membrane interactive AMPs to adopt this structure (Table 5.2), ranging from bombolitin II, a toxin from the bumblebee *Megabombus pennsylvanicus* [195], to colicin E1, a toxin from the bacterium *E. coli* [181, 196].

5.7
Amyloidogenic Mechanisms

Amyloid can be described as a non-covalent oligomer of extended β-sheets that laterally self-assemble to yield fibrils [197], and this protein quaternary structure is most strongly associated with proteins and peptides that cause human neurodegenerative disorders [198]. Recent evidence suggests that the neurotoxicity of some of these molecules such as Aβ40 and Aβ42, the causative agents of Alzheimer's disease (AD), primarily derives from their role as AMPs and their ability to

Table 5.3 AMPs that use amyloid structure in their membrane interactions.

AMP	Key references
α-AMPs	
LL-37	[114]
Magainin 2	[199]
Melittin	[199]
VP1	[179, 190, 200]
Plantaracin	[201]
Temporins B and L	[202, 203]
Other AMPs	
Aβ40 and Aβ42	[204]
Eosinophil cationic protein	[205]
Lactoferrin	[206]
Dermaseptin PD-3-7	[207]
Dermaseptin S9	[208, 209]
Protegrin-1	[210]
Sakacin P	[199]
Microcin E492	[211]

permeabilize bacterial membranes via amyloid mediated mechanisms (Table 5.3) [132]. These observations led to the demonstration that a number of established α-CAMPs are able to adopt amyloid structure [132, 133] of which the best characterized are those detected in frogs, including magainin and temporins from the *Rana* genus (Table 5.3). Experimental studies have shown that in the highly anisotropic environment of the membrane interface, anionic lipid induces these peptides to undergo a series of conformational changes. These changes are accompanied by self-associations, which results in the production of amyloid fibrils via intermediate states populated by various species of soluble peptide oligomers [133]. It is now widely accepted that amyloid fibrils *per se* are non-toxic [212] and, in general, it is these soluble oligomeric species that play active roles in the function of α-CAMPs [133]. For example, it is believed that anionic lipid in the bacterial membrane promotes the formation of soluble α-helical and/or β-sheet oligomers by temporin B and temporin L, which then mediate membrane permeabilization and inactivation of the target organism [202]. Similar antibacterial mechanisms appear to be used by other α-CAMPs known to form amyloid, including LL-37 from humans [114] and plantaracin A from bacteria [201]. Based on this work, a model has been presented to describe amyloid formation by α-CAMPs (Figure 5.8) and according to this model, monomers of these peptides, which are generally unstructured in aqueous solution, adopt α-helical structures at the membrane interface that lead to membrane insertion and self-assembly to form α-helical multimers. These α-helical species then undergo conformational conversions to form β-sheet multimers that can further aggregate to generate the β-structures associated with amyloid fibrils (Figure 5.8). High propensities for self-association

5.7 Amyloidogenic Mechanisms

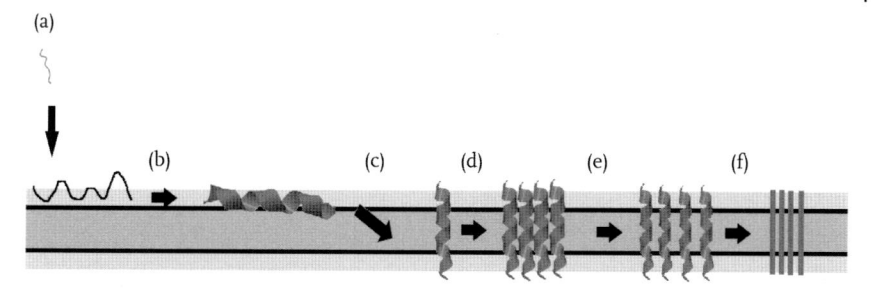

Figure 5.8 Putative schematic representation for amyloid formation by amyloidic α-CAMPs (adapted from [133]). In this scheme, the unfolded peptide (a) is attracted to the membrane via electrostatic associations between its cationic amino acid residues and anionic lipids in the bilayer. In response to the presence of these lipids and the ansiotropic environment of the membrane environment, the peptide then undergoes conformational changes, which leads to the adoption of an α-helical structure (b) accompanied by membrane insertion (c). At this juncture, anionic lipids that are associated with peptide molecules may become re-orientated to accommodate the amphiphilic characteristics of these molecules as described by the "leaky slit" model, for example (Figure 5.9). The alignment (d) and oligomerization (e) of inserted peptide molecules is then promoted by the presence of anionic lipid in the membrane via its ability to neutralize the cationic charge of the peptide. These oligomers subsequently adopt β-sheet conformations, which then leads to the formation of β-state oligomers (f) that can adopt higher levels of aggregation and form amyloid fibers that may also incorporate membrane lipid.

and conformational ambivalence to support environment-driven interconversion between secondary structures are key to the process of amyloid formation [133]. A number of theoretical techniques to identify regions of protein primary structure with these propensities have been developed and, using these techniques, it was predicted that many α-CAMPs from amphibians had the potential to form amyloid fibrils [202].

Currently, the mechanisms of bacterial membrane permeabilization used by amyloidogenic α-CAMPs are largely uncharacterized, and major unknowns are the structures and biophysical characteristics of the toxic oligomers that these peptides form. It has been hypothesized that these oligomers would be transient and although they may form channel-like structures, they are likely to vary in both size and morphology, including both circular, "pore-like," and linear arrangements of peptides, as described by the "leaky slit" model shown in Figure 5.9. It has also been hypothesized that for some α-AMPs, these oligomers may incorporate lipids, showing similarities to the toroidal pore mechanism of membrane permeabilization. A general requirement for the formation of these oligomers appears to be that the peptide segments forming them possess amphiphilic characteristics in order to concomitantly accommodate interaction with the membrane hydrophobic core and the hydrophilic head-group regions. Nonetheless, there appears to be no sequence homology between α-CAMPs that form these oligomers and, indeed, it

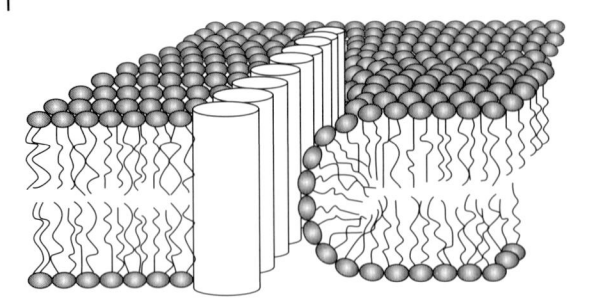

Figure 5.9 "Leaky slit" model for membrane permeabilization by toxic pre-amyloid oligomers of α-CAMPs (adapted from [133]). This model proposes that the simplest toxic arrangement for these oligomers is where peptide molecules form a linear amphiphilic array. In this arrangement, the hydrophobic face of the peptide molecules faces into the hydrophobic core of the bilayer arrangement. However, clearly, the opposing hydrophilic face of the peptide molecules cannot form a similar hydrophobic seal with the bilayer arrangement of membrane lipids. In response, these lipids are forced to adopt an arrangement with highly positive curvature so that their hydrophilic head-group region is able to interact with the hydrophilic face of the peptide oligomer. This arrangement shows similarities to the toroidal pore mechanism of membrane permeabilization proposed for host defense peptides and would be highly permeable to solutes thus making repair by the cell difficult.

has been suggested that peptide secondary structure is not a major factor in the antibacterial action of amyloidogenic AMPs. The only stringent requirements for the action of these peptides appear to relate to fibril formation, namely that the pre-amyloid oligomers that they form must be sufficiently long to span the width of the bilayer, which is generally achievable for peptides of around 20 amino acid residues or more [133, 202]. For shorter peptides, dimerization may be required as has been suggested for the 13-residue α-AMP, temporin L [213].

5.8
Discussion

This chapter has shown that in order to exert their antimicrobial action, CAMPs migrate through the microbial envelope to interact with the CM of these organisms. En route, these peptides may undergo conformational changes in the presence of the Gram-negative outer membrane or interact with lipid components of this membrane that act as receptors for these peptides. In other cases, CAMPs exhibit conformational changes in the presence of the microbial CM; however, in all cases, these peptides interact with this membrane to inactivate the target organism. The ability of CAMPs to partition into the microbial CM is influenced by a number of factors, including the presence of receptors, the $\Delta\psi$, and membrane fluidity along with the presence of sterols, anionic lipids and lipid-specific domains.

5.8 Discussion

The partitioning of CAMPs into the microbial CM can lead to a variety of membrane-permeabilizing mechanisms and a number of these models to describe mechanisms for α-CAMPs have been reviewed here. Some α-CAMPs use modes of action that involve pore or channel formation in the bilayer that are believed to kill bacteria via the induction of wholesale membrane lysis. The major examples of these latter models include the formation of amyloid-associated channels, barrel-stave pores, and variants of toroidal pores. These variants include the carpet mechanism, which can be considered as an extreme case of the toroidal pore model, and disordered toroidal pores, which, given the recent literature devoted to their discussion, seem set to become major established pore models. Transient forms of many of these pore models also appear to facilitate the translocation of α-CAMPs that kill bacteria by essentially non-membranolytic mechanisms. However, a number of the models described here would seem to more describe effects of α-CAMPs on membrane lipid organization that form a part of the models of pore and channel formation described above. For example, membrane thinning is considered to be an element of the toroidal pore model and membrane thickening appears in extensions to the aggregate model. Indeed, these effects on membrane organization appear to be generic features of the pore-forming mechanisms used by α-AMPs [127]. More recently, it has been suggested that the tilted peptide mechanism may contribute to the process of channel formation utilized by amyloidogenic α-CAMPs while the interfacial activity model would appear to constitute an element of disordered pore formation [6].

This chapter has clearly shown that a number of the models described here exhibit overlap in the processes involved in their descriptions, particularly in the case of the toroidal pore model and its variants. It is also noticeable that many of these latter models have been proposed based on the membrane interactions of magainins and their analogs. As recently postulated, it seems possible that this overlap may be explained in some cases by different studies presenting what were essentially different views of the same mechanism of membrane permeabilization used by an α-AMP [122]. Indeed, Wimley [122] also raised the possibility that there may be only one fundamental model for the membrane interactions of α-CAMPs and that the many different reported models are only manifestations of this fundamental model under specific conditions. Consistent with this view, several authors gave suggested that the route for intracellular action and/or membrane interaction by α-AMPs may be a multicomponent paradigm with many of the models described in Table 5.1 potentially accessible to these peptides. For example, the SHM model proposes that the toroidal pore model, the carpet mechanism, and the translocation of α-AMPs by transient pores are interconnected and form a network of potential pathways for α-AMPs to exert their antibacterial action [126]. More recent work extended the SHM model and suggested that most of the major pore-forming models involved in the ability of α-AMPs permeabilize and translocate membranes (Table 5.1) were interconnected to form numerous pathways potentially available to these peptides [120]. Moreover, these latter authors suggested that models such as the barrel-stave pore, toroidal pore, and disordered toroidal pore should be viewed as extreme states with mixed varieties of these models and interconversion between alternative states likely. Clearly, this latter scenario could give rise to the notion that there are multiple models for the membrane interactions involved in the antibacterial action of α-CAMPs when these perceived models are in fact different facets of a multicomponent paradigm. However, as to whether a fundamental model for the membrane interactions of α-CAMPs exists, that is a question yet to be definitively answered by research.

Although there is some debate as to the details of the models that describe the membrane interaction of α-AMPs, these models are useful reference frameworks and have utility in broader contexts. For example, these models have helped elucidate the membrane interactive mechanisms of other classes of biologically active peptides, which show structural and functional similarities to α-AMPs [132, 185], such as cell-penetrating peptides [102, 103]. More importantly, models described in this chapter have helped in the development of therapeutic drugs and their delivery. For example, buforin II translocation across the membrane has been shown to accommodate cargo addition, making it a promising candidate for development as a cell-penetrating peptide able to carry a wide range of macromolecules into living cells. A number of other AMPs have been shown able to serve as cell-penetrating peptides, and thereby the potential to deliver a range of therapeutic molecules in the treatment of cancer and other diseases [102]. Based on the pH-dependent ability of clavanin A to lyse membranes, a number of potential antimicrobial and anticancer agents have been designed that are inactive at neutral pH but at lower pH, whist benign towards normal eukaryotic cells, show toxicity to microbial cells and cancer cells [214]. Of major therapeutic importance, the paradigm of Aβ40 and Aβ42 as amyloidogenic AMPs has illuminated the currently held hypothesis that amyloid-mediated pore/channel formation in the neuronal membrane by these peptides may underlie their neurotoxicity and the pathogenesis of AD [215]. Based on this hypothesis, extensive searches for channel-blocking drugs are currently being conducted [215–218] that may help alleviate a devastating disease, which is the major contributor to dementia and predicted to affect around 42 million people worldwide by 2020 [219].

References

1 Yeagle, P.L. (2012) Introduction to lipid bilayers, in *Structure of Biological Membranes* (ed. P.L. Yeagle), CRC Press, Boca Raton, FL, pp. 1–7.

2 Haney, E.F., Nathoo, S., Vogel, H.J., and Prenner, E.J. (2010) Induction of non-lamellar lipid phases by antimicrobial peptides: a potential link to mode of action. *Chemistry and Physics of Lipids*, **163**, 82–93.

3 Teixeira, V., Feio, M.J., and Bastos, M. (2012) Role of lipids in the interaction of antimicrobial peptides with membranes. *Progress in Lipid Research*, **51**, 149–177.

4 Brogden, K.A. (2005) Antimicrobial peptides: pore formers or metabolic inhibitors in bacteria? *Nature Reviews Microbiology*, **3**, 238–250.

5 Dennison, S.R., Wallace, J., Harris, F., and Phoenix, D.A. (2005) Amphiphilic alpha-helical antimicrobial peptides and their structure/function relationships. *Protein and Peptide Letters*, **12**, 31–39.

6 Nguyen, L.T., Haney, E.F., and Vogel, H.J. (2011) The expanding scope of antimicrobial peptide structures and their modes of action. *Trends in Biotechnology*, **29**, 464–472.

7 Yeaman, M.R. and Yount, N.Y. (2003) Mechanisms of antimicrobial peptide action and resistance. *Pharmacological Reviews*, **55**, 27–55.

8 Giuliani, A., Pirri, G., and Nicoletto, S.F. (2007) Antimicrobial peptides: an overview of a promising class of therapeutics. *Central European Journal of Biology*, **2**, 1–33.

9 Nicolas, P. (2009) Multifunctional host defense peptides: intracellular-targeting antimicrobial peptides. *FEBS Journal*, **276**, 6483–6496.

10 Marcos, J.F. and Gandia, M. (2009) Antimicrobial peptides: to membranes and beyond. *Expert Opinion on Drug Discovery*, **4**, 659–671.

11 Dai, T., Huang, Y.-Y., and Hamblin, M.R. (2009) Photodynamic therapy for localized infections – state of the art. *Photodiagnosis and Photodynamic Therapy*, **6**, 170–188.

12 Grün, C.H. (2003) Structure and biosynthesis of fungal α-glucans, PhD Dissertation, University of Utrecht.

13 Papo, N., Oren, Z., Pag, U., Sahl, H.G., and Shai, Y. (2002) The consequence of sequence alteration of an amphipathic alpha-helical antimicrobial peptide and its diastereomers. *The Journal of Biological Chemistry*, **277**, 33913–33921.

14 McPhee, J.B., Tamber, S., Brazas, M.D., Lewenza, S., and Hancock, R.E.W. (2009) Antibiotic resistance due to reduced uptake, in *Antimicrobial Drug Resistance: Principles and Practice for the Clinical Bench* (ed. D.L. Mayers), Humana Press, Totowa, NJ, pp. 97–110.

15 Papo, N. and Shai, Y. (2005) Host defense peptides as new weapons in cancer treatment. *Cellular and Molecular Life Sciences*, **62**, 784–790.

16 Bucki, R., Namiot, D.B., Namiot, Z., Savage, P.B., and Janmey, P.A. (2008) Salivary mucins inhibit antibacterial activity of the cathelicidin-derived LL-37 peptide but not the cationic steroid CSA-13. *The Journal of Antimicrobial Chemotherapy*, **62**, 329–335.

17 Nishikawa, H. and Kitani, S. (2011) Gangliosides inhibit bee venom melittin cytotoxicity but not phospholipase A_2-induced degranulation in mast cells. *Toxicology and Applied Pharmacology*, **252**, 228–236.

18 Fadnes, B., Rekdal, O., and Uhlin-Hansen, L. (2009) The anticancer activity of lytic peptides is inhibited by heparan sulfate on the surface of the tumor cells. *BMC Cancer*, **9**, 183.

19 De Brucker, K., Cammue, B.P.A., and Thevissen, K. (2011) Apoptosis-inducing antifungal peptides and proteins. *Biochemical Society Transactions*, **39**, 1527–1532.

20 Batoni, G., Maisetta, G., Brancatisano, F.L., Esin, S., and Campa, M. (2011) Use of antimicrobial peptides against microbial biofilms: advantages and limits. *Current Medicinal Chemistry*, **18**, 256–279.

21 Cos, P., Tote, K., Horemans, T., and Maes, L. (2010) Biofilms: an extra hurdle for effective antimicrobial therapy. *Current Pharmaceutical Design*, **16**, 2279–2295.

22 Epand, R.M. and Epand, R.F. (2010) Biophysical analysis of membrane-targeting antimicrobial peptides: membrane properties and the design of peptides specifically targeting Gram-negative bacteria. *Advances in Molecular and Cellular Microbiology*, **18**, 116–127.

23 Peschel, A. and Collins, L.V. (2001) Staphylococcal resistance to antimicrobial peptides of mammalian and bacterial origin. *Peptides*, **22**, 1651–1659.

24 Laverty, G., Gorman, S.P., and Gilmore, B.F. (2011) The potential of antimicrobial peptides as biocides. *International Journal of Molecular Sciences*, **12**, 6566–6596.

25 Giacometti, A., Cirioni, O., Del Prete, M.S., Barchiesi, F., Fortuna, M., Drenaggi, D., and Scalise, G. (2000) *In vitro* activities of membrane-active peptides alone and in combination with clinically used antimicrobial agents against *Stenotrophomonas maltophilia*.

Antimicrobial Agents and Chemotherapy, **44**, 1716–1719.

26 Iwasaki, T., Saido-Sakanaka, H., Asaoka, A., Taylor, D., Ishibashi, J., and Yamakawa, M. (2007) *In vitro* activity of diastereomeric antimicrobial peptides alone and in combination with antibiotics against methicillin-resistant *Staphylococcus aureus* and *Pseudomonas aeruginosa*. *Journal of Insect Biotechnology and Sericology*, **76**, 25–29.

27 Giacometti, A., Cirioni, O., Kamysz, W., D'Amato, G., Silvestri, C., Licci, A., Nadolski, P., Riva, A., Łukasiak, J., and Scalise, G. (2005) *In vitro* activity of MSI-78 alone and in combination with antibiotics against bacteria responsible for bloodstream infections in neutropenic patients. *International Journal of Antimicrobial Agents*, **26**, 235–240.

28 Liu, Y.F., Han, F.F., Xie, Y.G., and Wang, Y.Z. (2011) Comparative antimicrobial activity and mechanism of action of bovine lactoferricin-derived synthetic peptides. *Biometals*, **24**, 1069–1078.

29 Giacometti, A., Cirioni, O., Kamysz, W., Silvestri, C., Licci, A., D'Amato, G., Nadolski, P., Riva, A., Lukasiak, J., and Scalise, G. (2005) *In vitro* activity and killing effect of uperin 3.6 against Gram-positive cocci isolated from immunocompromised patients. *Antimicrobial Agents and Chemotherapy*, **49**, 3933–3936.

30 Jiang, Y., Yi, X., Li, M., Wang, T., Qi, T., and She, X. (2012) Antimicrobial activities of recombinant mouse β-defensin 3 and its synergy with antibiotics. *Journal of Materials Science. Materials in Medicine*, **23**, 1723–1728.

31 Harris, M.R. and Coote, P.J. (2010) Combination of caspofungin or anidulafungin with antimicrobial peptides results in potent synergistic killing of *Candida albicans* and *Candida glabrata in vitro*. *International Journal of Antimicrobial Agents*, **35**, 347–356.

32 Tanida, T., Okamoto, T., Ueta, E., Yamamoto, T., and Osaki, T. (2006) Antimicrobial peptides enhance the candidacidal activity of antifungal drugs by promoting the efflux of ATP from *Candida* cells. *The Journal of Antimicrobial Chemotherapy*, **57**, 94–103.

33 Hancock, R.E.W. and Patrzykat, A. (2002) Clinical development of cationic antimicrobial peptides: from natural to novel antibiotics. *Current Drug Targets – Infectious Disorders*, **2**, 79–83.

34 Brandenburg, K., Andra, J., Garidel, P., and Gutsmann, T. (2011) Peptide-based treatment of sepsis. *Applied Microbiology and Biotechnology*, **90**, 799–808.

35 Rivas, L., Roman Luque-Ortega, J., Fernandez-Reyes, M., and Andreu, D. (2010) Membrane-active peptides as anti-infectious agents. *Journal of Applied Biomedicine*, **8**, 159–167.

36 Asaduzzaman, S.M. and Sonomoto, K. (2009) Lantibiotics: diverse activities and unique modes of action. *Journal of Bioscience and Bioengineering*, **107**, 475–487.

37 Vassiliadis, G., Destoumieux-Garzon, D., and Peduzzi, J. (2011) *Class II Microcins*, Springer, New York.

38 Zhao, M. (2011) Lantibiotics as probes for phosphatidylethanolamine. *Amino Acids*, **41**, 1071–1079.

39 Jose, F.M., Mónica, G., Eleonora, H., Lourdes, C., and Alberto, M. (2012) Antifungal peptides: exploiting non-lytic mechanisms and cell penetration properties, in *Small Wonders: Peptides for Disease Control* (eds K. Rajasekaran, J.W. Cary, J.M. Jaynes, and E. Montesinos), American Chemical Society, Washington, DC, pp. 337–357.

40 Hale, J.D.F. and Hancock, R.E.W. (2007) Alternative mechanisms of action of cationic antimicrobial peptides on bacteria. *Expert Review of Anti-Infective Therapy*, **5**, 951–959.

41 Chang, T.-W., Lin, Y.-M., Wang, C.-F., and Liao, Y.-D. (2012) Outer membrane lipoprotein Lpp is Gram-negative bacterial cell surface receptor for cationic antimicrobial peptides. *The Journal of Biological Chemistry*, **287**, 418–428.

42 Dawson, R.M. and Liu, C.-Q. (2009) Cathelicidin peptide SMAP-29: comprehensive review of its properties and potential as a novel class of antibiotics. *Drug Development Research*, **70**, 481–498.

43 Mendez-Samperio, P. (2010) The human cathelicidin hCAP18/LL-37: a multifunctional peptide involved in mycobacterial infections. *Peptides*, **31**, 1791–1798.

44 Cederlund, A., Gudmundsson, G.H., and Agerberth, B. (2011) Antimicrobial peptides important in innate immunity. *FEBS Journal*, **278**, 3942–3951.

45 Kamimori, H., Hall, K., Craik, D.J., and Aguilar, M.-I. (2005) Studies on the membrane interactions of the cyclotides kalata B1 and kalata B6 on model membrane systems by surface plasmon resonance. *Analytical Biochemistry*, **337**, 149–153.

46 Burman, R., Strömstedt, A.A., Malmsten, M., and Göransson, U. (2011) Cyclotide–membrane interactions: defining factors of membrane binding, depletion and disruption. *Biochimica et Biophysica Acta – Biomembranes*, **1808**, 2665–2673.

47 Pranting, M., Loov, C., Burman, R., Goransson, U., and Andersson, D.I. (2010) The cyclotide cycloviolacin O2 from *Viola odorata* has potent bactericidal activity against Gram-negative bacteria. *The Journal of Antimicrobial Chemotherapy*, **65**, 1964–1971.

48 Gran, L., Sletten, K., and Skjeldal, L. (2008) Cyclic peptides from *Oldenlandia affinis* DC. Molecular and biological properties. *Chemistry & Biodiversity*, **5**, 2014–2022.

49 Wilmes, M., Cammue, B.P.A., Sahl, H.-G., and Thevissen, K. (2011) Antibiotic activities of host defense peptides: more to it than lipid bilayer perturbation. *Natural Product Reports*, **28**, 1350–1358.

50 Aerts, A.M., Francois, I.E.J.A., Cammue, B.P.A., and Thevissen, K. (2008) The mode of antifungal action of plant, insect and human defensins. *Cellular and Molecular Life Sciences*, **65**, 2069–2079.

51 Thevissen, K., Kristensen, H.-H., Thomma, B.P.H.J., Cammue, B.P.A., and Francois, I.E.J.A. (2007) Therapeutic potential of antifungal plant and insect defensins. *Drug Discovery Today*, **12**, 966–971.

52 van Dijk, A., Veldhuizen, E.J.A., and Haagsman, H.P. (2008) Avian defensins. *Veterinary Immunology and Immunopathology*, **124**, 1–18.

53 Arnett, E. and Seveau, S. (2011) The multifaceted activities of mammalian defensins. *Current Pharmaceutical Design*, **17**, 4254–4269.

54 Sharma, M. (2011) Plant defensins: novel antimicrobial peptides. *Vegetos*, **24**, 126–135.

55 Schneider, T. and Sahl, H.-G. (2010) Lipid II and other bactoprenol-bound cell wall precursors as drug targets. *Current Opinion in Investigational Drugs*, **11**, 157–164.

56 Shai, Y. (1999) Mechanism of the binding, insertion and destabilization of phospholipid bilayer membranes by alpha-helical antimicrobial and cell non-selective membrane-lytic peptides. *Biochimica et Biophysica Acta*, **1462**, 55–70.

57 Epand, R.M. and Epand, R.F. (2011) Bacterial membrane lipids in the action of antimicrobial agents. *Journal of Peptide Science*, **17**, 298–305.

58 Strömstedt, A.A., Ringstad, L., Schmidtchen, A., and Malmsten, M. (2010) Interaction between amphiphilic peptides and phospholipid membranes. *Current Opinion in Colloid & Interface Science*, **15**, 467–478.

59 Lohner, K. and Prenner, E.J. (1999) Differential scanning calorimetry and X-ray diffraction studies of the specificity of the interaction of antimicrobial peptides with membrane-mimetic systems. *Biochimica et Biophysica Acta – Biomembranes*, **1462**, 141–156.

60 Ratledge, C. and Wilkinson, S.G. (1988) *Microbial Lipids*, vol. 1, Academic Press, London.

61 van Meer, G., Voelker, D.R., and Feigenson, G.W. (2008) Membrane lipids: where they are and how they behave. *Nature Reviews Molecular Cell Biology*, **9**, 112–124.

62 Wang, G., Li, X., and Wang, Z. (2009) APD2: the updated antimicrobial peptide database and its application in peptide design. *Nucleic Acids Research*, **37**, D933–D937.

63 Wang, G., Li, X., and Zasloff, M. (2010) A database view of naturally occurring antimicrobial peptides: nomenclature, classification and amino acid sequence analysis. *Advances in Molecular and Cellular Microbiology*, **18**, 1–21.

64 Pelegrini, P.B., Perseghini del Sarto, R., Silva, O.N., Franco, O.L., and Grossi-de-Sa, M.F. (2011) Antibacterial peptides from plants: what they are and how they probably work. *Biochemistry Research International*, **2011**, 250349.

65 Fjell, C.D., Hiss, J.A., Hancock, R.E.W., and Schneider, G. (2012) Designing antimicrobial peptides: form follows function. *Nature Reviews Drug Discovery*, **11**, 37–51.

66 Harris, M., Mora-Montes, H.M., Gow, N.A.R., and Coote, P.J. (2009) Loss of mannosylphosphate from *Candida albicans* cell wall proteins results in enhanced resistance to the inhibitory effect of a cationic antimicrobial peptide via reduced peptide binding to the cell surface. *Microbiology*, **155**, 1058–1070.

67 Bradshaw, J.P. (2003) Cationic antimicrobial peptides – issues for potential clinical use. *Biodrugs*, **17**, 233–240.

68 Zwaal, R.F.A., Comfurius, P., and Bevers, E.M. (2005) Surface exposure of phosphatidylserine in pathological cells. *Cellular and Molecular Life Sciences*, **62**, 971–988.

69 Fadeel, B. and Xue, D. (2009) The ins and outs of phospholipid asymmetry in the plasma membrane: roles in health and disease. *Critical Reviews in Biochemistry and Molecular Biology*, **44**, 264–277.

70 Frolov, V.A., Shnyrova, A.V., and Zimmerberg, J. (2011) Lipid polymorphisms and membrane shape. *Cold Spring Harbor Perspectives in Biology*, **3**, a004747.

71 Chen, R. and Mark, A.E. (2011) The effect of membrane curvature on the conformation of antimicrobial peptides: implications for binding and the mechanism of action. *European Biophysics Journal with Biophysics Letters*, **40**, 545–553.

72 Pasupuleti, M., Schmidtchen, A., and Malmsten, M. (2012) Antimicrobial peptides: key components of the innate immune system. *Critical Reviews in Biotechnology*, **32**, 143–171.

73 Thevissen, K., Warnecke, D.C., Francois, I.E., Leipelt, M., Heinz, E., Ott, C., Zahringer, U., Thomma, B.P., Ferket, K.K., and Cammue, B.P. (2004) Defensins from insects and plants interact with fungal glucosylceramides. *The Journal of Biological Chemistry*, **279**, 3900–3905.

74 Thevissen, K., Ferket, K.K., Francois, I.E., and Cammue, B.P. (2003) Interactions of antifungal plant defensins with fungal membrane components. *Peptides*, **24**, 1705–1712.

75 Lee, D.G., Kim, P.I., Park, Y., Park, S.C., Woo, E.R., and Hahm, K.S. (2002) Antifungal mechanism of SMAP-29 (1–18) isolated from sheep myeloid mRNA against *Trichosporon beigelii*. *Biochemical and Biophysical Research Communications*, **295**, 591–596.

76 Frantz, J.-F., Desbat, B., and Dufourc, E.J. (2009) Selectivity of cateslytin for fungi: the role of acidic lipid-ergosterol membrane fluidity in antimicrobial action. *FASEB Journal*, **23**, 3692–3701.

77 Mason, A.J., Marquette, A., and Bechinger, B. (2007) Zwitterionic phospholipids and sterols modulate antimicrobial peptide-induced membrane destabilization. *Biophysical Journal*, **93**, 4289–4299.

78 Li, J.F., Zhang, J., Xu, X.Z., Han, Y.Y., Cui, X.W., Chen, Y.Q., and Zhang, S.Q. (2012) The antibacterial peptide ABP-CM4: the current state of its production and applications. *Amino Acids*, **42**, 2393–2402.

79 Zhang, J., Wu, X., and Zhang, S.-Q. (2008) Antifungal mechanism of antibacterial peptide, ABP-CM4, from *Bombyx mori* against *Aspergillus niger*. *Biotechnology Letters*, **30**, 2157–2163.

80 Qi, X., Zhou, C., Li, P., Xu, W., Cao, Y., Ling, H., Ning Chen, W., Ming Li, C., Xu, R., Lamrani, M., Mu, Y., Leong, S.S.J., Wook Chang, M., and Chan-Park, M.B. (2010) Novel short antibacterial and antifungal peptides with low

cytotoxicity: efficacy and action mechanisms. *Biochemical and Biophysical Research Communications*, **398**, 594–600.

81 Benincasa, M., Scocchi, M., Pacor, S., Tossi, A., Nobili, D., Basaglia, G., Busetti, M., and Gennaro, R. (2006) Fungicidal activity of five cathelicidin peptides against clinically isolated yeasts. *The Journal of Antimicrobial Chemotherapy*, **58**, 950–959.

82 Soltani, S., Keymanesh, K., and Sardari, S. (2007) In silico analysis of antifungal peptides: determining the lead template sequence of potent antifungal peptides. *Expert Opinion on Drug Discovery*, **2**, 837–847.

83 Hancock, R.E.W. and Chapple, D.S. (1999) Peptide antibiotics. *Antimicrobial Agents and Chemotherapy*, **43**, 1317–1323.

84 Jenssen, H., Hamill, P., and Hancock, R.E.W. (2006) Peptide antimicrobial agents. *Clinical Microbiology Reviews*, **19**, 491–511.

85 Lopez-Garcia, B., Marcos, J.F., Abad, C., and Perez-Paya, E. (2004) Stabilisation of mixed peptide/lipid complexes in selective antifungal hexapeptides. *Biochimica et Biophysica Acta – Biomembranes*, **1660**, 131–137.

86 Espenshade, P.J. and Hughes, A.L. (2007) Regulation of sterol synthesis in eukaryotes. *Annual Review of Genetics*, **41**, 401–427.

87 Henriksen, J., Rowat, A.C., Brief, E., Hsueh, Y.W., Thewalt, J.L., Zuckermann, M.J., and Ipsen, J.H. (2006) Universal behavior of membranes with sterols. *Biophysical Journal*, **90**, 1639–1649.

88 Hsueh, Y.-W., Chen, M.-T., Patty, P.J., Code, C., Cheng, J., Frisken, B.J., Zuckermann, M., and Thewalt, J. (2007) Ergosterol in POPC membranes: physical properties and comparison with structurally similar sterols. *Biophysical Journal*, **92**, 1606–1615.

89 Sood, R. and Kinnunen, P. (2009) Cholesterol, lanosterol, and ergosterol attenuate the membrane association of LL-37(W27F) and temporin L. *FEBS Journal*, **276**, 345–345.

90 Glukhov, E., Stark, M., Burrows, L.L., and Deber, C.M. (2005) Basis for selectivity of cationic antimicrobial peptides for bacterial versus mammalian membranes. *The Journal of Biological Chemistry*, **280**, 33960–33967.

91 Brender, J.R., McHenry, A.J., and Ramamoorthy, A. (2012) Does cholesterol play a role in the bacterial selectivity of antimicrobial peptides? *Frontiers in Immunology*, **3**, 195.

92 Verly, R.M., Rodrigues, M.A., Daghastanli, K.R.P., Denadai, A.M.L., Cuccovia, I.M., Bloch, C., Jr, Frezard, F., Santoro, M.M., Pilo-Veloso, D., and Bemquerer, M.P. (2008) Effect of cholesterol on the interaction of the amphibian antimicrobial peptide DD K with liposomes. *Peptides*, **29**, 15–24.

93 Wu, G., Wu, H., Fan, X., Zhao, R., Li, X., Wang, S., Ma, Y., Shen, Z., and Xi, T. (2010) Selective toxicity of antimicrobial peptide S-thanatin on bacteria. *Peptides*, **31**, 1669–1673.

94 Schmidtchen, A., Ringstad, L., Kasetty, G., Mizuno, H., Rutland, M.W., and Malmsten, M. (2011) Membrane selectivity by W-tagging of antimicrobial peptides. *Biochimica et Biophysica Acta – Biomembranes*, **1808**, 1081–1091.

95 Matsuzaki, K., Sugishita, K., Fujii, N., and Miyajima, K. (1995) Molecular basis for membrane selectivity of an antimicrobial peptide, Magainin-2. *Biochemistry*, **34**, 3423–3429.

96 M'Baye, G., Mely, Y., Duportail, G., and Klymchenko, A.S. (2008) Liquid ordered and gel phases of lipid bilayers: fluorescent probes reveal close fluidity but different hydration. *Biophysical Journal*, **95**, 1217–1225.

97 Raghuraman, H. and Chattopadhyay, A. (2007) Melittin: a membrane-active peptide with diverse functions. *Bioscience Reports*, **27**, 189–223.

98 Hall, K., Lee, T.-H., and Aguilar, M.-I. (2011) The role of electrostatic interactions in the membrane binding of melittin. *Journal of Molecular Recognition*, **24**, 108–118.

99 Raghuraman, H. and Chattopadhyay, A. (2004) Interaction of melittin with membrane cholesterol: a fluorescence approach. *Biophysical Journal*, **87**, 2419–2432.

100 Wessman, P., Morin, M., Reijmar, K., and Edwards, K. (2010) Effect of alpha-helical peptides on liposome structure: a comparative study of melittin and alamethicin. *Journal of Colloid and Interface Science*, **346**, 127–135.

101 Dekruijff, B. (1990) Cholesterol as a target for toxins. *Bioscience Reports*, **10**, 127–130.

102 Splith, K. and Neundorf, I. (2011) Antimicrobial peptides with cell-penetrating peptide properties and vice versa. *European Biophysics Journal*, **40**, 387–397.

103 Henriques, S.T., Melo, M.N., and Castanho, M.A. (2006) Cell-penetrating peptides and antimicrobial peptides: how different are they? *The Biochemical Journal*, **399**, 1–7.

104 Cho, J.H., Sung, B.H., and Kim, S.C. (2009) Buforins: histone H2A-derived antimicrobial peptides from toad stomach. *Biochimica et Biophysica Acta – Biomembranes*, **1788**, 1564–1569.

105 Xie, Y., Fleming, E., Chen, J.L., and Elmore, D.E. (2011) Effect of proline position on the antimicrobial mechanism of buforin II. *Peptides*, **32**, 677–682.

106 Lan, Y., Ye, Y., Kozlowska, J., Lam, J.K.W., Drake, A.F., and Mason, A.J. (2010) Structural contributions to the intracellular targeting strategies of antimicrobial peptides. *Biochimica et Biophysica Acta – Biomembranes*, **1798**, 1934–1943.

107 Smith, V.J., Desbois, A.P., and Dyrynda, E.A. (2010) Conventional and unconventional antimicrobials from fish, marine invertebrates and micro-algae. *Marine Drugs*, **8**, 1213–1262.

108 van Kan, E.J.M., Demel, R.A., Breukink, E., van der Bent, A., and de Kruijff, B. (2002) Clavanin permeabilizes target membranes via two distinctly different pH-dependent mechanisms. *Biochemistry*, **41**, 7529–7539.

109 van Kan, E.J.M., Ganchev, D.N., Snel, M.M.E., Chupin, V., van der Bent, A., and de Kruijff, B. (2003) The peptide antibiotic clavanin A interacts strongly and specifically with lipid bilayers. *Biochemistry*, **42**, 11366–11372.

110 Steinstraesser, L., Kraneburg, U., Jacobsen, F., and Al-Benna, S. (2011) Host defense peptides and their antimicrobial-immunomodulatory duality. *Immunobiology*, **216**, 322–333.

111 Jenssen, H. and Hancock, R.E.W. (2010) Therapeutic potential of HDPs as immunomodulatory agents, in *Antimicrobial Peptides: Methods and Protocols* (eds A. Giuliani and A.C. Rinaldi), Humana Press, Totowa, NJ, pp. 329–347.

112 Lee, C.-C., Sun, Y., Qian, S., and Huang, H.W. (2011) Transmembrane pores formed by human antimicrobial peptide LL-37. *Biophysical Journal*, **100**, 1688–1696.

113 Burton, M.F. and Steel, P.G. (2009) The chemistry and biology of LL-37. *Natural Product Reports*, **26**, 1572–1584.

114 Sood, R., Domanov, Y., Pietiainen, M., Kontinen, V.P., and Kinnunen, P.K.J. (2008) Binding of LL-37 to model biomembranes: insight into target vs host cell recognition. *Biochimica et Biophysica Acta – Biomembranes*, **1778**, 983–996.

115 Nijnik, A. and Hancock, R.E.W. (2009) The roles of cathelicidin LL-37 in immune defences and novel clinical applications. *Current Opinion in Hematology*, **16**, 41–47.

116 Wimley, W.C. and Hristova, K. (2011) Antimicrobial peptides: successes, challenges and unanswered questions. *The Journal of Membrane Biology*, **239**, 27–34.

117 Kobayashi, S., Chikushi, A., Tougu, S., Imura, Y., Nishida, M., Yano, Y., and Matsuzaki, K. (2004) Membrane translocation mechanism of the antimicrobial peptide buforin 2. *Biochemistry*, **43**, 15610–15616.

118 Pokorny, A. and Almeida, P.F.F. (2004) Kinetics of dye efflux and lipid flip-flop induced by delta-lysin in phosphatidylcholine vesicles and the mechanism of graded release by amphipathic, alpha-helical peptides. *Biochemistry*, **43**, 8846–8857.

119 Yoneyama, F., Imura, Y., Ohno, K., Zendo, T., Nakayama, J., Matsuzaki, K., and Sonomoto, K. (2009) Peptide-lipid huge toroidal pore, a new antimicrobial

mechanism mediated by a lactococcal bacteriocin, lacticin Q. *Antimicrobial Agents and Chemotherapy*, **53**, 3211–3217.
120 Sengupta, D., Leontiadou, H., Mark, A.E., and Marrink, S.J. (2008) Toroidal pores formed by antimicrobial peptides show significant disorder. *Biochimica et Biophysica Acta – Biomembranes*, **1778**, 2308–2317.
121 Matsuzaki, K. (1998) Magainins as paradigm for the mode of action of pore forming polypeptides. *Biochimica et Biophysica Acta*, **1376**, 391–400.
122 Wimley, W.C. (2010) Describing the mechanism of antimicrobial peptide action with the interfacial activity model. *ACS Chemical Biology*, **5**, 905–917.
123 Gregory, S.M., Pokorny, A., and Almeida, P.F.F. (2009) Magainin 2 revisited: a test of the quantitative model for the all-or-none permeabilization of phospholipid vesicles. *Biophysical Journal*, **96**, 116–131.
124 Bechinger, B. and Lohner, K. (2006) Detergent-like actions of linear amphipathic cationic antimicrobial peptides. *Biochimica et Biophysica Acta – Biomembranes*, **1758**, 1529–1539.
125 Chen, Y., Guarnieri, M.T., Vasil, A.I., Vasil, M.L., Mant, C.T., and Hodges, R.S. (2007) Role of peptide hydrophobicity in the mechanism of action of alpha-helical antimicrobial peptides. *Antimicrobial Agents and Chemotherapy*, **51**, 1398–1406.
126 Zasloff, M. (2002) Antimicrobial peptides of multicellular organisms. *Nature*, **415**, 389–395.
127 Lohner, K. (2009) New strategies for novel antibiotics: peptides targeting bacterial cell membranes. *General Physiology and Biophysics*, **28**, 105–116.
128 Pag, U., Oedenkoven, M., Sass, V., Shai, Y., Shamova, O., Antcheva, N., Tossi, A., and Sahl, H.-G. (2008) Analysis of *in vitro* activities and modes of action of synthetic antimicrobial peptides derived from an α-helical "sequence template". *The Journal of Antimicrobial Chemotherapy*, **61**, 341–352.
129 Mattila, J., Sabatini, K., and Kinnunen, P.K.J. (2008) Oxidized phospholipids as potential molecular targets for antimicrobial peptides. *Biochimica et Biophysica Acta – Biomembranes*, **1778**, 2041–2050.
130 Chan, D.I., Prenner, E.J., and Vogel, H.J. (2006) Tryptophan- and arginine-rich antimicrobial peptides: structures and mechanisms of action. *Biochimica et Biophysica Acta – Biomembranes*, **1758**, 1184–1202.
131 Dennison, S.R., Harris, F., and Phoenix, D.A. (2005) Are oblique orientated alpha-helices used by antimicrobial peptides for membrane invasion? *Protein and Peptide Letters*, **12**, 27–29.
132 Harris, F., Dennison, S.R., and Phoenix, D.A. (2012) Amyloidic host defence peptides: a new paradigm to investigate neurodegenerative diseases? *FASEB Journal*, **26**, 1776–1781.
133 Kinnunen, P.K.J. (2009) Amyloid formation on lipid membrane surfaces. *The Open Biology Journal*, **2**, 163–175.
134 Huang, Y., Huang, J., and Chen, Y. (2010) Alpha-helical cationic antimicrobial peptides: relationships of structure and function. *Protein & Cell*, **1**, 143–152.
135 Ehrenstein, G. and Lecar, H. (1977) Electrically gated ionic channels in lipid bilayers. *Quarterly Reviews of Biophysics*, **10**, 1–34.
136 Leitgeb, B., Szekeres, A., Manczinger, L., Vagvolgyi, C., and Kredics, L. (2007) The history of alamethicin: a review of the most extensively studied peptaibol. *Chemistry & Biodiversity*, **4**, 1027–1051.
137 Duclohier, H. (2010) Antimicrobial peptides and peptaibols, substitutes for conventional antibiotics. *Current Pharmaceutical Design*, **16**, 3212–3223.
138 Marsh, D. (2009) Orientation and peptide-lipid interactions of alamethicin incorporated in phospholipid membranes: polarized infrared and spin-label EPR spectroscopy. *Biochemistry*, **48**, 729–737.
139 Ludtke, S.J., He, K., Heller, W.T., Harroun, T.A., Yang, L., and Huang, H. (1996) Membrane pores induced by magainin. *Biochemistry*, **35**, 13723–13728.
140 Matsuzaki, K., Murase, O., Fujii, N., and Miyajima, K. (1996) An antimicrobial peptide, magainin 2, induced rapid

flip-flop of phospholipids coupled with pore formation and peptide translocation. *Biochemistry*, **35**, 11361–11368.
141 Zasloff, M. (1987) Magainins, a class of antimicrobial peptides from *Xenopus* skin: isolation, characterization of two active forms, and partial cDNA sequence of a precursor. *Proceedings of the National Academy of Sciences of the United States of America*, **84**, 5449–5453.
142 Tossi, A., Sandri, L., and Giangaspero, A. (2000) Amphipathic, alpha-helical antimicrobial peptides. *Biopolymers*, **55**, 4–30.
143 Shai, Y. and Oren, Z. (2001) From "carpet" mechanism to de-novo designed diastereomeric cell-selective antimicrobial peptides. *Peptides*, **22**, 1629–1641.
144 Kobayashi, S., Takeshima, K., Park, C.B., Kim, S.C., and Matsuzaki, K. (2000) Interactions of the novel antimicrobial peptide buforin 2 with lipid bilayers: proline as a translocation promoting factor. *Biochemistry*, **39**, 8648–8654.
145 Iwatani, S., Zendo, T., and Sonomoto, K. (2011) *Class IId or Linear and Non-Pediocin-Like Bacteriocins*, Springer, New York.
146 Matsuzaki, K. (1999) Why and how are peptide–lipid interactions utilized for self-defense? Magainins and tachyplesins as archetypes. *Biochimica et Biophysica Acta – Biomembranes*, **1462**, 1–10.
147 Powers, J.P.S., Tan, A., Ramamoorthy, A., and Hancock, R.E.W. (2005) Solution structure and interaction of the antimicrobial polyphemusins with lipid membranes. *Biochemistry*, **44**, 15504–15513.
148 Almeida, P.F. and Pokorny, A. (2009) Mechanisms of antimicrobial, cytolytic, and cell-penetrating peptides: from kinetics to thermodynamics. *Biochemistry*, **48**, 8083–8093.
149 Steiner, H., Hultmark, D., Engstrom, A., Bennich, H., and Boman, H.G. (1981) Sequence and specificity of two anti-bacterial proteins involved in insect immunity. *Nature*, **292**, 246–248.
150 Gregory, S.M., Cavenaugh, A., Journigan, V., Pokorny, A., and Almeida, P.F.F. (2008) A quantitative model for the all-or-none permeabilization of phospholipid vesicles by the antimicrobial peptide cecropin A. *Biophysical Journal*, **94**, 1667–1680.
151 Verdon, J., Girardin, N., Lacombe, C., Berjeaud, J.-M., and Héchard, Y. (2009) δ-Hemolysin, an update on a membrane-interacting peptide. *Peptides*, **30**, 817–823.
152 Leontiadou, H., Mark, A.E., and Marrink, S.J. (2006) Antimicrobial peptides in action. *Journal of the American Chemical Society*, **128**, 12156–12161.
153 Rzepiela, A.J., Sengupta, D., Goga, N., and Marrink, S.J. (2010) Membrane poration by antimicrobial peptides combining atomistic and coarse-grained descriptions. *Faraday Discussions*, **144**, 431–443.
154 Mihajlovic, M. and Lazaridis, T. (2010) Antimicrobial peptides bind more strongly to membrane pores. *Biochimica et Biophysica Acta – Biomembranes*, **1798**, 1494–1502.
155 Dempsey, C.E. (1990) The actions of melittin on membranes. *Biochimica et Biophysica Acta*, **1031**, 143–161.
156 Campagna, S., Saint, N., Molle, G., and Aumelas, A. (2007) Structure and mechanism of action of the antimicrobial peptide piscidin. *Biochemistry*, **46**, 1771–1778.
157 Woo, H.J. and Wallqvist, A. (2011) Spontaneous buckling of lipid bilayer and vesicle budding induced by antimicrobial peptide magainin 2: a coarse-grained simulation study. *The Journal of Physical Chemistry B*, **115**, 8122–8129.
158 Pouny, Y., Rapaport, D., Mor, A., Nicolas, P., and Shai, Y. (1992) Interaction of antimicrobial Dermaseptin and its fluorescently labeled analogs with phospholipid membrabes. *Biochemistry*, **31**, 12416–12423.
159 Mor, A., Nguyen, V.H., Delfour, A., Miglioresamour, D., and Nicolas, P.

(1991) Isolation, amino-acid sequence, and synthesis of dermaseptin, a novel antimicrobial peptide of amphibian skin. *Biochemistry*, **30**, 8824–8830.

160 Chen, Y.X., Vasil, A.I., Rehaume, L., Mant, C.T., Burns, J.L., Vasil, M.L., Hancock, R.E.W., and Hodges, R.S. (2006) Comparison of biophysical and biologic properties of alpha-helical enantiomeric antimicrobial peptides. *Chemical Biology & Drug Design*, **67**, 162–173.

161 Carotenuto, A., Malfi, S., Saviello, M.R., Campiglia, P., Gomez-Monterrey, I., Mangoni, M.L., Gaddi, L.M.H., Novellino, E., and Grieco, P. (2008) A different molecular mechanism underlying antimicrobial and hemolytic actions of temporins A and L. *Journal of Medicinal Chemistry*, **51**, 2354–2362.

162 Mangoni, M.L. (2006) Temporins, anti-infective peptides with expanding properties. *Cellular and Molecular Life Sciences*, **63**, 1060–1069.

163 Amiche, M. and Galanth, C. (2011) Dermaseptins as models for the elucidation of membrane-acting helical amphipathic antimicrobial peptides. *Current Pharmaceutical Biotechnology*, **12**, 1184–1193.

164 Aisenbrey, C. and Bechinger, B. (2011) Membrane insertion of polypeptides and their interactions with other biomacromolecules, in *Thermodynamics – Kinetics of Dynamic Systems* (ed. J.C.M. Piraján), InTech, New York, pp. 381–402.

165 Bechinger, B. (2009) Rationalizing the membrane interactions of cationic amphipathic antimicrobial peptides by their molecular shape. *Current Opinion in Colloid & Interface Science*, **14**, 349–355.

166 Lazarev, V.N. and Govorun, V.M. (2010) Antimicrobial peptides and their use in medicine. *Applied Biochemistry and Microbiology*, **46**, 803–814.

167 Miteva, M., Andersson, M., Karshikoff, A., and Otting, G. (1999) Molecular electroporation: a unifying concept for the description of membrane pore formation by antibacterial peptides, exemplified with NK-lysin. *FEBS Letters*, **462**, 155–158.

168 Sang, Y. and Blecha, F. (2009) Porcine host defense peptides: expanding repertoire and functions. *Developmental and Comparative Immunology*, **33**, 334–343.

169 Weissenhorn, W., Hinz, A., and Gaudin, Y. (2007) Virus membrane fusion. *FEBS Letters*, **581**, 2150–2155.

170 Brasseur, R., Lorge, P., Goormaghtigh, E., Ruysschaert, J.M., Espion, D., and Burny, A. (1988) The mode of insertion of the paramyxovirus F1 N-terminus into lipid matrix, an initial step in host cell/virus fusion. *Virus Genes*, **1**, 325–332.

171 Charloteaux, B., Lorin, A., Brasseur, R., and Lins, L. (2009) The "tilted peptide theory" links membrane insertion properties and fusogenicity of viral fusion peptides. *Protein and Peptide Letters*, **16**, 718–725.

172 Lins, L., Decaffmeyer, M., Thomas, A., and Brasseur, R. (2008) Relationships between the orientation and the structural properties of peptides and their membrane interactions. *Biochimica et Biophysica Acta – Biomembranes*, **1778**, 1537–1544.

173 Harris, F., Daman, A., Wallace, J., Dennison, S.R., and Phoenix, D.A. (2006) Oblique orientated alpha-helices and their prediction. *Current Protein & Peptide Science*, **7**, 529–537.

174 Brasseur, R. (2000) Tilted peptides: a motif for membrane destabilization (hypothesis). *Molecular Membrane Biology*, **17**, 31–40.

175 Thomas, A. and Brasseur, R. (2006) Tilted peptides: the history. *Current Protein & Peptide Science*, **7**, 523–527.

176 Taylor, A. and Sansom, M.S.P. (2010) Studies on viral fusion peptides: the distribution of lipophilic and electrostatic potential over the peptide determines the angle of insertion into a membrane. *European Biophysics Journal with Biophysics Letters*, **39**, 1537–1545.

177 Harris, F., Dennison, S., and Phoenix, D.A. (2006) The prediction of hydrophobicity gradients within membrane interactive protein

alpha-helices using a novel graphical technique. *Protein and Peptide Letters*, **13**, 595–600.

178 Harris, F., Wallace, J., and Phoenix, D.A. (2000) Use of hydrophobic moment plot methodology to aid the identification of oblique orientated alpha-helices. *Molecular Membrane Biology*, **17**, 201–207.

179 Dennison, S.R., Dante, S., Hauss, T., Brandenburg, K., Harris, F., and Phoenix, D.A. (2005) Investigations into the membrane interactions of m-calpain domain V. *Biophysical Journal*, **88**, 3008–3017.

180 Dennison, S.R., Morton, L.H.G., Brandenburg, K., Harris, F., and Phoenix, D.A. (2006) Investigations into the ability of an oblique alpha-helical template to provide the basis for design of an antimicrobial anionic amphiphilic peptide. *FEBS Journal*, **273**, 3792–3803.

181 Lins, L., El Kirat, K., Charloteaux, B., Flore, C., Stroobant, V., Thomas, A., Dufrene, Y., and Brasseur, R. (2007) Lipid-destabilizing properties of the hydrophobic helices H8 and H9 from colicin E1. *Molecular Membrane Biology*, **24**, 419–430.

182 Pillot, T., Lins, L., Goethals, M., Vanloo, B., Baert, J., Vandekerckhove, J., Rosseneu, M., and Brasseur, R. (1997) The 118–135 peptide lot the human prion protein forms amyloid fibrils and induces liposome fusion. *Journal of Molecular Biology*, **274**, 381–393.

183 Fantini, J., Carlus, D., and Yahi, N. (2011) The fusogenic tilted peptide (67–78) of alpha-synuclein is a cholesterol binding domain. *Biochimica et Biophysica Acta – Biomembranes*, **1808**, 2343–2351.

184 Pillot, T., Goethals, M., Vanloo, B., Talussot, C., Brasseur, R., Vandekerckhove, J., Rosseneu, M., and Lins, L. (1996) Fusogenic properties of the C-terminal domain of the Alzheimer beta-amyloid peptide. *The Journal of Biological Chemistry*, **271**, 28757–28765.

185 Joanne, P., Nicolas, P., and El Amri, C. (2009) Antimicrobial peptides and viral fusion peptides: how different they are? *Protein and Peptide Letters*, **16**, 743–750.

186 Gao, B., Sherman, P., Luo, L., Bowie, J., and Zhu, S. (2009) Structural and functional characterization of two genetically related meucin peptides highlights evolutionary divergence and convergence in antimicrobial peptides. *FASEB Journal*, **23**, 1230–1245.

187 Dennison, S.R., Harris, F., Bhatt, T., Singh, J., and Phoenix, D.A. (2009) The effect of C-terminal amidation on the efficacy and selectivity of antimicrobial and anticancer peptides. *Molecular and Cellular Biochemistry*, **332**, 43–50.

188 Zhu, S.Y., Aumelas, A., and Gao, B. (2011) Convergent evolution-guided design of antimicrobial peptides derived from influenza A virus hemagglutinin. *Journal of Medicinal Chemistry*, **54**, 1091–1095.

189 Fernandez, D.I., Gehman, J.D., and Separovic, F. (2009) Membrane interactions of antimicrobial peptides from Australian frogs. *Biochimica et Biophysica Acta – Biomembranes*, **1788**, 1630–1638.

190 Brandenburg, K., Harris, F., Dennison, S., Seydel, U., and Phoenix, D. (2002) Domain V of m-calpain shows the potential to form an oblique-orientated alpha-helix, which may modulate the enzyme's activity via interactions with anionic lipid. *European Journal of Biochemistry*, **269**, 5414–5422.

191 Cross, K.J., Langley, W.A., Russell, R.J., Skehel, J.J., and Steinhauer, D.A. (2009) Composition and functions of the influenza fusion peptide. *Protein and Peptide Letters*, **16**, 766–778.

192 Dennison, S.R., Harris, F., and Phoenix, D.A. (2007) The interactions of aurein 1.2 with cancer cell membranes. *Biophysical Chemistry*, **127**, 78–83.

193 Marcotte, I., Wegener, K.L., Lam, Y.H., Chia, B.C.S., de Planque, M.R.R., Bowie, J.H., Auger, M., and Separovic, F. (2003) Interaction of antimicrobial peptides from Australian amphibians with lipid membranes. *Chemistry and Physics of Lipids*, **122**, 107–120.

194 Dennison, S.R., Morton, L.H.G., Harris, F., and Phoenix, D.A. (2007) Antimicrobial properties of a lipid interactive alpha-helical peptide VP1

against *Staphylococcus aureus* bacteria. *Biophysical Chemistry*, **129**, 279–283.

195 Javkhlantugs, N., Naito, A., and Ueda, K. (2011) Molecular dynamics simulation of bombolitin II in the dipalmitoylphosphatidylcholine membrane bilayer. *Biophysical Journal*, **101**, 1212–1220.

196 Ho, D., and Merrill, A.R. (2009) Evidence for the amphipathic nature and tilted topology of helices 4 and 5 in the closed state of the colicin E1 channel. *Biochemistry*, **48**, 1369–1380.

197 Greenwald, J. and Riek, R. (2010) Biology of amyloid: structure, function, and regulation. *Structure*, **18**, 1244–1260.

198 Lee, S.-J., Lim, H.-S., Masliah, E., and Lee, H.-J. (2011) Protein aggregate spreading in neurodegenerative diseases: problems and perspectives. *Neuroscience Research*, **70**, 339–348.

199 Zhao, H.X., Jutila, A., Nurminen, T., Wickstrom, S.A., Keski-Oja, J., and Kinnunen, P.K.J. (2005) Binding of endostatin to phosphatidylserine-containing membranes and formation of amyloid-like fibers. *Biochemistry*, **44**, 2857–2863.

200 Shanmugam, G., Phambu, N., and Polavarapu, P.L. (2011) Unusual structural transition of antimicrobial VP1 peptide. *Biophysical Chemistry*, **155**, 104–108.

201 Zhao, H., Sood, R., Jutila, A., Bose, S., Fimland, G., Nissen-Meyer, J., and Kinnunen, P.K.J. (2006) Interaction of the antimicrobial peptide pheromone Plantaricin A with model membranes: implications for a novel mechanism of action. *Biochimica et Biophysica Acta – Biomembranes*, **1758**, 1461–1474.

202 Mahalka, A.K. and Kinnunen, P.K.J. (2009) Binding of amphipathic alpha-helical antimicrobial peptides to lipid membranes: lessons from temporins B and L. *Biochimica et Biophysica Acta – Biomembranes*, **1788**, 1600–1609.

203 Sood, R., Domanov, Y., and Kinnunen, P.K.J. (2007) Fluorescent temporin B derivative and its binding to liposomes. *Journal of Fluorescence*, **17**, 223–234.

204 Soscia, S.J., Kirby, J.E., Washicosky, K.J., Tucker, S.M., Ingelsson, M., Hyman, B., Burton, M.A., Goldstein, L.E., Duong, S., Tanzi, R.E., and Moir, R.D. (2010) The Alzheimer's disease-associated amyloid beta-protein is an antimicrobial peptide. *PLoS ONE*, **5**, e9505.

205 Torrent, M., Odorizzi, F., Nogues, M.V., and Boix, E. (2010) Eosinophil cationic protein aggregation: identification of an N-terminus amyloid prone region. *Biomacromolecules*, **11**, 1983–1990.

206 Nilsson, M.R. and Dobson, C.M. (2003) *In vitro* characterization of lactoferrin aggregation and amyloid formation. *Biochemistry*, **42**, 375–382.

207 Gossler-Schofberger, R., Hesser, G., Muik, M., Wechselberger, C., and Jilek, A. (2009) An orphan dermaseptin from frog skin reversibly assembles to amyloid-like aggregates in a pH-dependent fashion. *FEBS Journal*, **276**, 5849–5859.

208 Nicolas, P., and El Amri, C. (2009) The dermaseptin superfamily: a gene-based combinatorial library of antimicrobial peptides. *Biochimica et Biophysica Acta – Biomembranes*, **1788**, 1537–1550.

209 Auvynet, C., El Amri, C., Lacombe, C., Bruston, F., Bourdais, J., Nicolas, P., and Rosenstein, Y. (2008) Structural requirements for antimicrobial versus chemoattractant activities for dermaseptin S9. *FEBS Journal*, **275**, 4134–4151.

210 Jang, H.J.H., Arce, F.T., Mustata, M., Ramachandran, S., Capone, R., Nussinov, R., and Lal, R. (2011) Antimicrobial protegrin-1 forms amyloid-like fibrils with rapid kinetics suggesting a functional link. *Biophysical Journal*, **100**, 1775–1783.

211 Bieler, S., Estrada, L., Lagos, R., Baeza, M., Castilla, J., and Soto, C. (2005) Amyloid formation modulates the biological activity of a bacterial protein. *The Journal of Biological Chemistry*, **280**, 26880–26885.

212 Butterfield, S.M. and Lashuel, H.A. (2010) Amyloidogenic protein membrane interactions: mechanistic insight from model systems. *Angewandte Chemie International Edition*, **49**, 5628–5654.

213 Zhao, H.X. and Kinnunen, P.K.J. (2002) Binding of the antimicrobial peptide temporin L to liposomes assessed by Trp fluorescence. *The Journal of Biological Chemistry*, **277**, 25170–25177.

214 Harris, F., Dennison, S.R., Singh, J., and Phoenix, D.A. (2011) On the selectivity and efficacy of defense peptides with respect to cancer cells. *Medicinal Research Reviews*, doi: 10.1002/med.20252

215 Kawahara, M., Ohtsuka, I., Yokoyama, S., Kato-Negishi, M., and Sadakane, Y. (2011) Membrane incorporation, channel formation, and disruption of calcium homeostasis by Alzheimer's beta-amyloid protein. *International Journal of Alzheimer's Disease*, **2011**, 304583.

216 Arispe, N., Diaz, J.C., and Simakova, O. (2007) A beta ion channels. Prospects for treating Alzheimer's disease with A beta channel blockers. *Biochimica et Biophysica Acta – Biomembranes*, **1768**, 1952–1965.

217 Diaz, J.C., Simakova, O., Jacobson, K.A., Arispe, N., and Pollard, H.B. (2009) Small molecule blockers of the Alzheimer A beta calcium channel potently protect neurons from A beta cytotoxicity. *Proceedings of the National Academy of Sciences of the United States of America*, **106**, 3348–3353.

218 Kawahara, M. (2010) Neurotoxicity of beta-amyloid protein: oligomerization, channel formation and calcium dyshomeostasis. *Current Pharmaceutical Design*, **16**, 2779–2789.

219 Ballard, C., Gauthier, S., Corbett, A., Brayne, C., Aarsland, D., and Jones, E. (2011) Alzheimer's disease. *Lancet*, **377**, 1019–1031.

6
Selectivity and Toxicity of Oncolytic Antimicrobial Peptides

Summary

Cancer is a major cause of premature death and there is an urgent need for new anticancer agents with novel mechanisms of action. In this chapter, we review recent studies on a group of peptides that show much promise in this regard, exemplified by the α-helical antimicrobial peptides, cecropins from arthropods, and magainins and aureins from amphibians. While these molecules have established roles as antimicrobial factors and modulators of innate immune systems, it is becoming increasingly clear that they show potent anticancer activity. Generally, α-helical anticancer peptides (α-ACPs) exhibit selectivity for cancer and microbial cells primarily due to their elevated levels of negative membrane surface charge as compared to non-cancerous eukaryotic cells. The anticancer activity of α-ACPs normally occurs at micromolar levels, but is not accompanied by significant levels of hemolysis or toxicity to other mammalian cells. Structure–function studies have established that architectural features of α-ACPs such as amphiphilicity levels and hydrophobic arc size are of major importance for the ability of these peptides to invade cancer cell membranes. In the vast majority of cases the mechanisms underlying such killing involves disruption of mitochondrial membrane integrity and/or that of the plasma membrane of the target tumor cells. Moreover, these mechanisms do not appear to proceed via receptor-mediated routes, but are thought to be effected in most cases by the carpet/toroidal pore model and variants. Usually, these membrane interactions lead to loss of membrane integrity and cell death utilizing apoptotic and necrotic pathways. It is concluded that that α-ACPs are major contenders in the search for new anticancer drugs, underlined by the fact that a number of these peptides have been patented in this capacity.

6.1
Introduction

Globally, cancer is now the third leading cause of death, and it has been projected that by 2030 there will be around 26 million new cancer cases and 17 million

Antimicrobial Peptides, First Edition. David A. Phoenix, Sarah R. Dennison, and Frederick Harris.
© 2013 Wiley-VCH Verlag GmbH & Co. KGaA. Published 2013 by Wiley-VCH Verlag GmbH & Co. KGaA.

cancer deaths on an annual basis [1]. The disease is initiated by a series of cumulative genetic and epigenetic changes that occur in normal cells, and is characterized by a number of specific behaviors at the cellular and molecular level. Cancer cells provide their own growth signals, ignore growth-inhibitory signals, avoid cell death, replicate without limit, sustain angiogenesis, and invade tissues through basement membranes and capillary walls. In addition, the immune system fails to eliminate cancer cells due to the immunosuppressive effects mediated by tumor cells and tumor-infiltrating host cells [2–5]. It has been predicted that preventive measures, coordinated on a global scale, provide the only feasible approach to slowly and, ultimately, reverse the world-wide increase in cancer; however, at the moment, cancer management is the only available therapeutic option [1]. Currently, the strategies of choice for cancer management focus on the conventional cytotoxic treatments of radiation therapy (RT), which is relatively precise and used to achieve local control of cancers, and chemotherapy (CT), which exerts a systemic effect and is used in a broad array of cancer treatments [6–10]. However, both forms of therapy suffer from low therapeutic indices and a broad spectrum of severe side-effects, with delayed neurotoxicity deriving from both RT and CT becoming a crucial issue in cancer treatment [11, 12]. In the case of CT, side-effects are exacerbated by the fact that the majority of drugs currently used in this form of treatment display little or no selectivity for cancer cells over non-cancerous cells [6]. Major examples of these drugs include the very commonly used alkylating agents, temozolomide and carmustinine [13], and the mitotic inhibitor, paclitaxel [14].

Another major limitation to the successful treatment of cancer with both CT and RT is the development of resistance, which is currently an important medical problem and can be due to a number of factors. Anticancer drugs can show poor penetration of tumors and hypoxic cells in the center of these growths are essentially in a growth-arrested state, which makes them much less susceptible to conventional anticancer drugs [15]. There is growing evidence that autophagy, which allows a cell to respond to changing environmental conditions such as nutrient deprivation, may play an important role in conferring cancer cells with resistance to established anticancer therapies [16]. In addition, cancer cells can develop multidrug resistance, which makes these cells resistant not only to the drug of treatment, but also to a variety of other unrelated compounds. Indeed, it is generally observed in the case of patients who relapse after CT that they exhibit tumors that are more resistant to this regime than the primary tumor [6]. A variety of mechanisms are believed to endow cancer cells with multidrug resistance, including an increased expression of drug-detoxifying enzymes and drug transporters and/or an increased ability to repair DNA damage and defects in the cellular machinery that mediate apoptosis [17–19]. A major factor in the onset of multidrug resistance is the over expression of the MDR1 gene, which causes resistance to a broad spectrum of drugs by transporting these compounds out of the cell before they can interact with their intracellular targets [19–21].

An alternative to cytotoxic therapies is immunotherapy, which aims to manipulate the immune system to create a hostile environment for cancer cells in the body [22]. Essentially, cancer immunotherapy involves treatment and/or prevention with a variety of vaccines, including peptide vaccines based on T- and B-cell epitopes [23–25], DNA vaccines [26], and vaccination using whole tumor cells [27], immunotoxins [28], dendritic cells [29], viral vectors [30], antibodies [31], and adoptive transfer of T cells to harness the body's own immune system towards the targeting of cancer cells for destruction [32]. However, immunotherapies are associated with problems such as adverse toxicity, reverse autoimmunity, poor tissue penetration, and the easy clearance of immunotherapeutic agents [27, 33].

The associated toxicities and the limited success of traditional cancer treatments in maximizing cure rates have issued a clear mandate for the development of innovative therapeutic strategies to combat the disease. In response, there have been concerted efforts to identify new cancer biomarkers [34] and to develop the targeted delivery of therapeutic agents to cancers [6, 35–38], such as drugs that target cancer cell mitochondria [39], hybrid tubulin-targeting compounds [40], and antiangiogenics [41]. However, it is becoming increasingly clear that cancer cells are highly heterogeneous and that they exhibit deregulation in multiple cellular signaling pathways, which strongly limits the potential of cancer treatments that use specific agents or inhibitors that target only one biological event [2, 3]. In order to overcome this limitation, anticancer treatments involving various combinations of conventional cytotoxic therapies, immunotherapy, and targeted strategies using multiple agents with distinct targets are being developed, although in many cases such treatment can be associated with unacceptable dose-related toxicity [42–52].

Another response to the mandate for the development of innovative anticancer strategies has been to identify and develop peptides with therapeutically useful anticancer potential [53–55]. A major focus of research into this area has been on defense peptides, which exhibit potent toxicity to cancer cells, including those with multidrug resistance, and can in some cases exhibit selectivity for these cells over non-cancerous cells [54, 56–77]. Nonetheless, the mechanisms underpinning the selectivity and toxicity of these peptides to cancer cells are, as yet, not fully understood, and we review our current understanding in this area in this chapter.

6.2
Peptide-Based Factors That Contribute to the Anticancer Action of Anticancer Peptides

Defense peptides that function as anticancer peptides (ACPs) have been identified in a diverse array of organisms [54, 56–65], with recent examples reported in plants [72, 73], arthropods [66, 74], fish [75], amphibians [54, 76, 77], and mammals [68], but it is only over the last decade that there has been a concerted effort investigate the structure–function relationships that underpin their action against cancer cells [67, 69–71]. One major approach used to gain insight into these structure–function relationships has been the statistical analysis of physicochemical data and correlation with LD_{50} (lethal dose, 50%) for peptides [62, 67, 71, 78–80] contained in databases of defense peptides [81–84], such as the APD2 [85, 86] and AMSDb databases (http://www.bbcm.units.it/~tossi/antimic.html). The results of these analyses were generally consistent with those of previous work on antibacterial efficacy (Chapter 2) [62]. Pair correlations between sequence length and net charge as well as sequence length and hydrophobicity showed that these structural characteristics, along with the secondary structure of ACPs, vary widely. In general, these analyses showed that ACPs vary in length between five and 45 residues, possess net charges from −1 to +11, and exhibit a range of secondary structures, including extended structures (E-ACPs); α-helical conformations (α-ACPs), β-sheet (β-ACPs), and other cysteine motif configurations; and mixed α-helical and β-sheet structures [67], although many reported ACPs are still of unknown structure [67]. Since these latter analyses, a number of anionic ACPs have been reported in plants across a variety of families [73], especially cyclotides [69, 70], which generally possess cysteine motif structures and are listed in the CyBase [81] and PhytAMP [82] databases. These anionic peptides are generally around 30 residues in length and possess a net negative charge of −1 [87].

A recent analysis of ACPs showed that these peptides can be broadly divided into two major subgroups [71]. The first group (Tables 6.1 and 6.2), includes peptides that show little evidence of selectivity, and are toxic to both cancerous and non-cancerous cells (ACP_T), with examples including mastopran from wasps and other toxins from arthropods, which are α-ACP_T peptides [66, 74], along with cycloviolacin O2 and other cyclotides, which are plant β-ACP_T peptides [69, 70]. The second group (Tables 6.1 and 6.2) includes peptides that possess toxicity to cancer cells, but are ineffective against non-cancerous cells and erythrocytes (ACP_{AO}), and are exemplified by lasioglossins from bees, which are α-ACP_{AO} peptides [66, 96], along with lactoferricin from mammals [68] and hepcidin TH2-3 from fish [75, 105], which are β-ACP_{AO} peptides.

Information provided by ACP databases is of great utility when analyzing the structure–function relationships of these peptides. However, for in-depth analysis of these relationships, it is necessary to relate the structural characteristics of ACPs to their toxicity and selectivity towards cancer cells of similar type and under the same conditions [116]. For most analyses of databases, this is not generally achievable because their listed LD_{50} values are generated against different cell types using differing protocols [67]. In response, we investigated the structure–function relationships of ACPs using a database published by Owen [83], which compromised around 160 endogenous and synthetic α-ACPs that were uniformly tested for toxicity and selectivity towards cancer cells from breast adenocarcinoma (MCF7), colon adenocarcinoma (SW480), melanoma (BMKC), lung large cell carcinoma (H1299), cervical epithelial carcinoma (HeLaS3), and prostate adenocarcinoma (PC3) [83]. It was found that these peptides could be divided into three datasets: one that

6.2 Peptide-Based Factors That Contribute to the Anticancer Action of Anticancer Peptides

Table 6.1 Representative α-ACPs.

Selectivity	Peptide	Origins	References
α-ACP$_{AO}$	Cecropins	*Hyalophora cecropia* and *Musca domestica*	[59, 88]
	Magainins	*Xenopus laevis*	[59]
	Aureins	*Litoria aurea* and *Litoria raniformis*	[76, 77]
	Citropins,	*Litoria citropa*	[89, 90]
	Gaegurins	*Rana rugosa*	[91, 92]
	Polybia-MP1	*Polybia paulista*	[93, 94]
	Epinecidin-1	*Epinephelus coloides*	[95]
	Lasioglossins	*Lasioglossum laticeps*	[96]
	NK-2	Porcine	[97]
	Buforin IIb	Amphibian	[98, 99]
	CB1a	Moth	[100]
α-ACP$_T$	Melittin	*Apis mellifura*	[63]
	Temporin L	*Rana temporaria*	[101]
	Temporin-1DRa	*Rana draytonii*	[102]
	BMAP-27	*Bos taurus*	[59]
	BMAP 28		
	LL-37	*Homo sapiens*	[103]

α-ACP$_{AO}$ and α-ACP$_T$ peptides from a range of species. Included are the host organisms and key references. For synthetic α-ACPs, the generic name of the source peptide host is given and is non-italicized.

included α-ACP$_{AO}$ peptides, one that contained α-ACP$_T$ peptides, and a third that was formed by α-helical peptides with inactivity against eukaryotic cells (α-ACP$_I$) [79, 80]. Moreover, as described in Chapter 4, these peptides had the added advantage that the spatial regularity of the residues forming their α-helical secondary structure supports the quantification of a range of structure-dependent parameters that can be related to their selectivity and toxicity to cancer cells [62, 117–122]. Using the database of Owen [83], we performed a cluster analysis of all ACPs in the database, and showed that they formed 21 clusters based on similarities in net charge, hydrophobicity, and amphiphilicity [80], which are generally accepted as being key determinants in the action of ACPs [62, 71]. It was found that that no single combination of these structural properties was optimal for efficacy of the α-ACP$_{AO}$ and α-ACP$_T$ peptides, and given that peptides of the α-ACP$_I$ dataset were distributed across these clusters, this indicated that knowledge of net charge, hydrophobicity, and amphiphilicity alone are not sufficient for the prediction of anticancer activity [80]. In another study, we investigated the role of C-terminal amidation in the anticancer action of ACPs, which is the most common posttranslational modification observed for these peptides [79], exemplified by α-ACPs from amphibian skin secretions [123, 124]. C-terminal amidation has the effect of increasing both the cationicity and α-helical propensity of ACPs [62], and to perform this investigation, we utilized a dataset of peptides extracted from the database of Owen [83], which included non-amidated α-ACP$_{AO}$ and α-ACP$_T$

Table 6.2 Representative β-ACPs and E-ACPs.

Selectivity	Peptide	Source	References
β-ACP$_{AO}$	Lactoferricin B	*Bos taurus*	[104]
	Hepcidin TH2-3	*Orechromis mossambicus*	[105]
β-ACP$_{AO}$ (cyclotides)	Cycloviolacin O2	*Viola odorata*	[72, 106]
	Varv A and varv F	*Viola arvensis*	[106]
	Varv A, varv E, and vitri A	*Viola tricolor*	[72, 107]
	Vibi D, vibi E, vibi G, and vibi H	*Viola biflora*	[72, 108]
	Psyle A–psyle F	*Psychotria leptothyrsa*	[109]
	MCoCC-1 and MCoCC-2	*Momordica cochinchinensis*	[110]
β-ACP$_T$	HNP-1, HNP-2, and HNP-3	*Homo sapiens*	[111]
	Gomesin	*Acanthoscurria gomesiana*	[58, 112]
	Tachyplesin 1	*Tachypleus tridentatus*	[59]
E-ACP$_{AO}$	ChBac3.4	*Capra hirca*	[113]
E-ACP$_T$	PR-39	Porcine	[59, 114]
	Indolicidin	Bovine	[83, 115]

β-ACP$_{AO}$ and β-ACP$_T$ peptides with a number of E-ACP$_{AO}$ and E-ACP$_T$ peptides from a range of species. Included are the host organisms and key references. For synthetic β-ACPs and β-ACPs, the generic name of the source peptide host is given and is non-italicized.

peptides along with their C-terminally amidated isoforms. Using a range of theoretical techniques, we showed that the acquisition of a C-terminal amide group had no apparent effect on the ability of α-ACPs to discriminate between non-cancerous cells and cancer cells. In the same study, we also showed that the C-terminal amidation of α-ACP$_{AO}$ and α-ACP$_T$ peptides had a variable effect on the levels of toxicity exhibited by these peptides to cancer cells. These toxicity levels remained either unaffected or showed up to 10-fold increases or decreases in magnitude, clearly showing that C-terminal amidation cannot be taken as a predictor of peptide anticancer activity [79]. To elucidate the role of amphiphilicity in the anticancer action of ACPs, we performed extended hydrophobic moment plot analysis of the α-ACP$_{AO}$, α-ACP$_T$, and α-ACP$_I$ datasets, and these analyses predicted that over half of the α-ACP$_{AO}$ and α-ACP$_T$ peptides would be active at the interface. This result was consistent with the view that amphiphilicity is a key driver of the membrane interactions of α-ACPs and suggested that this structural characteristic most likely plays a role in the efficacy of these peptides rather than their selectivity. This analysis also predicted that only 14% of α-ACP$_{AO}$ peptides compared to 45% of α-ACP$_T$ peptides were candidates to possess the hydrophobicity gradients associated with tilted α-helix formation (Chapter 4) [117]. These results suggested that the absence of this tilted structure may support the cancer cell selectivity of α-ACPs, although it was also predicted by these analyses that α-ACP$_I$ peptides would form also tilted α-helices along with surface-active α-helices, clearly showing that other factors are involved in determining the efficacy and selectivity of α-ACPs [80]. These individual studies provided some insight into the structure–function relationships driving the anticancer action of α-ACPs; however, to obtain a clearer view of these relationships, in 2011 [71], we undertook a major study in which statistical comparisons were performed between peptides of the α-ACP$_{AO}$, α-ACP$_T$, and α-ACP$_I$ datasets in respect of residue composition, sequence length, net positive charge, hydrophobicity, and amphiphilicity (Table 6.3).

6.2.1
Sequence Length

In our study of 2011 [71], sequence lengths were determined for peptides in the α-ACP$_{AO}$, α-ACP$_T$, and α-ACP$_I$ datasets, and were found to range between four and 38 residues (Table 6.3), which is comparable to that found in previous studies [62] and would include the bulk of ACPs currently listed in the APD2 database [85, 86]. The medians for sequence length in our analysis were found to follow the rank order: α-ACP$_T$ (sequence length = 15 residues) > α-ACP$_I$ (sequence length = 12 residues) > α-ACP$_{AO}$ (sequence length = 11 residues). While these differences were statistically relevant they are small in terms of residue variation and it seems unlikely that these variations in sequence length would play a key role in either the toxicity or selectivity of α-ACPs. This is supported by a number of reports indicating that there was no clear correlation between this structural characteristic and the toxicity of defense peptides [62, 121, 125]. However, a recent study has suggested that sequence length may be a factor in deciding which modes of membrane interaction are available to ACPs in order to exert their anticancer activity. This study suggested that shorter peptides are less likely to use oblique orientation models of insertion and are more likely to use carpet-type mechanisms of membrane lysis while peptides with over 20 residues, sufficient to span the bilayer, may tend to adopt pore-forming mechanisms of membrane permeation (Chapter 5) [116].

6.2.2
Net Positive Charge

The possession of a net positive charge is fundamental to the anticancer action of most ACPs [67, 68], and in our study of 2011 [71] this structural characteristic was determined for peptides across the α-ACP$_{AO}$, α-ACP$_T$, and α-ACP$_I$ datasets. The net positive charge of these peptides was found to range from 0 to +12 (Table 6.3)

Table 6.3 Summary statistics for the structural characteristics off α-ACPs.

Property	α-ACP$_I$			α-ACP$_{AO}$			α-ACP$_T$		
	Minimum	Maximum	Median	Minimum	Maximum	Median	Minimum	Maximum	Median
No. of residues	4	19	12	5	22	11	5	38	15
Net charge	0	6	4	1	7	4	0	12	5
<H_0>	−0.59	0.21	−0.01	−0.46	0.13	−0.05	−0.80	0.51	−0.18
Arc size (°)	80	120	200	180	280	180	60	260	180
<μ_H>	0.49	0.78	0.71	0.53	0.78	0.73	0.33	1.05	0.74

Summarized above is a range of statistics for structural properties of peptides in the α-ACP$_I$, α-ACP$_{AO}$, and α-ACP$_T$ datasets presented by Owen [83]. Listed are the minimum, maximum, and median values of sequence length, total net charge, mean hydrophobicity <H>, mean amphiphilicity <μ_H>, and hydrophobic arc θ, all of which were determined as described in Harris et al. [71].

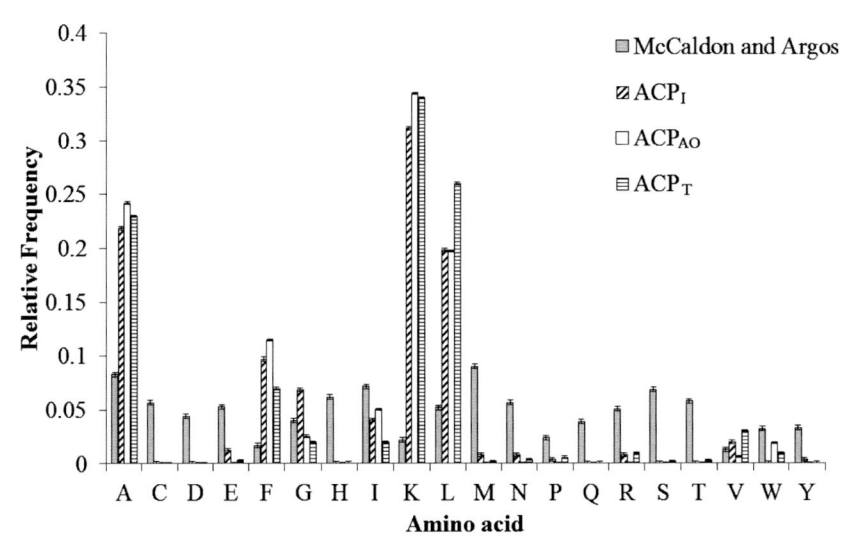

Figure 6.1 Residue distributions of α-ACPs. Overall amino acid residue compositions of peptides in the α-ACP$_I$, α-ACP$_{AO}$, and α-ACP$_T$ datasets presented by Owen [83]. Also shown are the overall residue compositions of peptides in the McCaldon and Argos dataset, which is a random sample of oligopeptides [126]. Residue frequencies were computed as described in [71] and these data used to identify differences in these frequencies, both within and between peptides of the McCaldon and Argos, α-ACP$_I$, α-ACP$_{AO}$, and α-ACP$_T$ datasets.

[67]. To investigate the role of net positive charge in α-ACPs further, we compared the distribution of charged residues across the α-ACP$_{AO}$, α-ACP$_T$, and α-ACP$_I$ datasets (Figure 6.1) [126]. The ACPs show no significant difference in net charge, indicating this was not a key factor in selectivity. These comparisons showed that glutamic acid and aspartic acid were found to be present in these datasets at levels over 10-fold lower than in the case of the McCaldon and Argos dataset, indicating that they do not generally play a major role in the selectivity of toxicity of α-ACPs to cancer cells. In the case of the few anionic ACPs that have been reported, the role of glutamic acid and aspartic acid residues in the anticancer action of these peptides is poorly understood [73]. Roles that have been characterized include DNA binding for ACPs with antiproliferative activity against cancer cells [127] and lipid interaction for these peptides with anticancer membranolytic activity [70]. In the case of the positively charged residues, arginine and lysine, striking differences were observed between the distributions of these residues in peptides in the α-ACP$_{AO}$, α-ACP$_T$, and α-ACP$_I$ datasets and those in the McCaldon and Argos dataset (Figure 6.1). Compared to the latter dataset, the level of arginine residues in the α-ACP$_{AO}$, α-ACP$_T$, and α-ACP$_I$ datasets was found to be around 10-fold lower, which suggests that arginine may not generally play a significant role in the selectivity and toxicity of α-ACPs to cancer cells. In contrast, lysine was found to be

strongly represented in the α-ACP$_{AO}$, α-ACP$_T$, and α-ACP$_I$ datasets with a relative frequency greater than 0.31, which is over 20-fold higher than that of the residue in McCaldon and Argos dataset, and this strongly suggested that lysine was important to the action of α-ACPs. Taken together, these observations led to the conclusion that lysine is both the basic residue most preferred by α-ACPs and is by far the biggest contributor to the net positive charge of these peptides.

The finding that α-ACPs have a preference for a particular positively charged residue is not without precedent and other classes of ACPs show a strong bias towards either lysine or arginine. For example, this latter residue is exclusively expressed in many β-ACPs such as protegrin-1, which is a porcine β-ACP$_T$ peptide [116], PR-39, which is a porcine E-ACP$_T$, and ChBac3.4, which is a goat E-ACP$_{AO}$ [113]. The reasons underlying such preferences are not immediately apparent, but they are clearly related to differences in the nature of the positively charged side-chains possessed by these residues. A major difference between these side-chains is that in the case of lysine its positive charge is localized to a primary amine group, whereas in the case of arginine it is dispersed over a guanidinium group [128, 129]. This delocalization of charge gives the arginine side-chain a stronger ability than that of lysine to engage in electrostatic interactions and it has been suggested that this ability gives the former residue a higher affinity for some anionic membrane components as compared to lysine [130]. Indeed, it is generally recognized that due to these differences in charge distribution, arginine has a high affinity for both anionic and zwitterionic membranes, whereas lysine has strong preference for anionic membranes [131]. A clear implication from this latter observation is that lysine residues may play a role in the cancer cell selectivity of α-ACPs with a preference for this residue, given the anionic nature of cancer cell membranes. Another major difference between the side-chains of lysine and arginine is the fact that the former side-chain is significantly more hydrophobic than the latter side-chain [132], and taken with the differences in their charge characteristics, these observations clearly suggest that differences in the hydrophobicity and/or amphiphilicity of the lysine and arginine side-chains could be a factor in the preference of ACPs for a given cationic residue. For example, the positive charge, hydrophobicity, and amphiphilicity of these side-chains underpin the ability of both lysine and arginine to interact with membranes via the snorkeling mechanism. Using this mechanism, the charged moieties of these residue side-chains engage in electrostatic interactions with the membrane lipid head-group region while the long apolar regions of these side-chains snorkel, or extend, into the acyl chain region of the bilayer, thereby enabling the parent peptide to penetrate more deeply into the membrane [133, 134]. However, the more hydrophobic side-chain of lysine is known to enhance the snorkeling ability of this residue compared to that of arginine [135] and it has previously been suggested that snorkeling by the former residue may be involved in mechanisms that contribute to the toxicity of α-ACPs to cancer cells [62, 136]. It may therefore be that the preference of α-ACPs for lysine enhances their ability to insert into cancer cell membranes and thereby the efficacy of their anticancer action. However, our analysis of 2011 [71] showed that comparable levels of lysine were present in the α-ACP$_{AO}$, α-ACP$_T$, and α-ACP$_I$ datasets, which indicated that factors other than the occurrence of this residue are involved in the selectivity and toxicity of α-ACPs for cancer cells.

Histidine is often described as an aromatic residue due to its possession of imidazole side-chain and at physiological pH, the pK_a of this side-chain is around 6, which means that the residue is uncharged [137]. However, at low pH, histidine becomes fully charged and while the residue does not normally make a significant contribution to the net positive charge of ACPs, it may be considered to be a moderately hydrophilic, basic residue [138]. The residue was found to be present at only very low levels in the α-ACP$_{AO}$, α-ACP$_T$, and α-ACP$_I$ datasets (relative frequencies below 0.005) (Figure 6.1). This represented levels of the residue that were in excess of 10-fold lower than those in the McCaldon and Argos dataset (Figure 6.1), which suggested that histidine does not generally play a major role in the selectivity or toxicity of α-ACPs to cancer cells. However, based on the fact that the surface of membranes from both cancer and microbial cells generally exhibits a pH lower than the pH of 7.2–7.4 found in the bulk phase of normal tissues and organs [139, 140], it has been suggested that pH-dependent properties of histidine may feature in the anticancer action of some ACPs. Based on this suggestion, histidine-containing peptides have been designed that function as pH-dependent α-ACP$_{AO}$ and α-ACP$_T$ peptides against both cancer cell lines and *in vivo* tumors [141–144]. As an example, it has been recently shown that a number of designed histidine-rich ACP$_{AO}$ peptides, such as [D]-H$_6$L$_9$, exhibit pH-dependent anticancer activity against various cell lines, including human prostate carcinoma (CL1 and 22RV1) [145] and human synovial sarcoma (SW982) [146]. Moreover, when these peptides were administered to mice, both intratumorally and via systemic inoculation, they were found to produce large reductions in the tumor growth of various xenografts derived from CL1, 22RV1, and SW982 cells [145, 146]. As an added advantage, these peptides were found to induce a significant decrease in vasculature within and around cancerous tissue as compared to untreated tumors, and it was suggested that [D]-H$_6$L$_9$ and the other ACPs tested may possess antiangiogenic properties [145, 146].

6.2.3
Hydrophobicity

In relation to hydrophobicity, this structural characteristic can be taken as a measure of the affinity of α-ACP peptides for the apolar core region of the bilayer [119, 120] and we showed in a previous study that a number of α-ACPs are candidates for tilted α-helix formation. These peptides have a strong propensity to locate part of their structure in this core region [80] and are characterized by an asymmetric distribution of hydrophobicity along the α-helical long axis [117]. More usually though, in the database analyzed [71], hydrophobic residues show an approximately constant distribution along the α-helical long axis of α-ACPs [147]. The distribution of hydrophobicity is commonly measured as the mean hydrophobicity of the parent α-helix, $<H>$, or by the angle, θ, which subtended its major hydrophobic arc [62]. Accordingly, in our study of 2011 [71], we quantified this structural characteristic for peptides across the α-ACP$_{AO}$, α-ACP$_T$, and α-ACP$_I$

datasets using both θ and $<H>$, and it was found that hydrophobic arc sizes ranged between $\theta = 60°$ and $260°$ while $<H>$ ranged between -0.8 and 0.51 (Table 6.3). These ranges in θ and $<H>$ are comparable to those previously reported for α-ACPs [62], and suggested that there is a wide variation in the affinity of these peptides for the membrane apolar core region. Statistical analysis revealed that there was no significant difference in the medians of either θ or $<H>$ for peptides of the datasets, which clearly suggested that hydrophobicity does not play a major role in the selectivity of α-ACPs for cancerous cells over non-cancerous cells. However, this result could indicate that a threshold value of this structural characteristic may be necessary to drive the interaction of these peptides with the hydrophobic core of target cell membranes and thereby their toxicity to eukaryotic cells in general. Supporting this suggestion, a number of studies have suggested that threshold hydrophobicites are required to drive the membrane interactions of α-AMPs, many of which are known to serve as α-ACPs [148–150], including the designed peptide, V13K [151, 152]. A recent major study on a series of α-ACPs based on V13K showed that there was a correlation between the hydrophobicity of these peptides and their anticancer action, with a threshold hydrophobicity observed for efficient anticancer action against cells from human cervix carcinoma cells (HeLa), melanoma (A375), colorectal carcinoma cells (SW1116), breast adenocarcinoma (MCF-7), lung carcinoma (H1299), rhabdomyosarcoma, lung carcinoma cells (A549), and murine melanoma cells (B16). Moreover, these studies also found that this threshold bounded a narrow window in hydrophobicity in which the toxicity and selectivity of these peptides for cancer cells were optimal [151].

Hydrophobicity is most often measured as $<H>$ and clearly, it is a function of the residue composition of the parent α-ACPs (Chapter 4). As revealed by our study of 2011 [71], residues that make positive contributions to $<H>$ vary widely in their individual levels of hydrophobicity. A number of weakly and moderately hydrophobic residues were either absent or present at very low levels in the α-ACP_{AO}, α-ACP_T, and α-ACP_I datasets, levels that were generally in excess of tenfold lower than those in the McCaldon and Argos dataset (Figure 6.1). Taken together, these observations suggested that these residues, including cysteine, proline, glycine, methionine, and tyrosine, played no major role in the anticancer action of α-ACPs. In general, the function of these residues in the anticancer action of ACPs is poorly understood, although in a number of cases it has been shown that they play important roles in the action of individual or small groups of α-ACPs against cancer cells and several recent examples are described below.

Cysteine was absent from each of our datasets and it has been established that it occurs rarely in α-ACPs [85, 86], primarily due to its very low α-helix-forming propensity [153]. Indeed, the only occurrence of cysteine as a free residue in ACPs currently listed in the APD2 database [85, 86] is OEP3121, a host defense peptide from the earthworm, which is not well characterized [154]. Cysteine most usually occurs in β-ACPs [56, 59] such as the β-hairpin molecule of gomesin, a spider β-ACPT peptide [58], where these residues participate in cysteine bonds or disulfide bridges due to the ability of their thiol side-chains to form covalent bonds with other cysteine side-chains (Chapter 2) [137].

Proline showed a very low frequency in our datasets (relative frequency less than 0.005) and is generally found at relatively low levels in α-ACPs (Figure 6.1) [85, 86], primarily due to the residue's very low α-helix-forming propensity [153]. The amino acid side-chain of proline is unable to donate an amide hydrogen bond, which causes steric interference to α-helix formation, resulting in a break or kink in the helical structure [155]. Internal prolines have been observed in a number of α-ACP$_{AO}$ peptides [85], such as buforin IIb [98], along with α-ACP$_T$ peptides, such as melittin [63]. In the case of the latter peptide, it has been shown that a kink in its backbone due to presence of proline gives melittin a flexibility that is important for efficient membrane interaction [156] and hence its anticancer action [59, 156], which has been suggested could also be the case for other ACPs [71]. In addition, proline-rich ACPs have been identified in a number of organisms [59, 113] such as the cyclic phakellistatins from sea sponges [53], which have potent activity against various cancer cells, including those from murine fibrosarcoma (WEHI-164) [157, 158] and leukemia (P388) [53] while incorporation therefore unusual when present may have structural significance.

Glycine lacks a side-chain [137], which gives the residue high conformational flexibility [153] and, as a consequence, it resembles proline by its ability to act as a breaker of α-helical structure [155]. Glycine was found to be twice as abundant in the ACP$_I$ dataset (relative frequency 0.075) than the α-ACP$_{AO}$ and α-ACP$_T$ datasets (relative frequency less than 0.025) (Figure 6.1), which implied that the inactivity of peptides in the α-ACP$_I$ dataset may, in some cases, be related to a reduced propensity for α-helix formation. Nonetheless, the conformational flexibility of the residue was recently shown to be important to the anticancer action of SK84, an insect ACP$_{AO}$, which was found to possess multiple glycine residues in its N-terminal domain. The flexibility of this domain appeared to facilitate the formation of an elastic structure in cancer cell membranes thereby contributing to bilayer disruption and the action of SK84 against cells from human leukemia (THP-1), liver cancer (HepG2), and breast cancer (MCF-7) [159]. Glycine is also known to be an efficient N-capping agent, which is important to the propensity of peptides to form an α-helical structure, and it has previously been suggested that the residue might serve such a purpose in α-ACPs [62]. Most recently, glycine capping has been used to design a series of highly efficient α-ACP$_{AO}$ peptides G(IIKK)nI-NH$_2$ with activity against microbial and tumor cells [160]. Interestingly, these peptides appeared to be one of the first attempts to design ACPs whose anticancer action incorporated the concept of imperfect amphiphilicity where the apolar face of the peptide is perturbed by the introduction of a cationic residue to enhance membrane interaction, previously predicted to serve as a basis for the design of novel antimicrobial peptides (AMPs) (Chapter 4) [161].

Tryptophan was found to have a low frequency of occurrence in the α-ACP$_{AO}$, α-ACP$_T$, and α-ACP$_I$ datasets (relative frequency less than 0.025). Nonetheless, the residue was found to be present in the α-ACP$_{AO}$ and α-ACP$_T$ datasets, but is absent from the α-ACP$_I$ dataset, which could provide evidence of relevance to the anticancer action of α-ACPs. Consistent with this suggestion, there is evidence indicating that tryptophan may play a role in the ability of ACPs to partition into cancer cell membranes. The residue possesses a side-chain with aromatic characteristics [137] due to the presence of π-ring systems with a significant quadruple moment, which gives rise to negatively charged regions that allow these aromatic residues to engage in cation–π contacts with positively charged species such as the basic residues described above [128]. In addition to cation–π contacts, tryptophan is also able to engage in hydrogen bonding and other electrostatic interactions with components of the lipid head-group region, which gives these residues a strong preference for an interfacial location [162–165]. It has previously been suggested that this preference may play a role in the orientation of ACPs in the cancer cell membrane [71], and in support of this suggestion, several studies have shown that there may be a preferential interaction between tryptophan residues and cholesterol located near the membrane–water interface, although this has been disputed [166]. Nonetheless, there is substantial evidence to support the view that the membrane interactions of some α-ACP$_{AO}$ and α-ACP$_T$ peptides may be mediated by the ability of their tryptophan residues to form a stable complex with cholesterol [156].

The most strongly hydrophobic residues, listed in descending rank order, are isoleucine, phenylalanine, valine, leucine, and alanine [138], and each of these residues was found to be strongly represented in the datasets of our 2011 study (Figure 6.1) [71]. Phenylalanine is the best characterized of these residues [167], and was found to be strongly represented in the α-ACP$_{AO}$, α-ACP$_T$, and α-ACP$_I$ datasets (relative frequencies between 0.075 and 0.125) (Figure 6.1), which clearly suggested functionality. It would seem that this functionality is related to the fact that, similarly to tryptophan, the side-chain of phenylalanine possesses aromatic characteristics due to the presence of π-ring systems [137]. However, in contrast to tryptophan, the side-chain of phenylalanine does not possess a hydrogen bond donor and has a weaker capacity for cation–π contacts, which taken with hydrophobic and lipophilic effects leads the residue to possess a strong preference for location in the hydrophobic core of the bilayer [164]. This preference appears to underpin the major contribution made by phenylalanine to the anticancer action of α-ACPs, which is enhancing the affinity of these peptides for the lipidic region of cancer cell membranes [62]. In addition, studies on defense peptides have suggested that phenylalanine may play other roles in the anticancer action of α-ACPs. In aureins, which function as both α-AMPs and α-ACPs, the residue appears to play a role in maintaining the surface topography of the peptide [167], whereas in the case of clavanins, which are α-AMPs, the residue appears to play a role in maintaining the conformational flexibility of these peptides [168]. Phenylalanine is overly represented in E-ACP$_T$ peptides such as prophenin PF2 [169] and PR-39 [59], and it may be that the residue contributes to the conformational flexibility of the extended structures possessed by these peptides.

Alanine, leucine, isoleucine, and valine were found to be represented in α-ACP$_{AO}$, α-ACP$_T$, and α-ACP$_I$ datasets at varying levels (relative frequencies between 0.26 and 0.06) (Figure 6.1), and it is generally accepted that, similarly to phenylalanine, the strong hydrophobicity of these residues is a major factor in the affinity of α-ACPs for cancer cell membranes and the anticancer action of these peptides

[62]. However, the occurrences of alanine and leucine (relative frequencies 0.21 and 0.26 respectively) in the our datasets were found to be much greater than those of valine and isoleucine (relative frequencies less than 0.06) (Figure 6.1), and a bias for the former two residues over other hydrophobic residues has previously been reported for α-ACPs [62] and α-AMPs [121]. It has been suggested that this bias may be related to differences in both the relative α-helix-forming propensities [71] and hydrophobicities of isoleucine, phenylalanine, valine, leucine, and alanine [160]. Nonetheless, the mutations of alanine and leucine for valine, isoleucine, and phenylalanine appear to be conservative substitutions as found in recent studies on aureins, which showed that homologs differing solely by the mutation of leucine to phenylalanine exhibit similar general profiles of activity when directed against the same range of human cancer cells, which included lung, colon, ovarian, prostate, and breast cancers [76]. Moreover, the mutation of leucine and alanine to isoleucine, valine, and phenylalanine has been identified in a number of homologous α-ACPs such as BMAP-27 and BMAP-28, which are bovine α-ACP$_T$ peptides, and gaegurin 5 and gaegurin 6, which are amphibian α-ACP$_{AO}$ peptides [59]. At present, the reasons underlying a preference for leucine and alanine by α-ACPs over other hydrophobic residues are unclear; however, taken together, the results presented here clearly imply that in some cases this preference is related to other factors in addition to the hydrophobic characteristics of these former two residues.

6.2.4
Amphiphilicity

As described above, examination of the residue composition of α-ACPs [71] clearly suggested that amphiphilicity is important to the anticancer action of these peptides (Figure 6.1). It is generally recognized that amphiphilicity is a major determinant in the ability of peptides partition into membranes, which themselves are amphiphilic, and may be defined as the ordered spatial segregation of hydrophilic and hydrophobic residues within the structure of the peptide [120]. A number of ACPs possess tertiary amphiphilicity where residues that are distal in the primary structure of the peptide are brought together in its final three-dimensional to form amphiphilic sites, exemplified by the cystine knot structures of plant cyclotides (Figure 6.2) [171]. These β-ACP$_T$ peptides are cyclic molecules, and cyclization brings together clusters of charged and hydrophobic residues that are located proximal to the N- and C-termini of these peptides in their precursor, linear form [172]. Recent studies have shown that the juxtaposition of these residue clusters on the surface of cyclotides is a key determinant in the ability of these peptides to partition into membranes and hence their anticancer activity [69, 70]. However, the majority of ACPs demonstrate secondary amphiphilicity, which arises when the folding of these peptides leads to residue segregation and secondary structures with opposing polar and apolar faces [120] (Figure 6.3). Typical examples of this form of amphiphilicity are provided by β-ACP peptides such as gomesin [58] along with magainin 2, which is an amphibian α-ACP$_{AO}$ (Figure 6.4) [173], and can be

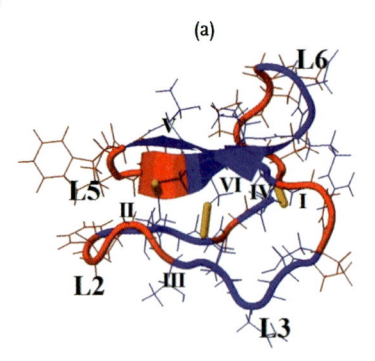

Figure 6.2 Tertiary amphiphilicity of ACPs: structures for Möbius cyclotides (a), which include kalata B2, cycloviolacin O24, and vhl-2, and bracelet cyclotides (b), which include cycloviolacins Y4 and Y5 (adapted from [170]). The core of these structures is the "cystine knot" motif formed by conserved cysteine residues, which is indicated above in both structures as I–V. Residues between these cysteine residues form backbone loops that define the general surface topology of these peptides. Loops 2, 3, 5, and 6 are indicated above in both structures as L2, L3, L4, and L6. Both structural types exhibit hydrophobic patches on their molecular surface, which is centered on L5 and L6 in Möbius peptides (a) and on L2 and L3 in bracelet peptides (b). For both structures, hydrophilic and charged regions are indicated in red.

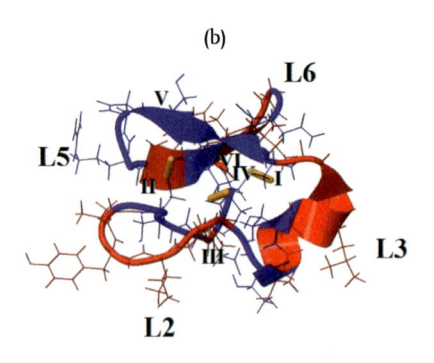

taken as structurally representative of the peptides included in the three datasets of our 2011 study [80]. It is well established that secondary amphiphilicity can be quantified by a number techniques. The hydrophobic moment ($<\mu_H>$) is by far the most commonly employed [119, 120, 174] (Chapter 4) concept even though it was developed by Eisenberg et al. [175], almost 30 years ago. Using values of $<\mu_H>$ as a measure of secondary amphiphilicity it was found that values of this parameter ranged between 0.33 and 1.05 for peptides in the α-ACP$_I$, α-ACP$_{AO}$, and α-ACP$_T$ datasets (Table 6.3) [71]. This range is in close agreement with previous work [62] and is consistent with the view that α-ACPs are active at the interface – generally accepted as a common step in the anticancer mechanisms of most known ACPs. Statistical analysis revealed that there was no significant difference in the medians of $<\mu_H>$ for peptides of these datasets, although as previously reported for AMPs (Chapter 2) it is likely there is an optimal range in $<\mu_H>$ values for the anticancer action of ACP peptides.

6.3
Membrane-Based Factors That Contribute to the Anticancer Action of ACPs

It is generally accepted that ACPs exert their anticancer action via bilayer interaction involving membranolytic mechanisms and/or membrane translocation to utilize intracellular sites of action. In most cases, ACPs internalized to the cancer cell interior appear to interact with the mitochondrial membrane to initiate

6.3 Membrane-Based Factors That Contribute to the Anticancer Action of ACPs

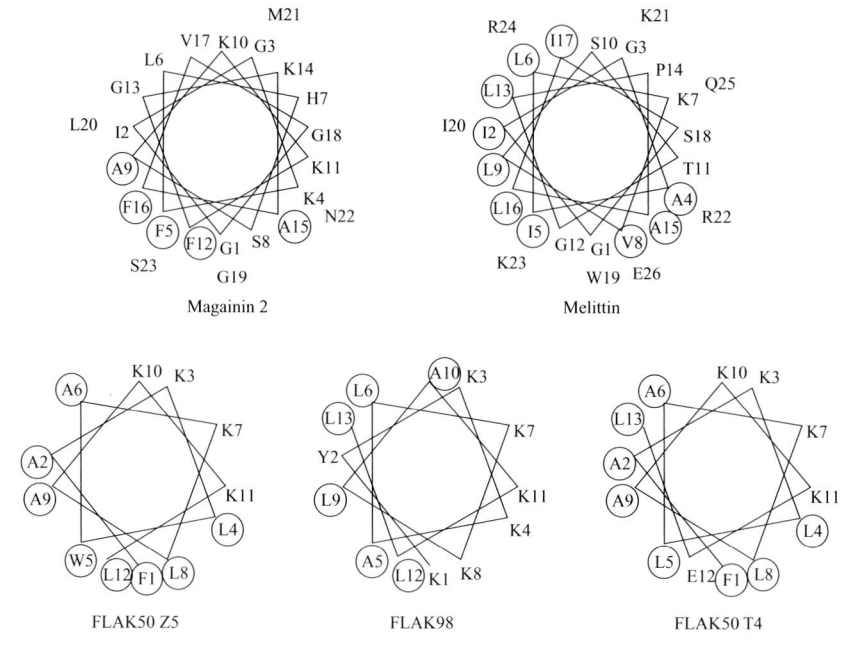

Figure 6.3 Secondary amphiphilicity of ACPs: peptides from the α-ACP$_I$, α-ACP$_{AO}$, and α-ACP$_T$ datasets when represented as α-helical wheels, which are two-dimensional axial projections of these peptides, assuming an angular periodicity of 100°. The examples shown include magainin 2 (Figure 6.4) and FLAK50 Z5, which are endogenous and synthetic members of the α-ACP$_{AO}$ dataset, respectively; melittin and FLAK98, which are endogenous and synthetic members of the α-ACP$_T$ dataset, respectively; and FLAK50 T4, which is a synthetic member of the α-ACP$_I$ dataset [80]. The amphiphilicity of these α-ACPs can be clearly seen by the segregation of hydrophilic residues and hydrophobic residues (circled). The sequences of these examples were obtained from Owen [83].

apoptotic mechanisms of cell death, although in a few cases specific intracellular targets have been reported for these peptides [59, 60, 68]. For example, PR-39, which is a porcine ACP$_T$ peptide, appears to translocate the cytoplasmic membrane of target cancer cells to interact with proteins containing SH3-binding motifs, which leads to the suppression of the invasive activity of these cells and the prevention of metastasis [67]. Most recently, an anionic ACP, *Cr*-ACP1, has been reported in plants, which appeared to enter target cancer cells and kill these cells by inhibiting their ability to proliferate as demonstrated for human colon carcinoma (HCT15) and epidermoid cancers (Hp2). The peptide appeared to function as an α-ACP$_{AO}$ peptide and exert its anti-proliferative activity *via* DNA binding, leading to cell cycle arrest and the induction of apoptosis [127].

Clearly, a fundamental requirement for the anticancer action of ACPs is the ability to target and interact with the membranes of cancer cells, and several

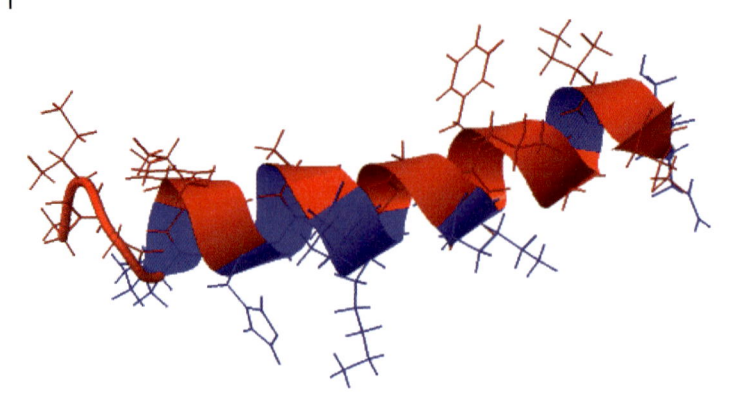

Figure 6.4 Secondary amphiphilicity of magainin 2, an amphibian α-ACP$_{AO}$ [173], which is clearly shown by the segregation of hydrophobic residues (blue) and hydrophilic residues (red) about the α-helical long axis. The sequence of the peptide was taken from [173], and its structural amphiphilicity can be taken as representative of that shown by peptides in the α-ACP$_{AO}$, ACP$_T$, and α-ACP$_I$ datasets [80].

generic properties of these membranes have been identified, which differ to those of non-cancerous cells and appear able to attenuate the selectivity and toxicity of ACPs for the former cells over the latter [59, 67, 71]. As described above, the extracellular environment of tumors is relatively acidic compared to that of normal healthy tissues and is able to facilitate the anticancer action of histidine-rich ACPs. This acidification is primarily due to the accumulation of lactic acid, which is produced during aerobic and anaerobic glycolysis, but is inefficiently cleared due to the inadequate vasculature found in growing tumors [176, 177]. As another example, the membranes of many cancer cells, such as lymphomas, lung carcinomas, and neural tumors, exhibit increased membrane fluidity in relation to non-cancerous cells [67] and it has been suggested that this increased membrane fluidity may enhance the anticancer activity of α-ACP$_{AO}$ peptides by augmenting their ability to destabilize the membranes of target cells [59]. Interestingly, it has also been reported that cells from soft tissue tumors such as hepatoma exhibit decreased membrane fluidity relative to their healthy counterparts [67] and it may be that in these cases, altered membrane fluidity provides cancer cells with protection from the activity of ACPs. As a further example, the selectivity of α-ACP$_{AO}$ peptides may be influenced by the high negative transmembrane potential ($\Delta\psi$) associated with cancer cell membranes relative to non-cancerous cells with disturbance of this transmembrane potential by the membranolytic anticancer mechanisms of these peptides, resulting in the loss of electrolytes and cell death [64, 178]. It has also been shown that in relation to non-cancerous cells, cancer cells tend to have more abundant microvilli, effectively increasing their outer surface area, and it has been suggested that this morphological difference should allow the binding of larger numbers of ACPs per cell [179]. Support for this suggestion came from studies on the anticancer action of the moth α-ACP$_{AO}$ peptide, cecropin B, when directed against leukemia cells and stomach carcinoma (AGS). Based on scanning electron microscopy, it was suggested that interaction of cecropin B with the large numbers of irregular microvilli present on the surface of these cells served to localize the peptide proximal to the membrane, thereby enhancing its selectivity and membranolytic toxicity for cancer cells [180, 181].

In addition to these generic membrane-based factors, a number of specific molecules and classes of molecules have been identified that are constituents of the eukaryotic membrane, and that appear able to attenuate the selectivity and toxicity of ACPs with respect to cancer cells, including receptors, cholesterol, and a variety of anionic membrane components, such as glycoproteins, glycolipids, proteoglycans (PGs), and phospholipids [56, 62, 67, 71].

6.3.1
Membrane Receptors

It is generally believed that chiral molecular recognition is not a prerequisite for the anticancer activity of ACPs and that their action against cancer cells involves membrane interaction via non-receptor-mediated pathways [56, 182]. Strongly supporting this belief, recent studies on kalata B1 and gomesin (Table 6.2) showed that substitution of the entire complement of L-amino acids forming these peptides with the corresponding D-amino acids produced molecules with comparable levels of anticancer activity and membrane interaction [58, 112, 183]. Similar results have been presented for a number of other ACPs [64, 89, 184], including melittin, cecropin, magainin, and analogs of these peptides (Table 6.1) [183, 185–190]. Nonetheless, it has increasingly been reported that the activity of L-enantiomers and D-enantiomers of defined peptides towards the same target cells can vary widely [186, 191–193], with examples including HNP-1, LfB, and derivatives of these peptides (Table 6.2) [194–196]. These reports clearly suggest that the recognition of stereospecific membrane receptors may underpin the ability of some ACPs to target cancer cells, thereby enhancing the selectivity and toxicity of these peptides for these cells [62, 197]. Indeed, it is worthy of note that in their antimicrobial capacity some defense peptides have been shown to exhibit varying levels of selectivity for specific lipids in their membrane interactions. For example, bacterial lantibiotics such as nisin bind lipid II [198] while cinnamycin and duramycin bind PE [199]. Other PE-binding peptides include cyclotides [70, 200] and the Aβ peptides, which are best known for their role in Alzheimer's disease [201], but have recently been shown to function as anionic AMPs [202].

6.3.2
Cholesterol

Cholesterol is a major sterol of eukaryotic cell membranes [203], and it has been previously suggested that this compound may generally offer protection to eukaryotic membranes from the lytic action of some α-ACPs by changing membrane fluidity and thereby reducing the ability of these peptides to partition into the membrane [65]. Consistent with this suggestion, it was found that increasing levels of membrane cholesterol, thereby decreasing membrane fluidity, inhibited the lytic ability of a number of α-ACPs towards membranes of non-cancerous eukaryotic cells and their lipid mimics, including cecropins [204, 205], and magainins [206, 207]. Based on these and other results, it has also been proposed that the presence of cholesterol may make an important contribution to the general inability of ACP_{AO} peptides to disrupt erythrocyte membranes [62], which inherently contain high levels of the sterol [208]. It has also been demonstrated that increasing levels of cholesterol, along with other sterols, can decrease the activity of ACP_T peptides against non-cancerous cells and erythrocytes, thereby reducing the toxicity of these peptides to these cells [208–211]. Examples of these peptides include kalata B1, kalata B2, and cycloviolacin O2 [212]; melittin and temporin L, which is an amphibian α-ACP_T peptide [101]; and LL-37, which is a human α-ACP_T peptide [103]. In relation to cancerous cells, it has recently been demonstrated that elevated levels of cholesterol-rich lipid rafts are present in breast and prostate cancer cell lines [213]. Studies on melittin have shown that it is able to associate with membrane cholesterol [211, 214, 215] while other ACP_T peptides form complexes with membrane cholesterol [156, 216]. Based on these studies, it was predicted that the association of ACPs with cholesterol-rich lipid rafts in the cancer cell membrane may decrease the toxicity of these peptides to cancer cells [71]. However, it has also been suggested that the formation of cholesterol-rich lipid rafts may deplete the bulk cancer membrane of the sterol, leading to regions with increased fluidity that contribute to an increased susceptibility to the action of some ACPs and thereby increasing the toxicity of these peptides [67]. A clear implication from these studies is that the presence of cholesterol-rich lipid rafts in the cancer cell membrane may be a key factor in differentiating the activity of both ACP_T and ACP_{AO} peptides against cancer cells [56, 59].

6.3.3
Anionic Membrane Components

The major determinant in the selectivity and toxicity of ACPs for cancer cells over non-cancerous cells appears to be the over-representation of anionic membrane components on the surface of the former cells compared to the latter. The increased levels of these anionic molecules gives cancer cells a net negative charge, which allows positively charged ACP_{AO} and ACP_T peptides to target and bind these cells with increased efficacy compared to non-cancerous cells [56, 59], whose outer surface is electrically neutral [217, 218]. The electrical neutrality of these latter membranes is believed to be a major contributor to the inability of ACP_{AO} peptides to target non-cancerous cells, thereby protecting the host from the action of these peptides. It has also been shown that the electrical neutrality of non-cancerous cell membranes helps to provide these cells with some protection from the action of ACP_T peptides although, clearly, this protective effect is limited and other factors support the ability of these latter peptides to kill these cells [219].

It is also well established that electrostatically driven interactions with anionic components of the cancer cell membrane is fundamental to the selectivity and toxicity of ACP_{AO} and ACP_T peptides for these cells [54, 56, 57, 59–63, 220]. In the case of ACPs that translocate to the cytostolic compartment of cancer cells to preferentially interact with mitochondrial membranes, it has been suggested that these interactions may involve the high levels of anionic lipid expressed by these membranes such as cardiolipin [67]. It has also been shown that the membranes of mitochondria in a variety of melanoma express significant amounts of phosphatidylserine (PS), which could also contribute to the ability of ACPs to interact with these membranes [67]. However, most research into the interaction of ACP_{AO} and ACP_T peptides with anionic components of cancer cell membranes have focused on those expressed in the cytoplasmic membrane of these cells, and have identified a number molecules and classes of molecules that appear to be involved in these interactions, including glycoproteins, glycolipids, PGs, and phospholipids.

6.3.4
Glycoproteins and Glycolipids

Glycoproteins and glycolipids represent a major source of negative charge on the surface of human cells by virtue of their glycosylated side-chains [221–223] whose exposed terminal positions are occupied by sialic acids, which are anionic sugar residues [224]. There is some evidence to suggest that the interaction of sialic acids with ACP_T peptides can reduce the toxicity of these peptides by inhibiting their action against non-cancerous cells [225, 226] and erythrocytes [64]. In relation to cancer cells, it has been reported that changes in the glycosylation of glycoproteins and glycolipids, including the enhanced expression of their terminal sialic acids, is characteristic of a variety of cancers [227, 228]. As an example, recent studies have shown that levels of mucin 1, which is a glycoprotein, and cell surface sialylation are elevated in the vast majority of breast carcinomas and often in other cancers, such as ovarian, lung, colon, and pancreatic carcinoma [229]. The level of surface sialylation on cancer cells appears to correlate with the metastatic potential of different cancer cell types, serving a protective function against the host immune system, and facilitating the migration and invasion of these cells via modulation of adhesiveness [230–232].

It has previously been suggested that increased levels of sialic acids on glycoproteins and glycolipids may contribute to the selectivity of ACPs for cancerous cells over non-cancerous cells [62]. Strongly supporting this suggestion, several investigations have shown that enzymatic digestion of sialyl residues on the surface of cancer cells greatly reduces the ability of α-ACPs to target these cells [98, 233]. Examples of these peptides include BMAP27 and BMAP28 [59], and buforin IIB, which is a synthetic amphibian ACP_{AO} peptide [99]. Moreover, it has also been shown that gangliosides, which are glycosphingolipids that carry the sialyated side-chains of some glycolipids [221], are more abundant on the outer surface of cancer cells than non-cancerous cells [234, 235]. These glycosphingolipids were found to act as specific targets for buforin IIb, and were essential to both the selectivity of the peptide for cancer cells and its ability to translocate the membranes of these cells to attack intracellular targets [98, 99]. Nonetheless, it was recently reported that the anticancer activity of a peptide derived from human lactoferricin was unaffected by the removal of sialic acid moieties from rhabdosarcoma and it was concluded that other targets were preferred by the peptide for its activity against this tumor type [67]. Other studies have shown that gangliosides can inhibit the toxicity of melittin to non-cancerous cells while there is evidence to suggest that sialic acids may generally protect erythrocytes from the action of ACPs. In combination, these studies clearly show that sialyated components of the cancer cell membrane play an important although complex role in the selectivity and toxicity of ACP_{AO} peptide and α-ACP_T peptides for cancer cells.

6.3.5
Proteoglycans

PGs represent another source of negative charge on the surface of human cells due to their highly negatively charged glycosaminoglycan (GAG) side-chains [236], primarily in the form of chondroitin sulfate and heparan sulfate (HS), which consist of linear repeats of up to 100 highly sulfated disaccharide units [237]. The level of these PGs expressed on the cell surface along with the degree and the pattern of sulfation of their GAG side-chains has been shown to differ between cancer cells to non-cancerous cells. These differences can lead to the expression of PGs on the surface of cancerous cells that show much higher levels of negative charge than those of their non-cancerous counterparts [238–240]. Recent studies have suggested that the interaction of HS with cationic peptides may initiate the cell entry of these peptides [241], and a number of investigations have focused on the ability of α-ACP_{AO} and α-ACP_T peptides to bind to GAG. For example, magainin 2, was found not to bind HS, whereas strong binding was observed for LL-37, which is a human α-ACP_T peptide [242], and melittin [243] In the case of the latter peptide, interaction with HS was accompanied by a conformational change in melittin, which led to the adoption of a more α-helical structure, and it has been suggested that its interaction and HS is most likely a first step in the peptide's internalization into cancer cells [243, 244]. More recently, CB1a, which is a synthetic α-ACP_{AO} with potent toxicity against stomach carcinoma (AGS), leukemia (HL-60 and CCRF-CEM), and lung cancer cells (NCI-H520 and NCI-H661) was found to adopt an α-helical structure in the presence of heparin, an anionic structural analog of HS, and exhibited the ability to translocate across the cancer cell membrane. Based on these results, it was suggested the anticancer action of CB1a may show similarities to that of melittin and involve interaction with intracellular targets within cancer cells [100]. In a major recent study, the anticancer action of several ACPs was investigated with respect to the effects of varying GAG levels on the surface of target cancer cells. This study found that lactoferricin B (LfB), which is a bovine β-ACP_{AO} peptide [245, 246], and KW5, a synthetic α-ACP_T, displayed higher cytotoxic activity against melanoma cells (FEMX), which expressed a larger amount of GAGs compared to colon carcinoma cells (HT-29), which expressed much lower levels of these anionic PG side-chains [247]. Clearly, this result is in agreement with other studies, which have shown that increasing the level of negatively charged components on the surface of target cancer cells generally enhances the cytotoxic activity of defense peptides [97, 98, 191]. However, contrary to what may be expected, the study [247] also found that reducing the sulfation levels of GAGs expressed by the melanoma and carcinoma investigated had the effect of enhancing the cytotoxic activity of both LfB and KW5, indicating that in these cases, the anionic PG side-chains were in fact inhibiting the anticancer activity of some peptides. Further investigation into this inhibitory effect suggested that it depended on two major factors. Compared to the melanoma cell line, the carcinoma cell line expressed larger PGs, which may keep bound ACPs at a greater distance from the membrane. In addition, PGs of the carcinoma cell line expressed GAGs with much higher levels of HS than those of the melanoma cell line, and both LfB and KW5 primarily bound to HS [247]. Based on these combined data, it was suggested that that PGs expressed at higher levels by cancer cells may have an inhibitory effect on the lytic action of ACPs by binding these peptides and thereby restricting their access to the membrane. However, this effect is influenced by the size of the cell surface PGs along with the size, chemical nature, and sulfation levels of their GAGs [247]. More recent studies have suggested that the effect of HS on the anticancer action of ACPs also varies with the characteristics of these peptides when it was found that the anticancer activity of small membranolytic ACPs derived from LfB was either unaffected or increased by the presence of the proteoglycan [248].

Hyaluronan (hyaluronic acid or hyaluronate) is another anionic GAG that associates with PGs, but differs to all other GAGs in that it is non-sulfated and non-covalently linked to core proteins [237]. Many tumor cells express hyaluronan at levels that are much higher than those found on normal tissues [249, 250], and in a recent study, tachyplesin I, which is found in crabs, was shown to target and bind hyaluronan on carcinoma cells that overexpressed the GAG [251]. Using what appears to be an anticancer mechanism not previously reported, while bound to cancer cells, the peptide also bound the C1q component of complement in human serum, which led to the complement-mediated lysis of tachyplesin I-coated cancer cells [251]. Hyaluronan is also highly expressed on the surface of endothelial cells involved in neovascularization [249, 250], and it has been suggested that interactions between these GAGs and the peptide may contribute to the killing of cancer cells via complement-mediated destruction of tumor-associated vasculature [59].

6.3.6
Phospholipids

Phosphatidylserine is a negatively charged phospholipid that localizes exclusively in the inner membrane leaflet of normal cells but can be exposed on the surface of these cells by translocation to the outer leaflet in a range of pathological conditions including cancers [252]. It has been shown that cancer cells expose between 3- and 7-fold more PS than normal keratinocytes, including those from human ovarian carcinoma [253], gastric carcinoma [253], leukemias [97, 254], and various melanoma [255, 256]. More recently, it has been shown that different melanomas expose between 4- and 11-fold more PS than melanocytes where the level of the exposed phospholipid correlated with the progression of these tumors [257, 258]. Exposed PS has also been demonstrated on the surface of cells from primary cancer cell cultures, including glioblastoma and rhabdomyosarcoma [67, 257].

A number of studies have shown that the exposure of PS on the surface of cancer cells is required to facilitate the anticancer action of ACPs [59, 62, 71], as demonstrated for number of synthetic ACPs when it was shown that the colocalization of this exposed phospholipid and these peptides was essential for their anticancer activity against cells from human prostate carcinoma (CL1 and 22RV1) and human breast cancer (RFP-MDA-MB-231) [259]. More recent work has shown that the colocalization of PS on the surface of cancer cells with some ACPs is required to initiate novel anticancer mechanisms based on amyloidogenesis (Chapter 5) [71, 260]. As examples, this mechanism has been proposed to underpin the anticancer activity of eosinophil cationic antimicrobial protein [261], which may be considered a human ACP_{AO} [262], and is able to kill cells from human leukemia (HL-60), cervix adenocarcinoma (HeLa) [263], erythroleukemia (K562), and epidermoid carcinoma (A431) [262]. Amyloidogenesis has also been proposed to underpin the anticancer activity of plantarcin A [264], which is a bacterial α-ACP_T [265] able to kill human leukemic T cells (Jurkat) along with cells from murine neuroblastoma (N2A) and pheochromocytoma (PC12) [264, 265]. It is possible that similar mechanisms of anticancer action are utilized by many other ACPs based on the results of a recent theoretical study, which predicted that a large number of these peptides possessed the potential to form amyloid fibrils [266].

To gain insight into the role of PS in the anticancer action of ACPs, a recent study investigated the ability of aurein 1.2, an amphibian α-ACP_{AO} peptide [89, 90], to partition into model glioma and astrocytoma membranes in which the levels of this phospholipid were varied. These experiments showed that a strong positive correlation existed between the ability of the peptide to penetrate these membranes and their PS levels, clearly suggesting that the phospholipid was important to the peptides selectivity for toxicity to cancer cells [62, 136]. This suggestion was strongly supported by later work on aurein 1.2 and citropin 1.1, an amphibian α-ACP_{AO} peptide [116, 267], which investigated the interaction of the peptides with anionic phospholipids in model bilayers [268]. In response, several studies have investigated the variation of endogenous surface-exposed PS on cancer cells and its relation to the ability of ACPs to kill these cells, which included neuroblastoma, lymphoma, glioma, carcinoma and myeloma. It was found by these investigations that endogenous PS levels varied widely across cancer cells, with murine myeloma expressing levels of the phospholipid that were around 100-fold higher than those of human lung cancer cells [97, 191]. Paralleling the results of Dennison et al. [62,

136], these studies also demonstrated a strong positive correlation between the levels of PS found on the surface of these cancer cells and the cytotoxicity of several ACPs, including NK-2, a synthetic α-ACP$_{AO}$ of porcine derivation [97], and analogs of beetle defensins [191].

In addition to cancer cells, PS and other anionic phospholipids are also over represented on the surface of tumor-associated vascular endothelial cells [269], which are produced by tumor-induced angiogenesis and are essential for tumor survival and propagation [270]. It appears that the interaction of some ACPs with these anionic lipids can play a role in the death of the associated tumor by inactivating tumor-associated vascular endothelial cells, thereby directly inhibiting angiogenesis [63], with examples including human defensins [111], gomesin [112], RGD-tachyplesin 1 [271], and LfB [272]. In the case of this latter peptide, it appears that it is able to block angiogenesis not only by directly binding anionic phospholipids, but also indirectly by binding HS on the cancer cell surface, thereby blocking the induction of angiogenesis by preventing the docking of associated growth factors with their membrane receptors [272]. PS exposed on surface of tumor-associated vascular also appears to facilitate amyloid-based mechanisms of antiangiogenesis in the case of endostatin, an endogenous human ACP cleaved from collagen XVIII [273, 274]. Many more endogenous human protein fragments that inhibit angiogenesis are known and it has been suggested that these inhibitory effects may also be mediated by the ability of these peptides to form amyloid structures [275]. Based on these studies, it has been proposed that tumor-associated vasculature may provide another target for the anticancer action of ACPs [67].

6.4 Discussion

This chapter has reviewed evidence of differences in a number of membrane-based factors that contribute to the anticancer action of both α-ACPs and other ACPs. This evidence has shown that that some of these latter peptides may possess the ability to recognize membrane-based receptors, which could contribute to the selectivity and toxicity of ACP$_{AO}$ and ACP$_T$ peptides for cancer cells. Physical and electrochemical properties, such as the acidic extracellular environment, increased levels of microvilli, and higher transmembrane potentials that are associated with cancer cells, appear to contribute to the selectivity of ACP$_{AO}$ peptides for these cells over non-cancerous cells. These differences in membrane properties may also contribute to the toxicity of both these latter peptides and ACP$_T$ peptides for target eukaryotic cells, although this possibility does not appear to have been extensively researched. The net negative surface charge carried by cancer cells as compared to the overall neutral charge carried by non-cancerous cells appears to be the major factor in the anticancer action of ACPs. This negative surface charge facilitates the targeting and electrostatic binding of cancer cells by cationic ACPs, and, in general, these binding events appear to be a major determinant in the selectivity and toxicity of ACP_{AO} peptides for cancer cells and the toxicity of ACP_T peptides for these latter cells. Consistent with these observations, this chapter has shown that in relation to non-cancerous cells, a variety of anionic membrane-based moieties, including sialic acid residues, sulfated GAGs, and PS, are over-represented on the surface of cancer cells. Nonetheless, this chapter has also shown that these anionic moieties can exhibit widely differing levels of expression between cancer cell types and vary greatly in their structural characteristics, which under some circumstances can either reduce or not affect both the selectivity and toxicity of ACPs for cancer cells. Indeed, it appears that the structural characteristics of some anionic moieties on the surface of non-cancerous cells, which carry no net positive charge, enable these moieties to bind ACPs and inhibit their lytic action, thereby effectively influencing the selectivity and toxicity of ACPs for cancer cells.

Cholesterol is present in the membranes of both cancerous and non-cancerous cells, and this presence appears to have a variable effect on the selectivity and toxicity of ACPs for these cells. The presence of the sterol in the membranes of non-cancerous cells appears able to inhibit the action of some α-ACP_{AO} peptides against these cells via changes in membrane fluidity. Clearly, this ability can contribute to the lack of toxicity shown by these peptides for non-cancerous cells and effectively enhance their selectivity for cancer cells. The presence of cholesterol in the membranes of non-cancerous cells also appears able to inhibit the action of ACP_T peptides, thereby reducing their toxicity to these cells. There is evidence to suggest that the membranes of some cancerous cells contain cholesterol-rich lipid rafts that are able to associate with ACP_T peptides, and such associations may be a key factor in differentiating the activity of both ACP_T and ACP_{AO} peptides towards cancer cells. A clear implication from these studies is that the presence of cholesterol-rich lipid rafts in the cancer cell membrane may be a key factor in differentiating the activity of both ACP_T and ACP_{AO} peptides against cancer cells [56, 59].

This chapter has also reviewed evidence that differences in a number of peptide-based factors could contribute to the anticancer action of both α-ACPs and other ACPs. This evidence has shown that differences between α-ACP_{AO} and α-ACP_T peptides in relation to their residue composition, sequence length, net positive charge, and molecular architecture such as that inherent in amphiphilicity may each have some influence on the selectivity of the former peptides for cancer cells over non-cancerous cells. However, taken overall, this chapter strongly suggests that no single membrane-based or peptide-based factor, alone, is responsible for the anticancer action of ACPs. Rather, it would seem that, according to the case, varying contributions from each of a number of these factors determines the selectivity and toxicity shown by ACP_{AO} peptides for cancer cells along with the general toxicity shown to eukaryotic cells by ACP_T peptides. In a sense, it may be that the anticancer action of ACPs parallels the "lock and key" model postulated for enzyme activity [276] where the molecular architecture of the peptide has to support its binding and insertion into a membrane of a given composition. Such binding and insertion is clearly dependent on the membrane packing, but also the shape and physiochemical properties of the peptide. For example, as described above, net positive charge is a major component of the "key" in the anticancer action of many ACPs, which facilitates binding of the peptide to anionic moieties on the cancer cell surface. In some cases, it may be that when these two factors are not balanced, such as when the anionic moieties on the cancer cell surface bind a peptide too tightly or in insufficient quantities for anticancer action, the peptide "key" cannot fully engage or efficiently fit the "lock" formed from the lipid packing arrangement and anticancer action is not initiated. Currently, a number of studies are in progress, and are trying to improve the selectivity and toxicity of ACPs to cancer cells by varying parameters, such as length, charge, hydrophobicity, and residue composition [277]. However, as emphasized in this chapter, in the design of novel ACPs, the characteristics of not only these peptides, but also those of the target cancer cell membranes must be taken into consideration. Indeed, there is a wide variation in the characteristics of cancer cell membranes and it may be that the design of novel ACPs should be guided by a cancer cell-specific approach rather than the generic approach, which appears to be currently used.

References

1 Thun, M.J., DeLancey, J.O., Center, M.M., Jemal, A., and Ward, E.M. (2010) The global burden of cancer: priorities for prevention. *Carcinogenesis*, **31**, 100–110.

2 Kreeger, P.K. and Lauffenburger, D.A. (2010) Cancer systems biology: a network modeling perspective. *Carcinogenesis*, **31**, 2–8.

3 Marusyk, A. and Polyak, K. (2009) Tumor heterogeneity: causes and consequences. *Biochimica et Biophysica Acta*, **1805**, 105–117.

4 Brooks, S.A., Lomax-Browne, H.J., Carter, T.M., Kinch, C.E., and Hall, D.M. (2010) Molecular interactions in cancer cell metastasis. *Acta Histochemica*, **112**, 3–25.

5 Sung, S.Y. and Johnstone, P.A.S. (2007) Tumor microenvironment promotes cancer progression, metastasis, and therapeutic resistance. *Current Problems in Cancer*, **31**, 36–100.

6 Urruticoechea, A., Alemany, R., Balart, J., Villanueva, A., Vinals, F., and Capella, G. (2009) Recent advances in cancer therapy: an overview. *Current Pharmaceutical Design*, **16**, 3–10.

7 Lee, L.J. and Harris, J.R. (2009) Innovations in radiation therapy (RT) for breast cancer. *The Breast*, **18**, S103–S111.

8 Mallick, I. and Waldron, J.N. (2009) Radiation therapy for head and neck cancers. *Seminars in Oncology Nursing*, **25**, 193–202.

9 Aebi, S. (2006) Chemotherapy – overview and future perspectives. *The Breast*, **15**, S11–S14.

10 Verweij, J. and de Jonge, M.J.A. (2000) Achievements and future of chemotherapy. *European Journal of Cancer*, **36**, 1479–1487.

11 Schiff, D., Wen, P.Y., and van den Bent, M.J. (2009) Neurological adverse effects caused by cytotoxic and targeted therapies. *Nature Reviews. Clinical Oncology*, **6**, 596–603.

12 Soussain, C., Ricard, D., Fike, J.R., Mazeron, J.J., Psimaras, D., and Delattre, J. (2009) CNS complications of radiotherapy and chemotherapy. *Lancet*, **374**, 1639–1651.

13 Ralhan, R. and Kaur, J. (2007) Alkylating agents and cancer therapy. *Expert Opinion on Therapeutic Patents*, **17**, 1061–1075.

14 Fu, Y., Li, S., Zu, Y., Yang, G., Yang, Z., Luo, M., Jiang, S., Wink, M., and Efferth, T. (2009) Medicinal chemistry of

paclitaxel and its analogs. *Current Medicinal Chemistry*, **16**, 3966–3985.
15 Ruan, K., Song, G., and Ouyang, G. (2009) Role of hypoxia in the hallmarks of human cancer. *Journal of Cellular Biochemistry*, **107**, 1053–1062.
16 Livesey, K.M., Tang, D., Zeh, H.J., and Lotze, M.T. (2009) Autophagy inhibition in combination cancer treatment. *Current Opinion in Investigational Drugs*, **10**, 1269–1279.
17 Gillet, J. and Gottesman, M.M. (2010) Mechanisms of multidrug resistance in cancer. *Methods in Molecular Biology*, **596**, 47–76.
18 Baguley, B.C. (2010) Multidrug resistance in cancer. *Methods in Molecular Biology*, **596**, 1–14.
19 Liu, F.S. (2009) Mechanisms of chemotherapeutic drug resistance in cancer therapy – a quick review. *Taiwanese Journal of Obstetrics & Gynecology*, **48**, 239–244.
20 Goda, K., Bacso, Z., and Szabo, G. (2009) Multidrug resistance through the spectacle of P-glycoprotein. *Current Cancer Drug Targets*, **9**, 281–297.
21 Ozben, T. (2006) Mechanisms and strategies to overcome multiple drug resistance in cancer. *FEBS Letters*, **580**, 2903–2909.
22 Jorritsma, A., Schumacher, T.N.M., and Haanen, J. (2009) Immunotherapeutic strategies: the melanoma example. *Immunotherapy*, **1**, 679–690.
23 Mocellin, S., Pilati, P., and Nitti, D. (2009) Peptide-based anticancer vaccines: recent advances and future perspectives. *Current Medicinal Chemistry*, **16**, 4779–4796.
24 Mishra, S. and Sinha, S. (2009) Immunoinformatics and modeling perspective of T cell epitope-based cancer immunotherapy: a holistic picture. *Journal of Biomolecular Structure & Dynamics*, **27**, 293–305.
25 Kanduc, D. (2009) Epitopic peptides with low similarity to the host proteome: towards biological therapies without side effects. *Expert Opinion on Biological Therapy*, **9**, 45–53.
26 Bodles-Brakhop, A.M. and Draghia-Akli, R. (2008) DNA vaccination and gene therapy: optimization and delivery for cancer therapy. *Expert Review of Vaccines*, **7**, 1085–1101.
27 Copier, J., Dalgleish, A.G., Britten, C.M., Finke, L.H., Gaudernack, G., Gnjatic, S., Kallen, K., Kiessling, R., Schuessler-Lenz, M., Singh, H., Talmadge, J., Zwierzina, H., and Hakansson, L. (2009) Improving the efficacy of cancer immunotherapy. *European Journal of Cancer*, **45**, 1424–1431.
28 Mathew, M. and Verma, R.S. (2009) Humanized immunotoxins: a new generation of immunotoxins for targeted cancer therapy. *Cancer Science*, **100**, 1359–1365.
29 Ilett, E.J., Prestwich, R.J.D., and Melcher, A.A. (2010) The evolving role of dendritic cells in cancer therapy. *Expert Opinion on Biological Therapy*, **10**, 369–379.
30 Collins, S.A., Guinn, B.A., Harrison, P.T., Scallan, M.F., O'Sullivan, G.C., and Tangney, M. (2008) Viral vectors in cancer immunotherapy: which vector for which strategy? *Current Gene Therapy*, **8**, 66–78.
31 Huang, Z.Q. and Buchsbaum, D.J. (2009) Monoclonal antibodies in the treatment of pancreatic cancer. *Immunotherapy*, **1**, 223–239.
32 Eshhar, Z. (2010) Adoptive cancer immunotherapy using genetically engineered designer T-cells: first steps into the clinic. *Current Opinion in Molecular Therapeutics*, **12**, 55–63.
33 Elkord, E., Hawkins, R.E., and Stern, P.L. (2008) Immunotherapy for gastrointestinal cancer: current status and strategies for improving efficacy. *Expert Opinion on Biological Therapy*, **8**, 385–395.
34 Spitz, M.R. and Bondy, M.L. (2010) The evolving discipline of molecular epidemiology of cancer. *Carcinogenesis*, **31**, 127–134.
35 Fuchs, H. and Bachran, C. (2009) Targeted tumor therapies at a glance. *Current Drug Targets*, **10**, 89–93.
36 Sarkar, F.H. and Li, Y. (2009) Harnessing the fruits of nature for the development of multi-targeted cancer therapeutics. *Cancer Treatment Reviews*, **35**, 597–607.

37 Pallis, A.G., Serfass, L., Dziadziusko, R., van Meerbeeck, J.P., Fennell, D., Lacombe, D., Welch, J., and Gridelli, C. (2009) Targeted therapies in the treatment of advanced/metastatic NSCLC. *European Journal of Cancer*, **45**, 2473–2487.

38 Minniti, G., Muni, R., Lanzetta, G., Marchetti, P., and Enrici, R.M. (2009) Chemotherapy for glioblastoma: current treatment and future perspectives for cytotoxic and targeted agents. *Anticancer Research*, **29**, 5171–5184.

39 Pathania, D., Millard, M., and Neamati, N. (2009) Opportunities in discovery and delivery of anticancer drugs targeting mitochondria and cancer cell metabolism. *Advanced Drug Delivery Reviews*, **61**, 1250–1275.

40 Breen, E.C. and Walsh, J.J. (2010) Tubulin-targeting agents in hybrid drugs. *Current Medicinal Chemistry*, **17**, 609–639.

41 Li, Y. and Cozzi, P.J. (2009) Angiogenesis as a strategic target for prostate cancer therapy. *Medicinal Research Reviews*, **30**, 23–66.

42 Rodon, J., Perez, J., and Kurzrock, R. (2010) Combining targeted therapies: practical issues to consider at the bench and bedside. *Oncologist*, **15**, 37–50.

43 Cao, S., Cripps, A., and Wei, M.Q. (2010) New strategies for cancer gene therapy: progress and opportunities. *Clinical and Experimental Pharmacology and Physiology*, **37**, 108–114.

44 Gough, M.J. and Crittenden, M.R. (2009) Combination approaches to immunotherapy: the radiotherapy example. *Immunotherapy*, **1**, 1025–1037.

45 Zahorowska, B., Crowe, P.J., and Yang, J.-L. (2009) Combined therapies for cancer: a review of EGFR-targeted monotherapy and combination treatment with other drugs. *Journal of Cancer Research and Clinical Oncology*, **135**, 1137–1148.

46 Seufferlein, T., Ahn, J., Krndija, D., Lother, U., Adler, G., and von Wichert, G. (2009) Tumor biology and cancer therapy – an evolving relationship. *Cell Communication and Signaling*, **7**, 19.

47 Cirstea, D., Vallet, S., and Raje, N. (2009) Future novel single agent and combination therapies. *Cancer Journal*, **15**, 511–518.

48 Kamrava, M., Bernstein, M.B., Camphausen, K., and Hodge, J.W. (2009) Combining radiation, immunotherapy, and antiangiogenesis agents in the management of cancer: the Three Musketeers or just another quixotic combination? *Molecular BioSystems*, **5**, 1262–1270.

49 Sakamoto, J., Matsui, T., and Kodera, Y. (2009) Paclitaxel chemotherapy for the treatment of gastric cancer. *Gastric Cancer*, **12**, 69–78.

50 Klimm, B. and Engert, A. (2009) Combined modality treatment of Hodgkin's lymphoma. *Cancer Journal*, **15**, 143–149.

51 Magne, N., Deutsch, E., and Haie-Meder, C. (2008) Current data on radiochemotherapy and potential of targeted therapies for cervical cancers. *Cancer Radiothérapie*, **12**, 31–36.

52 Shannon, A.M. and Williams, K.J. (2008) Antiangiogenics and radiotherapy. *Journal of Pharmacy and Pharmacology*, **60**, 1029–1036.

53 Zheng, L.-H., Wang, Y.-J., Sheng, J., Wang, F., Zheng, Y., Lin, X.-K., and Sun, M. (2011) Antitumor peptides from marine organisms. *Marine Drugs*, **9**, 1840–1859.

54 Gomes, A., Bhattacharjee, P., Mishra, R., Biswas, A.K., Dasgupta, S.C., Giri, B., Debnath, A., Das Gupta, A., Das, T., and Gomes, A. (2010) Anticancer potential of animal venoms and toxins. *Indian Journal of Experimental Biology*, **48**, 93–103.

55 Udenigwe, C.C. and Aluko, R.E. (2012) Food protein-derived bioactive peptides: production, processing, and potential health benefits. *Journal of Food Science*, **77**, R11–R24.

56 Schweizer, F. (2009) Cationic amphiphilic peptides with cancer-selective toxicity. *European Journal of Pharmacology*, **625**, 190–194.

57 Smolarczyk, R., Cichon, T., and Szala, S. (2009) Peptides: a new class of anticancer drugs. *Postepy Higieny i Medycyny Doswiadczalnej*, **63**, 360–368.

58 Rodrigues, E.G., Dobroff, A.S., Taborda, C.P., and Travassos, L.R. (2009)

Antifungal and antitumor models of bioactive protective peptides. *Anais da Academia Brasileira de Ciencias*, **81**, 503–520.

59 Hoskin, D.W. and Ramamoorthy, A. (2008) Studies on anticancer activities of antimicrobial peptides. *Biochimica et Biophysica Acta*, **1778**, 357–375.

60 Bhutia, S.K. and Maiti, T.K. (2008) Targeting tumors with peptides from natural sources. *Trends in Biotechnology*, **26**, 210–217.

61 Slocinska, M., Marciniak, P., and Rosinski, G. (2008) Insects antiviral and anticancer peptides: new leads for the future? *Protein and Peptide Letters*, **15**, 578–585.

62 Dennison, S.R., Whittaker, M., Harris, F., and Phoenix, D.A. (2006) Anticancer alpha-helical peptides and structure–function relationships underpinning their interactions with tumour cell membranes. *Current Protein & Peptide Science*, **7**, 487–499.

63 Mader, J.S. and Hoskin, D.W. (2006) Cationic antimicrobial peptides as novel cytotoxic agents for cancer treatment. *Expert Opinion on Investigational Drugs*, **15**, 933–946.

64 Papo, N. and Shai, Y. (2005) Host defense peptides as new weapons in cancer treatment. *Cellular and Molecular Life Sciences*, **62**, 784–790.

65 Leuschner, C. and Hansel, W. (2004) Membrane disrupting lytic peptides for cancer treatments. *Current Pharmaceutical Design*, **10**, 2299–2310.

66 Heinen, T.E. and Gorini da Veiga, A.B. (2011) Arthropod venoms and cancer. *Toxicon*, **57**, 497–511.

67 Riedl, S., Zweytick, D., and Lohner, K. (2011) Membrane-active host defense peptides – challenges and perspectives for the development of novel anticancer drugs. *Chemistry and Physics of Lipids*, **164**, 766–781.

68 Al-Benna, S., Shai, Y., Jacobsen, F., and Steinstraesser, L. (2011) Oncolytic activities of host defense peptides. *International Journal of Molecular Sciences*, **12**, 8027–8051.

69 Burman, R., Herrmann, A., Tran, R., Kivela, J.-E., Lomize, A., Gullbo, J., and Goransson, U. (2011) Cytotoxic potency of small macrocyclic knot proteins: structure–activity and mechanistic studies of native and chemically modified cyclotides. *Organic & Biomolecular Chemistry*, **9**, 4306–4314.

70 Burman, R., Stromstedt, A.A., Malmsten, M., and Goransson, U. (2011) Cyclotide–membrane interactions: defining factors of membrane binding, depletion and disruption. *Biochimica et Biophysica Acta – Biomembranes*, **1808**, 2665–2673.

71 Harris, F., Dennison, S.R., Singh, J., and Phoenix, D.A. (2011) On the selectivity and efficacy of defense peptides with respect to cancer cells. *Medicinal Research Reviews*, doi: 10.1002/med.20252

72 Henriques, S.T. and Craik, D.J. (2010) Cyclotides as templates in drug design. *Drug Discovery Today*, **15**, 57–64.

73 Harris, F., Dennison, S.R., and Phoenix, D.A. (2019) Anionic antimicrobial peptides from eukaryotic organisms. *Current Protein Peptide Science*, **10**, 585–606.

74 Slaninová, J., Mlsová, V., Kroupová, H., Alán, L., Tůmová, T., Monincová, L., Borovičková, L., Fučík, V., and Čeřovský, V. (2012) Toxicity study of antimicrobial peptides from wild bee venom and their analogs toward mammalian normal and cancer cells. *Peptides*, **33**, 18–26.

75 Rajanbabu, V. and Chen, J.-Y. (2011) Applications of antimicrobial peptides from fish and perspectives for the future. *Peptides*, **32**, 415–420.

76 Rozek, T., Bowie, J.H., Wallace, J.C., and Tyler, M.J. (2000) The antibiotic and anticancer active aurein peptides from the Australian Bell Frogs *Litoria aurea* and *Litoria raniformis*. Part 2. Sequence determination using electrospray mass spectrometry. *Rapid Communications in Mass Spectrometry*, **14**, 2002–2011.

77 Rozek, T., Wegener, K.L., Bowie, J.H., Olver, I.N., Carver, J.A., Wallace, J.C., and Tyler, M.J. (2000) The antibiotic and anticancer active aurein peptides from the Australian Bell Frogs *Litoria aurea* and *Litoria raniformis* – the solution structure of aurein 1.2. *European Journal of Biochemistry*, **267**, 5330–5341.

78 Kaas, Q. and Craik, D.J. (2010) Analysis and classification of circular proteins in CyBase. *Biopolymers*, **94**, 584–591.

79 Dennison, S.R., Harris, F., Bhatt, T., Singh, J., and Phoenix, D.A. (2009) The effect of C-terminal amidation on the efficacy and selectivity of antimicrobial and anticancer peptides. *Molecular and Cellular Biochemistry*, **332**, 43–50.

80 Dennison, S.R., Harris, F., Bhatt, T., Singh, J., and Phoenix, D.A. (2010) A theoretical analysis of secondary structural characteristics of anticancer peptides. *Molecular and Cellular Biochemistry*, **333**, 129–135.

81 Wang, C.K.L., Kaas, Q., Chiche, L., and Craik, D.J. (2008) CyBase: a database of cyclic protein sequences and structures, with applications in protein discovery and engineering. *Nucleic Acids Research*, **36**, D206–D210.

82 Hammami, R., Ben Hamida, J., Vergoten, G., and Fliss, I. (2009) PhytAMP: a database dedicated to antimicrobial plant peptides. *Nucleic Acids Research*, **37**, D963–D968.

83 Owen, D.R. (2005) Short Bioactive Peptides, US Patent 6,875,744.

84 Owen, D.R. and Anonymous (2008) Methods for Use of Short Bioactive Peptides, US Patent 7,381,704.

85 Wang, G., Li, X., and Wang, Z. (2009) APD2: the updated antimicrobial peptide database and its application in peptide design. *Nucleic Acids Research*, **37**, D933–D937.

86 Wang, G., Li, X., and Zasloff, M. (2010) A database view of naturally occurring antimicrobial peptides: nomenclature, classification and amino acid sequence analysis. *Advances in Molecular and Cellular Microbiology*, **18**, 1–21.

87 Harris, F., Dennison, S.R., and Phoenix, D.A. (2011) Anionic antimicrobial peptides from eukaryotic organisms and their mechanisms of action. *Current Chemical Biology*, **5**, 142–153.

88 Jin, X., Mei, H., Li, X., Ma, Y., Zeng, A.-H., Wang, Y., Lu, X., Chu, F., Wu, Q., and Zhu, J. (2010) Apoptosis-inducing activity of the antimicrobial peptide cecropin of *Musca domestica* in human hepatocellular carcinoma cell line BEL-7402 and the possible mechanism. *Acta Biochimica et Biophysica Sinica*, **42**, 259–265.

89 Doyle, J., Brinkworth, C.S., Wegener, K.L., Carver, J.A., Llewellyn, L.E., Olver, I.N., Bowie, J.H., Wabnitz, P.A., and Tyler, M.J. (2003) nNOS inhibition, antimicrobial and anticancer activity of the amphibian skin peptide, citropin 1.1 and synthetic modifications – the solution structure of a modified citropin 1.1. *European Journal of Biochemistry*, **270**, 1141–1153.

90 Apponyi, M.A., Pukala, T.L., Brinkworth, C.S., Maselli, V.A., Bowie, J.H., Tyler, M.J., Booker, G.W., Wallace, J.C., John, A., Separovic, F., Doyle, J., and Llewellyn, L. (2004) Host-defence peptides of Australian anurans: structure, mechanism of action and evolutionary significance. *Peptides*, **25**, 1035–1054.

91 Lu, C.X., Nan, K.J., and Lei, Y. (2008) Agents from amphibians with anticancer properties. *Anti-Cancer Drugs*, **19**, 931–939.

92 Kim, S.K., Kim, S.S., Bang, Y.J., Kim, S.J., and Lee, B.J. (2003) *In vitro* activities of native and designed peptide antibiotics against drug sensitive and resistant tumor cell lines. *Peptides*, **24**, 945–953.

93 dos Santos Cabrera, M.P., Arcisio-Miranda, M., Gorjão, R., Leite, N.B., de Souza, B.M., Palma, M.S., Cury, R., Neto, J.R., and Procopio, J. (2009) Influence of the bilayer composition on the membrane-disruption effect of Polybia-MP1, a mastoparan peptide with antimicrobial and leukemic cell selectivity. *Biophysical Journal*, **96**, 156a–156a.

94 Wang, K.R., Zhang, B.Z., Zhang, W., Yan, J.X., Li, J., and Wang, R. (2008) Antitumor effects, cell selectivity and structure–activity relationship of a novel antimicrobial peptide polybia-MPI. *Peptides*, **29**, 963–968.

95 Lin, W.J., Chien, Y.L., Pan, C.Y., Lin, T.L., Chen, J.Y., Chiu, S.J., and Hui, C.F. (2009) Epinecidin-1, an antimicrobial peptide from fish (*Epinephelus coioides*) which has an antitumor effect like lytic peptides in human fibrosarcoma cells. *Peptides*, **30**, 283–290.

96 Cerovsky, V., Budesinsky, M., Hovorka, O., Cvacka, J., Voburka, Z., Slaninova, J., Borovickova, L., Fucik, V., Bednarova, L., Votruba, I., and Straka, J. (2009) Lasioglossins: three novel antimicrobial peptides from the venom of the eusocial bee *Lasioglossum laticeps* (Hymenoptera: Halictidae). *ChemBioChem*, **10**, 2089–2099.

97 Schroder-Borm, H., Bakalova, R., and Andra, J. (2005) The NK-lysin derived peptide NK-2 preferentially kills cancer cells with increased surface levels of negatively charged phosphatidylserine. *FEBS Letters*, **579**, 6128–6134.

98 Lee, H.S., Park, C.B., Kim, J.M., Jang, S.A., Park, I.Y., Kim, M.S., Cho, J.H., and Kim, S.C. (2008) Mechanism of anticancer activity of buforin IIb, a histone H2A-derived peptide. *Cancer Letters*, **271**, 47–55.

99 Cho, J.H., Sung, B.H., and Kim, S.C. (2009) Buforins: histone H2A-derived antimicrobial peptides from toad stomach. *Biochimica et Biophysica Acta*, **1788**, 1564–1569.

100 Wu, J.M., Jan, P.S., Yu, H.C., Haung, H.Y., Fang, H.J., Chang, Y.I., Cheng, J.W., and Chen, H.M. (2009) Structure and function of a custom anticancer peptide, CB1a. *Peptides*, **30**, 839–848.

101 Rinaldi, A.C., Mangoni, M.L., Rufo, A., Luzi, C., Barra, D., Zhao, H.X., Kinnunen, P.K.J., Bozzi, A., Di Giulio, A., and Simmaco, M. (2002) Temporin L: antimicrobial, haemolytic and cytotoxic activities, and effects on membrane permeabilization in lipid vesicles. *Biochemical Journal*, **368**, 91–100.

102 Conlon, J.M., Abdel-Wahab, Y.H.A., Flatt, P.R., Leprince, J., Vaudry, H., Jouenne, T., and Condamine, E. (2009) A glycine–leucine-rich peptide structurally related to the plasticins from skin secretions of the frog *Leptodactylus laticeps* (Leptodactylidae). *Peptides*, **30**, 888–892.

103 Wu, W.K., Sung, J.J., To, K.F., Yu, L., Li, H.T., Li, Z.J., Chu, K.M., Yu, J., and Cho, C.H. (2010) The host defense peptide LL-37 activates the tumor-suppressing bone morphogenetic protein signaling via inhibition of proteasome in gastric cancer cells. *Journal of Cellular Physiology*, **223**, 178–186.

104 Lizzi, A.R., Carnicelli, V., Clarkson, M.M., Di Giulio, A., and Oratore, A. (2009) Lactoferrin derived peptides: mechanisms of action and their perspectives as antimicrobial and antitumoral agents. *Mini-Reviews in Medicinal Chemistry*, **9**, 687–695.

105 Chen, J.Y., Lin, W.J., and Lin, T.L. (2009) A fish antimicrobial peptide, tilapia hepcidin TH2-3, shows potent antitumor activity against human fibrosarcoma cells. *Peptides*, **30**, 1636–1642.

106 Lindholm, P., Goransson, U., Johansson, S., Claeson, P., Gullbo, J., Larsson, R., Bohlin, L., and Backlund, A. (2002) Cyclotides: a novel type of cytotoxic agents. *Molecular Cancer Therapeutics*, **1**, 365–369.

107 Svangard, E., Goransson, U., Hocaoglu, Z., Gullbo, J., Larsson, R., Claeson, P., and Bohlin, L. (2004) Cytotoxic cyclotides from *Viola tricolor*. *Journal of Natural Products*, **67**, 144–147.

108 Herrmann, A., Burman, R., Mylne, J.S., Karlsson, G., Gullbo, J., Craik, D.J., Clark, R.J., and Göransson, U. (2008) The alpine violet, *Viola biflora*, is a rich source of cyclotides with potent cytotoxicity. *Phytochemistry*, **69**, 939–952.

109 Gerlach, S.L., Goransson, U., and Mondal, D. (2009) Monitoring the anti-cancer effects and chemosensitizing abilities of novel cyclotides from *Psychotria leptothyrsa*. *FASEB Journal*, **23**, 756–710.

110 Chan, L.Y., Wang, C.K.L., Major, J.M., Greenwood, K.P., Lewis, R.J., Craik, D.J., and Daly, N.L. (2009) Isolation and characterization of peptides from *Momordica cochinchinensis* seeds. *Journal of Natural Products*, **72**, 1453–1458.

111 Droin, N., Hendra, J.B., Ducoroy, P., and Solary, E. (2009) Human defensins as cancer biomarkers and antitumour molecules. *Journal of Proteomics*, **72**, 918–927.

112 Rodrigues, E.G., Dobroff, A.S.S., Cavarsan, C.F., Paschoalin, T.,

Nimrichter, L., Mortara, R.A., Santos, E.L., Fazio, M.A., Miranda, A., Daffre, S., and Travassos, L. (2008) Effective topical treatment of subcutaneous murine B16F10-Nex2 melanoma by the antimicrobial peptide gomesin. *Neoplasia*, **10**, 61–68.

113 Shamova, O., Orlov, D., Stegemann, C., Czihal, P., Hoffmann, R., Brogden, K., Kolodkin, N., Sakuta, G., Tossi, A., Sahl, H.G., Kokryakov, V., and Lehrer, R. (2009) ChBac3.4: a novel proline-rich antimicrobial peptide from goat leukocytes. *International Journal of Peptide Research and Therapeutics*, **15**, 31–42.

114 Catrina, S.B., Refai, E., and Andersson, M. (2009) The cytotoxic effects of the anti-bacterial peptides on leukocytes. *Journal of Peptide Science*, **15**, 842–848.

115 Johnstone, S.A., Gelmon, K., Mayer, L.D., Hancock, R.E., and Bally, M.B. (2000) *In vitro* characterization of the anticancer activity of membrane-active cationic peptides. I. Peptide-mediated cytotoxicity and peptide-enhanced cytotoxic activity of doxorubicin against wild-type and P-glycoprotein over-expressing tumor cell lines. *Anti-Cancer Drug Design*, **15**, 151–160.

116 Koszalka, P., Kamysz, E., Wejda, M., Kamysz, W., and Bigda, J. (2011) Antitumor activity of antimicrobial peptides against U937 histiocytic cell line. *Acta Biochimica Polonica*, **58**, 111–117.

117 Harris, F., Daman, A., Wallace, J., Dennison, S.R., and Phoenix, D.A. (2006) Oblique orientated alpha-helices and their prediction. *Current Protein & Peptide Science*, **7**, 529–537.

118 Harris, F., Dennison, S., and Phoenix, D.A. (2006) The prediction of hydrophobicity gradients within membrane interactive protein alpha-helices using a novel graphical technique. *Protein and Peptide Letters*, **13**, 595–600.

119 Phoenix, D.A. and Harris, F. (2002) The hydrophobic moment and its use in the classification of amphiphilic structures [review]. *Molecular Membrane Biology*, **19**, 1–10.

120 Phoenix, D.A., Harris, F., Daman, O.A., and Wallace, J. (2002) The prediction of amphiphilic alpha-helices. *Current Protein & Peptide Science*, **3**, 201–221.

121 Dennison, S.R., Wallace, J., Harris, F., and Phoenix, D.A. (2005) Amphiphilic alpha-helical antimicrobial peptides and their structure–function relationships. *Protein and Peptide Letters*, **12**, 31–39.

122 Harris, F., Wallace, J., and Phoenix, D.A. (2000) Use of hydrophobic moment plot methodology to aid the identification of oblique orientated alpha-helices. *Molecular Membrane Biology*, **17**, 201–207.

123 Jackway, R.J., Pukala, T.L., Donnellan, S.C., Sherman, P.J., Tyler, M.J., and Bowie, J.H. (2011) Skin peptide and cDNA profiling of Australian anurans: genus and species identification and evolutionary trends. *Peptides*, **32**, 161–172.

124 Conlon, J. (2011) Structural diversity and species distribution of host-defense peptides in frog skin secretions. *Cellular and Molecular Life Sciences*, **68**, 2303–2315.

125 Tossi, A., Sandri, L., and Giangaspero, A. (2000) Amphipathic, alpha-helical antimicrobial peptides. *Biopolymers*, **55**, 4–30.

126 McCaldon, P. and Argos, P. (1988) Oligopeptide biases in protein sequences and their use in predicting protein coding regions in nucleotide sequences. *Proteins*, **4**, 99–122.

127 Mandal, S.M., Migliolo, L., Das, S., Mandal, M., Franco, O.L., and Hazra, T.K. (2012) Identification and characterization of a bactericidal and proapoptotic peptide from *Cycas revoluta* seeds with DNA binding properties. *Journal of Cellular Biochemistry*, **113**, 184–193.

128 Dougherty, D.A. (2007) Cation–pi interactions involving aromatic amino acids. *Journal of Nutrition*, **137**, 1504S–1508S.

129 Chan, D.I., Prenner, E.J., and Vogel, H.J. (2006) Tryptophan- and arginine-rich antimicrobial peptides: structures and mechanisms of action. *Biochimica et Biophysica Acta*, **1758**, 1184–1202.

130 Fromm, J.R., Hileman, R.E., Caldwell, E.E.O., Weiler, J.M., and Linhardt, R.J. (1995) Differences in the interaction of heparin with arginine and lysine and the importance of these basic-amino-acids in the binding of heparin to acidic fibroblast growth-factor. *Archives of Biochemistry and Biophysics*, **323**, 279–287.

131 Yang, S.T., Shin, S.Y., Lee, C.W., Kim, Y.C., Hahm, K.S., and Kim, J.I. (2003) Selective cytotoxicity following Arg-to-Lys substitution in tritrpticin adopting a unique amphipathic turn structure. *FEBS Letters*, **540**, 229–233.

132 Mant, C.T., Kovacs, J.M., Kim, H.M., Pollock, D.D., and Hodges, R.S. (2009) Intrinsic amino acid side-chain hydrophilicity/hydrophobicity coefficients determined by reversed-phase high-performance liquid chromatography of model peptides: comparison with other hydrophilicity/hydrophobicity scales. *Biopolymers*, **92**, 573–595.

133 Sengupta, D., Smith, J.C., and Ullmann, G.M. (2008) Partitioning of amino-acid analogs in a five-slab membrane model. *Biochimica et Biophysica Acta*, **1778**, 2234–2243.

134 Nyholm, T.K.M., Ozdirekcan, S., and Killian, J.A. (2007) How protein transmembrane segments sense the lipid environment. *Biochemistry*, **46**, 1457–1465.

135 Johansson, A.C.V. and Lindahl, E. (2008) Position-resolved free energy of solvation for amino acids in lipid membranes from molecular dynamics simulations. *Proteins*, **70**, 1332–1344.

136 Dennison, S.R., Harris, F., and Phoenix, D.A. (2007) The interactions of aurein 1.2 with cancer cell membranes. *Biophysical Chemistry*, **127**, 78–83.

137 Nozaki, Y., Tanford, C., and Hirs, C.H.W. (1967) Examination of titration behavior. *Methods in Enzymology*, **11**, 715–734.

138 Eisenberg, D., Schwarz, E., Komaromy, M., and Wall, R. (1984) Analysis of membrane and surface protein sequences with the hydrophobic moment plot. *Journal of Molecular Biology*, **179**, 125–142.

139 Raghunand, N., He, X., van Sluis, R., Mahoney, B., Baggett, B., Taylor, C.W., Paine-Murrieta, G., Roe, D., Bhujwalla, Z.M., and Gillies, R.J. (1999) Enhancement of chemotherapy by manipulation of tumour pH. *British Journal of Cancer*, **80**, 1005–1011.

140 Engin, K., Leeper, D.B., Cater, J.R., Thistlethwaite, A.J., Tupchong, L., and McFarlane, J.D. (1995) Extracellular pH distribution in human tumors. *International Journal of Hyperthermia*, **11**, 211–216.

141 Tu, Z.G., Volk, M., Shah, K., Clerkin, K., and Liang, J.F. (2009) Constructing bioactive peptides with pH-dependent activities. *Peptides*, **30**, 1523–1528.

142 Tu, Z.G., Young, A., Murphy, C., and Liang, J.F. (2009) The pH sensitivity of histidine-containing lytic peptides. *Journal of Peptide Science*, **15**, 790–795.

143 Shai, Y. and Arik, M. (2009) Histidine-Containing Diasteromeric Peptides and Uses Thereof. US Patent 2009/0305954.

144 Steinstraesser, L., Hauk, J., Al-Benna, S., Langer, S., Ring, A., Kesting, M., Sudhoff, H., Becerikli, M., Kafferlein, H., and Jacobsen, F. (2012) Genotoxic and cytotoxic activity of host defense peptides against human soft tissue sarcoma in an *in vitro* model. *Drug and Chemical Toxicology*, **35**, 96–103.

145 Makovitzki, A., Fink, A., and Shai, Y. (2009) Suppression of human solid tumor growth in mice by intratumor and systemic inoculation of histidine-rich and pH-dependent host defense-like lytic peptides. *Cancer Research*, **69**, 3458–3463.

146 Steinstraesser, L., Hauk, J., Schubert, C., Al-Benna, S., Stricker, I., Hatt, H., Shai, Y., Steinau, H.U., and Jacobsen, F. (2011) Suppression of soft tissue sarcoma growth by a host defense-like lytic peptide. *PLoS ONE*, **6**, e18321.

147 Thomas, A. and Brasseur, R. (2006) Tilted peptides: the history. *Current Protein & Peptide Science*, **7**, 523–527.

148 Matsuzaki, K. (2009) Control of cell selectivity of antimicrobial peptides. *Biochimica et Biophysica Acta – Biomembranes*, **1788**, 1687–1692.

149 Glukhov, E., Burrows, L.L., and Deber, C.M. (2008) Membrane interactions of designed cationic antimicrobial peptides: the two thresholds. *Biopolymers*, **89**, 360–371.

150 Deber, C.M., Liu, L.P., Wang, C., Goto, N.K., and Reithmeier, R.A.F. (2002) The hydrophobicity threshold for peptide insertion into membranes. *Current Topics in Membranes*, **52**, 465–479.

151 Huang, Y.-B., Wang, X.-F., Wang, H.-Y., Liu, Y., and Chen, Y. (2011) Studies on mechanism of action of anticancer peptides by modulation of hydrophobicity within a defined structural framework. *Molecular Cancer Therapeutics*, **10**, 416–426.

152 Chen, Y.X., Mant, C.T., Farmer, S.W., Hancock, R.E.W., Vasil, M.L., and Hodges, R.S. (2005) Rational design of alpha-helical antimicrobial peptides with enhanced activities and specificity/therapeutic index. *Journal of Biological Chemistry*, **280**, 12316–12329.

153 Pace, C.N. and Scholtz, J.M. (1998) A helix propensity scale based on experimental studies of peptides and proteins. *Biophysical Journal*, **75**, 422–427.

154 Bilej, M., Prochazkova, P., Silerova, M., and Joskova, R. (2010) Earthworm immunity, in Invertebrate Immunity (ed. K. Soderhall), Springer, Berlin, pp. 66–79.

155 Richardson, J.S. (1981) The anatomy and taxonomy of protein structure. *Advances in Protein Chemistry*, **34**, 167–339.

156 Raghuraman, H. and Chattopadhyay, A. (2007) Melittin: a membrane-active peptide with diverse functions. *Bioscience Reports*, **27**, 189–223.

157 Napolitano, A., Rodriquez, M., Bruno, I., Marzocco, S., Autore, G., Riccio, R., and Gomez-Paloma, L. (2003) Synthesis, structural aspects and cytotoxicity of the natural cyclopeptides yunnanins A, C and phakellistatins 1, 10. *Tetrahedron*, **59**, 10203–10211.

158 Napolitano, A., Bruno, I., Riccio, R., and Gomez-Paloma, L. (2005) Synthesis, structure, and biological aspects of cyclopeptides related to marine phakellistatins 7–9. *Tetrahedron*, **61**, 6808–6815.

159 Lu, J. and Chen, Z.W. (2010) Isolation, characterization and anti-cancer activity of SK84, a novel glycine-rich antimicrobial peptide from *Drosophila virilis*. *Peptides*, **31**, 44–50.

160 Hu, J., Chen, C.X., Zhang, S.Z., Zhao, X.C., Xu, H., Zhao, X.B., and Lu, J.R. (2011) Designed antimicrobial and antitumor peptides with high selectivity. *Biomacromolecules*, **12**, 3839–3843.

161 Mihajlovic, M. and Lazaridis, T. (2010) Antimicrobial peptides bind more strongly to membrane pores. *Biochimica et Biophysica Acta – Biomembranes*, **1798**, 1494–1502.

162 Blaser, G., Sanderson, J.M., and Wilson, M.R. (2009) Free-energy relationships for the interactions of tryptophan with phosphocholines. *Organic & Biomolecular Chemistry*, **7**, 5119–5128.

163 Stopar, D., Spruijt, R.B., and Hemminga, M.A. (2006) Anchoring mechanisms of membrane-associated M13 major coat protein. *Chemistry and Physics of Lipids*, **141**, 83–93.

164 Norman, K.E. and Nymeyer, H. (2006) Indole localization in lipid membranes revealed by molecular simulation. *Biophysical Journal*, **91**, 2046–2054.

165 Petersen, F.N.R., Jensen, M.O., and Nielsen, C.H. (2005) Interfacial tryptophan residues: a role for the cation–pi effect? *Biophysical Journal*, **89**, 3985–3996.

166 Holt, A., de Almeida, R.F.M., Nyholm, T.K.M., Loura, L.M.S., Daily, A.E., Staffhorst, R., Rijkers, D.T.S., Koeppe, R.E., Prieto, M., and Killian, J. (2008) Is there a preferential interaction between cholesterol and tryptophan residues in membrane proteins? *Biochemistry*, **47**, 2638–2649.

167 Dennison, S.R., Harris, F., and Phoenix, D.A. (2009) A study on the importance of phenylalanine for aurein functionality. *Protein and Peptide Letters*, **16**, 1455–1458.

168 van Kan, E.J.M., Demel, R.A., van der Bent, A., and de Kruijff, B. (2003) The role of the abundant phenylalanines in the mode of action of the antimicrobial peptide clavanin. *Biochimica et*

Biophysica Acta – Biomembranes, **1615**, 84–92.

169 Pleskach, V.A., Aleshina, G.M., Artsybasheva, I.V., Shamova, O.V., Kozhukharova, I.V., Goilo, T.A., and Kokriakov, V.N. (2000) Cytotoxic and mitogenic effect of antimicrobial peptides from neutrophils on cultured cells. *Tsitologiia*, **42**, 228–234.

170 Wang, C.K., Colgrave, M.L., Ireland, D.C., Kaas, Q., and Craik, D.J. (2009) Despite a conserved cystine knot motif, different cyclotides have different membrane binding modes. *Biophysical Journal*, **97**, 1471–1481.

171 Cascales, L. and Craik, D.J. (2010) Naturally occurring circular proteins: distribution, biosynthesis and evolution. *Organic & Biomolecular Chemistry*, **8**, 5035–5047.

172 Gillon, A.D., Saska, I., Jennings, C.V., Guarino, R.F., Craik, D.J., and Anderson, M.A. (2008) Biosynthesis of circular proteins in plants. *Plant Journal*, **53**, 505–515.

173 Lehmann, J., Retz, M., Sidhu, S.S., Suttmann, H., Sell, M., Paulsen, F., Harder, J., Unteregger, G., and Stockle, M. (2006) Antitumor activity of the antimicrobial peptide Magainin II against bladder cancer cell lines. *European Urology*, **50**, 141–147.

174 Phoenix, D. and Harris, F. (2006) The multi-purpose amphiphilic alpha-helix – a historical perspective. *Current Protein & Peptide Science*, **7**, 471–472.

175 Eisenberg, D., Weiss, R.M., and Terwilliger, T.C. (1982) The helical hydrophobic moment – a measure of the amphiphilicity of a helix. *Nature*, **299**, 371–374.

176 Wojtkowiak, J.W., Verduzco, D., Schramm, K.J., and Gillies, R.J. (2011) Drug resistance and cellular adaptation to tumor acidic pH microenvironment. *Molecular Pharmaceutics*, **8**, 2032–2038.

177 Lunt, S.J., Chaudary, N., and Hill, R.P. (2009) The tumor microenvironment and metastatic disease. *Clinical & Experimental Metastasis*, **26**, 19–34.

178 Utsugi, T., Schroit, A.J., Connor, J., Bucana, C.D., and Fidler, I.J. (1991) Elevated expression of phosphatidylserine in the outer-membrane leaflet of human tumor-cells and recognition by activated human blood monocytes. *Cancer Research*, **51**, 3062–3066.

179 Zwaal, R.F.A. and Schroit, A.J. (1997) Pathophysiologic implications of membrane phospholipid asymmetry in blood cells. *Blood*, **89**, 1121–1132.

180 Chan, S.C., Hui, L., and Chen, H.M. (1998) Enhancement of the cytolytic effect of anti-bacterial cecropin by the microvilli of cancer cells. *Anticancer Research*, **18**, 4467–4474.

181 Chan, S.C., Yau, W.L., Wang, W., Smith, D.K., Sheu, F.S., and Chen, H.M. (1998) Microscopic observations of the different morphological changes caused by anti-bacterial peptides on *Klebsiella pneumoniae* and HL-60 leukemia cells. *Journal of Peptide Science*, **4**, 413–425.

182 Bechinger, B. and Lohner, K. (2006) Detergent-like actions of linear amphipathic cationic antimicrobial peptides. *Biochimica et Biophysica Acta – Biomembranes*, **1758**, 1529–1539.

183 Sando, L., Henriques, S.T., Foley, F., Simonsen, S.M., Daly, N.L., Hall, K.N., Gustafson, K.R., Aguilar, M.-I., and Craik, D.J. (2011) A synthetic mirror image of Kalata B1 reveals that cyclotide activity is independent of a protein receptor. *ChemBioChem*, **12**, 2456–2462.

184 Chia, B.C.S., Carver, J.A., Mulhern, T.D., and Bowie, J.H. (2000) Maculatin 1.1, an anti-microbial peptide from the Australian tree frog, *Litoria genimaculata* – solution structure and biological activity. *European Journal of Biochemistry*, **267**, 1894–1908.

185 Hetru, C., Letellier, L., Oren, Z., Hoffmann, J.A., and Shai, Y. (2000) Androctonin, a hydrophilic disulfide-bridged non-haemolytic anti-microbial peptide: a plausible mode of action. *Biochemical Journal*, **345**, 653–664.

186 Vunnam, S., Juvvadi, P., Rotondi, K.S., and Merrifield, R.B. (1998) Synthesis and study of normal, enantio, retro, and retroenantio isomers of cecropin A–melittin hybrids, their end group effects and selective enzyme

inactivation. *The Journal of Peptide Research*, **51**, 38–44.

187 Merrifield, R.B., Juvvadi, P., Andreu, D., Ubach, J., Boman, A., and Boman, H.G. (1995) Retro and retroenantio analogs of cecropin–melittin hybrids. *Proceedings of the National Academy of Sciences of the United States of America*, **92**, 3449–3453.

188 Maloy, W.L. and Kari, U.P. (1995) Structure–activity studies on magainins and other host-defense peptides. *Biopolymers*, **37**, 105–122.

189 Wade, D., Boman, A., Wahlin, B., Drain, C.M., Andreu, D., Boman, H.G., and Merrifield, R.B. (1990) All-D amino acid-containing channel-forming antibiotic peptides. *Proceedings of the National Academy of Sciences of the United States of America*, **87**, 4761–4765.

190 Bessalle, R., Kapitkovsky, A., Gorea, A., Shalit, I., and Fridkin, M. (1990) All-D-magainin–chirality, antimicrobial activity and proteolytic resistance. *FEBS Letters*, **274**, 151–155.

191 Iwasaki, T., Ishibashi, J., Tanaka, H., Sato, M., Asaoka, A., Taylor, D., and Yamakawa, M. (2009) Selective cancer cell cytotoxicity of enantiomeric 9-mer peptides derived from beetle defensins depends on negatively charged phosphatidylserine on the cell surface. *Peptides*, **30**, 660–668.

192 Oudhoff, M.J., Bolscher, J.G.M., Nazmi, K., Kalay, H., van't Hof, W., Amerongen, A.V.N., and Veerman, E.C.I. (2008) Histatins are the major wound-closure stimulating factors in human saliva as identified in a cell culture assay. *FASEB Journal*, **22**, 3805–3812.

193 Fehlbaum, P., Bulet, P., Chernysh, S., Briand, J.P., Roussel, J.P., Letellier, L., Hetru, C., and Hoffmann, J.A. (1996) Structure–activity analysis of thanatin, a 21-residue inducible insect defense peptide with sequence homology to frog skin antimicrobial peptides. *Proceedings of the National Academy of Sciences of the United States of America*, **93**, 1221–1225.

194 Wei, G., de Leeuw, E., Pazgier, M., Yuan, W.R., Zou, G.Z., Wang, J.F., Ericksen, B., Lu, W.Y., Lehrer, R.I., and Lu, W.Y. (2009) Through the looking glass, mechanistic insights from enantiomeric human defensins. *Journal of Biological Chemistry*, **284**, 29180–29192.

195 Ulvatne, H. and Vorland, L.H. (2001) Bactericidal kinetics of 3 lactoferricins against *Staphylococcus aureus* and *Escherichia coli*. *Scandinavian Journal of Infectious Diseases*, **33**, 507–511.

196 Vorland, L.H., Ulvatne, H., Andersen, J., Haukland, H.H., Rekdal, O., Svendsen, J.S., and Gutteberg, T.J. (1999) Antibacterial effects of lactoferricin B. *Scandinavian Journal of Infectious Diseases*, **31**, 179–184.

197 Nicolas, P. (2009) Multifunctional host defense peptides: intracellular-targeting antimicrobial peptides. *FEBS Journal*, **276**, 6483–6496.

198 Asaduzzaman, S.M. and Sonomoto, K. (2009) Lantibiotics: diverse activities and unique modes of action. *Journal of Bioscience and Bioengineering*, **107**, 475–487.

199 Zhao, M. (2011) Lantibiotics as probes for phosphatidylethanolamine. *Amino Acids*, **41**, 1071–1079.

200 Kamimori, H., Hall, K., Craik, D.J., and Aguilar, M.I. (2005) Studies on the membrane interactions of the cyclotides kalata B1 and kalata B6 on model membrane systems by surface plasmon resonance. *Analytical Biochemistry*, **337**, 149–153.

201 Querfurth, H.W. and LaFerla, F.M. (2010) Alzheimer's disease. *New England Journal of Medicine*, **362**, 329–344.

202 Soscia, S.J., Kirby, J.E., Washicosky, K.J., Tucker, S.M., Ingelsson, M., Hyman, B., Burton, M.A., Goldstein, L.E., Duong, S., Tanzi, R.E., and Moir, R.D. (2010) The Alzheimer's disease-associated amyloid beta-protein is an antimicrobial peptide. *PLoS ONE*, **5**, e9505.

203 Simons, K. and Ikonen, E. (2000) Cell biology – how cells handle cholesterol. *Science*, **290**, 1721–1726.

204 Steiner, H., Andreu, D., and Merrifield, R.B. (1988) Binding and action of cecropin and cecropin analogs – antibacterial peptides from insects. *Biochimica et Biophysica Acta*, **939**, 260–266.

205 Silvestro, L., Weiser, J.N., and Axelsen, P.H. (1997) Structure–function studies of cecropin A activity in lipid membranes. *Biophysical Journal*, **72**, TU332–TU332.

206 Wojcik, C., Sawicki, W., Marianowski, P., Benchaib, M., Czyba, J.C., and Guerin, J.F. (2000) Cyclodextrin enhances spermicidal effects of magainin-2-amide. *Contraception*, **62**, 99–103.

207 Matsuzaki, K., Sugishita, K., Fujii, N., and Miyajima, K. (1995) Molecular-basis for membrane selectivity of an antimicrobial peptide, magainin-2. *Biochemistry*, **34**, 3423–3429.

208 Raghuraman, H. and Chattopadhyay, A. (2005) Cholesterol inhibits the lytic activity of melittin in erythrocytes. *Chemistry and Physics of Lipids*, **134**, 183–189.

209 Sood, R. and Kinnunen, P.K.J. (2008) Cholesterol, lanosterol, and ergosterol attenuate the membrane association of LL-37(W27F) and temporin L. *Biochimica et Biophysica Acta*, **1778**, 1460–1466.

210 Maher, S. and McClean, S. (2008) Melittin exhibits necrotic cytotoxicity in gastrointestinal cells which is attenuated by cholesterol. *Biochemical Pharmacology*, **75**, 1104–1114.

211 Raghuraman, H. and Chattopadhyay, A. (2004) Interaction of melittin with membrane cholesterol: a fluorescence approach. *Biophysical Journal*, **87**, 2419–2432.

212 Burman, R., Svedlund, E., Felth, J., Hassan, S., Herrmann, A., Clark, R.J., Craik, D.J., Bohlin, L., Claeson, P., Goransson, U., and Gullbo, J. (2010) Evaluation of toxicity and antitumor activity of cycloviolacin O2 in mice. *Biopolymers*, **94**, 626–634.

213 Li, Y.C., Park, M.J., Ye, S.K., Kim, C.W., and Kim, Y.N. (2006) Elevated levels of cholesterol-rich lipid rafts in cancer cells are correlated with apoptosis sensitivity induced by cholesterol-depleting agents. *American Journal of Pathology*, **168**, 1107–1118.

214 Hall, K., Lee, T.H., and Aguilar, M.I. (2011) The role of electrostatic interactions in the membrane binding of melittin. *Journal of Molecular Recognition*, **24**, 108–118.

215 Wessman, P., Morin, M., Reijmar, K., and Edwards, K. (2010) Effect of alpha-helical peptides on liposome structure: a comparative study of melittin and alamethicin. *Journal of Colloid and Interface Science*, **346**, 127–135.

216 De Kruijff, B. (1990) Cholesterol as a target for toxins. *Bioscience Reports*, **10**, 127–130.

217 Lohner, K. and Prenner, E.J. (1999) Differential scanning calorimetry and X-ray diffraction studies of the specificity of the interaction of antimicrobial peptides with membrane-mimetic systems. *Biochimica et Biophysica Acta – Biomembranes*, **1462**, 141–156.

218 Zachowski, A. (1993) Phospholipids in animal eukaryotic membranes – transverse asymmetry and movement. *Biochemical Journal*, **294**, 1–14.

219 Yeaman, M.R. and Yount, N.Y. (2003) Mechanisms of antimicrobial peptide action and resistance. *Pharmacological Reviews*, **55**, 27–55.

220 Giuliani, A., Pirri, G., and Nicoletto, S.F. (2007) Antimicrobial peptides: an overview of a promising class of therapeutics. *Central European Journal of Biology*, **2**, 1–33.

221 Lopez, P.H.H. and Schnaar, R.L. (2009) Gangliosides in cell recognition and membrane protein regulation. *Current Opinion in Structural Biology*, **19**, 549–557.

222 Kobata, A. (2007) Glycoprotein glycan structures, in Comprehensive Glycoscience – From Chemistry to Systems Biology (eds J.P. Kamerling, G.-J. Boons, Y.C. Lee, A. Suzuki, N. Taniguchi, and A.G.J. Voragen), Elsevier, Oxford, pp. 39–72.

223 Yu, R.K., Yanagisawa, M., and Ariga, T. (2007) Glycosphingolipid structures, in Comprehensive Glycoscience – From Chemistry to Systems Biology (eds J.P. Kamerling, G.-J. Boons, Y.C. Lee, A. Suzuki, N. Taniguchi, and A.G.J. Voragen), Elsevier, Oxford, pp. 73–122.

224 Miyagi, T. and Yamaguchi, K. (2007) Sialic acids, in Comprehensive

Glycoscience – From Chemistry to Systems Biology (eds J.P. Kamerling, G.-J. Boons, Y.C. Lee, A. Suzuki, N. Taniguchi, and A.G.J. Voragen), Elsevier, Oxford, pp. 297–323..

225 Bucki, R., Namiot, D.B., Namiot, Z., Savage, P.B., and Janmey, P.A. (2008) Salivary mucins inhibit antibacterial activity of the cathelicidin-derived LL-37 peptide but not the cationic steroid CSA-13. *Journal of Antimicrobial Chemotherapy*, **62**, 329–335.

226 Nishikawa, H. and Kitani, S. (2011) Gangliosides inhibit bee venom melittin cytotoxicity but not phospholipase A_2-induced degranulation in mast cells. *Toxicology and Applied Pharmacology*, **252**, 228–236.

227 Varki, A. (2008) Sialic acids in human health and disease. *Trends in Molecular Medicine*, **14**, 351–360.

228 Wang, P.-H. (2005) Altered glycosylation in cancer: sialic acids and sialyltransferases. *Journal of Cancer Molecules*, **1**, 73–81.

229 Bafna, S., Kaur, S., and Batra, S.K. (2010) Membrane-bound mucins: the mechanistic basis for alterations in the growth and survival of cancer cells. *Oncogene*, **29**, 2893–2904.

230 Wang, F.-L., Cui, S.-X., Sun, L.-P., Qu, X.-J., Xie, Y.-Y., Zhou, L., Mu, Y.-L., Tang, W., and Wang, Y.-S. (2009) High expression of alpha 2,3-linked sialic acid residues is associated with the metastatic potential of human gastric cancer. *Cancer Detection and Prevention*, **32**, 437–443.

231 Litynska, A., Przybylo, M., Pochec, E., Hoja-Lukowicz, D., Ciolczyk, D., Laidler, P., and Gil, D. (2001) Comparison of the lectin-binding pattern in different human melanoma cell lines. *Melanoma Research*, **11**, 205–212.

232 Cazet, A., Julien, S., Bobowski, M., Krzewinski-Recchi, M.-A., Harduin-Lepers, A., Groux-Degroote, S., and Delannoy, P. (2010) Consequences of the expression of sialylated antigens in breast cancer. *Carbohydrate Research*, **345**, 1377–1383.

233 Risso, A., Zanetti, M., and Gennaro, R. (1998) Cytotoxicity and apoptosis mediated by two peptides of innate immunity. *Cellular Immunology*, **189**, 107–115.

234 Ohyama, C. (2008) Glycosylation in bladder cancer. *International Journal of Clinical Oncology/Japan Society of Clinical Oncology*, **13**, 308–313.

235 Fredman, P., Hedberg, K., and Brezicka, T. (2003) Gangliosides as therapeutic targets for cancer. *Biodrugs*, **17**, 155–167.

236 Schaefer, L. and Schaefer, R.M. (2010) Proteoglycans: from structural compounds to signaling molecules. *Cell and Tissue Research*, **339**, 237–246.

237 Taylor, K.R. and Gallo, R.L. (2006) Glycosaminoglycans and their proteoglycans: host-associated molecular patterns for initiation and modulation of inflammation. *FASEB Journal*, **20**, 9–22.

238 Koo, C.Y., Sen, Y.P., Bay, B.H., and Yip, G.W. (2008) Targeting heparan sulfate proteoglycans in breast cancer treatment. *Recent Patents on Anti-Cancer Drug Discovery*, **3**, 151–158.

239 Asimakopoulou, A.P., Theocharis, A.D., Tzanakakis, G.N., and Karamanos, N.K. (2008) The biological role of chondroitin sulfate in cancer and chondroitin-based anticancer agents. *In Vivo*, **22**, 385–389.

240 Iozzo, R.V. and Sanderson, R.D. (2011) Proteoglycans in cancer biology, tumour microenvironment and angiogenesis. *Journal of Cellular and Molecular Medicine*, **15**, 1013–1031.

241 Poon, G.M.K. and Gariepy, J. (2007) Cell-surface proteoglycans as molecular portals for cationic peptide and polymer entry into cells. *Biochemical Society Transactions*, **35**, 788–793.

242 Baranska-Rybak, W., Sonesson, A., Nowicki, R., and Schmidtchen, A. (2006) Glycosaminoglycans inhibit the antibacterial activity of LL-37 in biological fluids. *Journal of Antimicrobial Chemotherapy*, **57**, 260–265.

243 Klocek, G. and Seelig, J. (2008) Melittin interaction with sulfated cell surface sugars. *Biochemistry*, **47**, 2841–2849.

244 Goncalves, E., Kitas, E., and Seelig, J. (2006) Structural and thermodynamic aspects of the interaction between heparan sulfate and analogs of melittin. *Biochemistry*, **45**, 3086–3094.

245 Li, G.-H., Zhang, J.-H., Qu, M.-R., and You, J.-M. (2009) Lactoferricin: a lactoferrin-derived multifunctional antimicrobial peptide. *Zhongguo Shengwu Huaxue yu Fenzi Shengwu Xuebao*, **25**, 796–804.

246 Gifford, J.L., Hunter, H.N., and Vogel, H.J. (2005) Lactoferricin: a lactoferrin-derived peptide with antimicrobial, antiviral, antitumor and immunological properties. *Cellular and Molecular Life Sciences*, **62**, 2588–2598.

247 Fadnes, B., Rekdal, O., and Uhlin-Hansen, L. (2009) The anticancer activity of lytic peptides is inhibited by heparan sulfate on the surface of the tumor cells. *BMC Cancer*, **9**, 183.

248 Fadnes, B., Uhlin-Hansen, L., Lindin, I., and Rekdal, O. (2011) Small lytic peptides escape the inhibitory effect of heparan sulfate on the surface of cancer cells. *BMC Cancer*, **11**, 116.

249 Itano, N. and Kimata, K. (2008) Altered hyaluronan biosynthesis in cancer progression. *Seminars in Cancer Biology*, **18**, 268–274.

250 Stern, R. (2005) Hyaluronan metabolism: a major paradox in cancer biology. *Pathologie Biologie*, **53**, 372–382.

251 Chen, J., Xu, X.-M., Underhill, C.B., Yang, S., Wang, L., Chen, Y., Hong, S., Creswell, K., and Zhang, L. (2005) Tachyplesin activates the classic complement pathway to kill tumor cells. *Cancer Research*, **65**, 4614–4622.

252 Zwaal, R.F.A., Comfurius, P., and Bevers, E.M. (2005) Surface exposure of phosphatidylserine in pathological cells. *Cellular and Molecular Life Sciences*, **62**, 971–988.

253 Rao, L.V.M., Tait, J.F., and Hoang, A.D. (1992) Binding of annexin-V to a human ovarian-carcinoma cell-line (Oc-2008)–contrasting effects on cell-surface factor-VIIa/tissue factor activity and prothrombinase activity. *Thrombosis Research*, **67**, 517–531.

254 Connor, J., Bucana, C., Fidler, I.J., and Schroit, A.J. (1989) Differentiation expression of phosphatidylserine in mammalian plasma membranes–quantitative assessment of outer-leaflet lipid by prothrombinase complex formation. *Proceedings of the National Academy of Sciences of the United States of America*, **86**, 3184–3188.

255 Cichorek, M., Kozlowska, K., Witkowski, J.M., and Zarzeczna, M. (2000) Flow cytometric estimation of the plasma membrane diversity of transplantable melanomas, using annexin V. *Folia Histochemica et Cytobiologica*, **38**, 41–43.

256 Kirszberg, C., Lima, L.G., de Oliveira, A.D.S., Pickering, W., Gray, E., Barrowcliffe, T.W., Rumjanek, V.M., and Monteiro, R.Q. (2009) Simultaneous tissue factor expression and phosphatidylserine exposure account for the highly procoagulant pattern of melanoma cell lines. *Melanoma Research*, **19**, 301–308.

257 Riedl, S., Rinner, B., Asslaber, M., Schaider, H., Walzer, S., Novak, A., Lohner, K., and Zweytick, D. (2011) In search of a novel target–phosphatidylserine exposed by non-apoptotic tumor cells and metastases of malignancies with poor treatment efficacy. *Biochimica et Biophysica Acta – Biomembranes*, **1808**, 2638–2645.

258 Zweytick, D., Deutsch, G., Andra, J., Blondelle, S.E., Vollmer, E., Jerala, R., and Lohner, K. (2011) Studies on lactoferricin-derived *Escherichia coli* membrane-active peptides reveal differences in the mechanism of N-acylated versus nonacylated peptides. *Journal of Biological Chemistry*, **286**, 21266–21276.

259 Papo, N., Seger, D., Makovitzki, A., Kalchenko, V., Eshhar, Z., Degani, H., and Shai, Y. (2006) Inhibition of tumor growth and elimination of multiple metastases in human prostate and breast xenografts by systemic inoculation of a host defense-like lytic peptide. *Cancer Research*, **66**, 5371–5378.

260 Harris, F., Dennison, S.R., and Phoenix, D.A. (2012) Aberrant action of amyloidogenic host defense peptides: a new paradigm to investigate neurodegenerative disorders? *FASEB Journal*, **26**, 1776–1781.

261 Torrent, M., Odorizzi, F., Nogues, M.V., and Boix, E. (2010) Eosinophil cationic protein aggregation: identification of an

N-terminus amyloid prone region. *Biomacromolecules*, **11**, 1983–1990.

262 Bystrom, J., Amin, K., and Bishop-Bailey, D. (2011) Analysing the eosinophil cationic protein – a clue to the function of the eosinophil granulocyte. *Respiratory Research*, **12**, 10.

263 Navarro, S., Aleu, J., Jimenez, M., Boix, E., Cuchillo, C.M., and Nogues, M.V. (2008) The cytotoxicity of eosinophil cationic protein/ribonuclease 3 on eukaryotic cell lines takes place through its aggregation on the cell membrane. *Cellular and Molecular Life Sciences*, **65**, 324–337.

264 Zhao, H., Sood, R., Jutila, A., Bose, S., Fimland, G., Nissen-Meyer, J., and Kinnunen, P.K. (2006) Interaction of the antimicrobial peptide pheromone Plantaricin A with model membranes: implications for a novel mechanism of action. *Biochimica et Biophysica acta*, **1758**, 1461–1474.

265 Sand, S.L., Oppegard, C., Ohara, S., Iijima, T., Naderi, S., Blomhoff, H.K., Nissen-Meyer, J., and Sand, O. (2010) Plantaricin A, a peptide pheromone produced by *Lactobacillus plantarum*, permeabilizes the cell membrane of both normal and cancerous lymphocytes and neuronal cells. *Peptides*, **31**, 1237–1244.

266 Mahalka, A.K. and Kinnunen, P.K.J. (2009) Binding of amphipathic alpha-helical antimicrobial peptides to lipid membranes: lessons from temporins B and L. *Biochimica et Biophysica Acta – Biomembranes*, **1788**, 1600–1609.

267 Doyle, J., Llewellyn, L.E., Brinkworth, C.S., Bowie, J.H., Wegener, K.L., Rozek, T., Wabnitz, P.A., Wallace, J.C., and Tyler, M.J. (2002) Amphibian peptides that inhibit neuronal nitric oxide synthase – the isolation of lesueurin from the skin secretion of the Australian Stony Creek Frog *Litoria lesueuri*. *European Journal of Biochemistry*, **269**, 100–109.

268 Gehman, J.D., Luc, F., Hall, K., Lee, T.H., Boland, M.P., Pukala, T.L., Bowie, J.H., Aguilar, M.I., and Separovic, F. (2008) Effect of antimicrobial peptides from Australian tree frogs on anionic phospholipid membranes. *Biochemistry*, **47**, 8557–8565.

269 Ran, S., Downes, A., and Thorpe, P.E. (2002) Increased exposure of anionic phospholipids on the surface of tumor blood vessels. *Cancer Research*, **62**, 6132–6140.

270 Papetti, M. and Herman, I.M. (2002) Mechanisms of normal and tumor-derived angiogenesis. *American Journal of Physiology*, **282**, C947–C970.

271 Chen, Y.X., Xu, X.M., Hong, S.G., Chen, J.G., Liu, N.F., Underhill, C.B., Creswell, K., and Zhang, L.R. (2001) RGD-tachyplesin inhibits tumor growth. *Cancer Research*, **61**, 2434–2438.

272 Mader, J.S., Smyth, D., Marshall, J., and Hoskin, D.W. (2006) Bovine lactoferricin inhibits basic fibroblast growth factor- and vascular endothelial growth factor (165)-induced angiogenesis by competing for heparin-like binding sites on endothelial cells. *American Journal of Pathology*, **169**, 1753–1766.

273 Fu, Y., Tang, H., Huang, Y., Song, N., and Luo, Y. (2009) Unraveling the mysteries of endostatin. *IUBMB Life*, **61**, 613–626.

274 Zhao, H.X., Jutila, A., Nurminen, T., Wickstrom, S.A., Keski-Oja, J., and Kinnunen, P.K.J. (2005) Binding of endostatin to phosphatidylserine-containing membranes and formation of amyloid-like fibers. *Biochemistry*, **44**, 2857–2863.

275 Gebbink, M., Voest, E.E., and Reijerkerk, A. (2004) Do antiangiogenic protein fragments have amyloid properties? *Blood*, **104**, 1601–1605.

276 Fischer, E. (1894) Einfluss der Configuration auf die Wirkung der Enzyme. *Berichte der deutschen chemischen Gesellschaft*, **27**, 2985–2993.

277 Held-Kuznetsov, V., Rotem, S., Assaraf, Y.G., and Mor, A. (2009) Host-defense peptide mimicry for novel antitumor agents. *FASEB Journal*, **23**, 4299–4307.

Index

Numbers in **bold** refer to Tables; Numbers in *italics* refer to Figures

a

Aβ peptides   199
Aβ40 and Aβ42   89–91, 99–102, **163**
– amyloids   163, **164**, 168
Acinetobacter baumannii   119
acne vulgaris   41
Actinomycesnaes lundii   98
acyl chains   91, 151–153, 190
adaptive immune system   1–3, 13, 15, 17, 18, 89, 154
Aesculus hippocastanum   150
agouti signal peptide (ASP)   17
alanine   45, 61, 99, 194, 195
alginate   148
"all-or-none" model   157
alloferons   55
almethicin   **155**, 156
α-amylase   11
α-AMPs   2–4, 7, 115–118
– amphiphilicity   115, 116, *117*, 117–119, *119*, 134
– – classification   127–129, 133
– – qualitative methods   118–121
– – quantitative methods   121–133
– amyloid structure   **164**, 165, 166
– anticancer   192, 194, 195
– membrane interaction models   149, 154–161, 163, 164, 165, 167, 168
α-anionic AMPs (α-AAMPs)   88, **88**, 89, 100, 101
– epidermis   92, 93, *93*, *94*
α-anticancer peptides (α-ACPs)   181, 184–187, **188**, *189*, 189–202, 204–206
– α-ACP$_{AO}$ 184–187, **188**, *189*, 191–199, 202, 204–206
– α-ACP$_I$ 185, 187, **188**, *189*, 189–194, 196, *197*, *198*, 202, 204

– α-ACP$_T$ 184–187, **188**, *189*, 189–196, *197*, *198*, 200, 202, 206
α-cationic AMPs (α-CAMPs)   43, *44*, 44–51, 60, 61
– membrane interaction models   145, 149, 152, 154, **155**, 159–168
α-chymotrypsin   97
α-defensins   2, 3, 7, *8*, 10, 51, 52, 135
α-glucan   *146*, *147*
α-lactalbumin (αLA)   97, **98**
α-melanocyte stimulating hormone (α-MSH)   11, 12
α-synuclein   163
alternate ligand model   13, 14
Alzheimer's disease (AD)   89–91, 102, 163, 168, 199
amatoxins   101
amino acids   7, *45*, *165*, 166
– AAMPs   84, **84**, *88*
– amphiphilicity   115–117, 134
– – qualitative methods   119–120, *120*
– – quantitative methods   123–126, 129, 130
– anticancer   *189*, 192, 199
– CAMPs   42, 43, 45–47, 53
amino glycosides   10, 133
5-aminolevulinic acid (ALA)   41
AMPer database   **5**, 135
amphiphilicity   7, 115–134, 195, *196*, 196
– AAMPs   91, 93, *93*, *94*, 99, 101
– ACPs   181, 185–187, 190, 193, 195, *196*, *197*, *198*, 206
– CAMPs   42, 43, 49–51, 55, 61, 99
– membrane interaction models   158, 165, 166
– primary   116, 121

– secondary 116, 117, 119, 122, 195, 196, *197*, 198
– tertiary 116–118, 195, *196*
Amphipathic Index (AI) 124
AMSDb database **5**, 184
amyloid fibrils 164, 165, *165*
amyloidogenesis 89, 145, 149, 161, 163–168
– AAMPs 89, 91, 99
– anticancer 204, 205
angiotensin-converting enzyme (ACE) 90
anionic ACPs 184, 189, 197, 199–201
anionic AMPs (AAMPs) 3, 42, 47, 83–102
anionic β-defensins 95, 96
anionic lipids 54, 56, 60, 91
– membrane interaction models 145, 149–152, 161, 164–166
antibiotics 1, 2, 83, 84, 94, 102, 135, 148
– multidrug resistance 39–41, 102
anticancer peptides (ACPs) 181–196, 205–207
– ACP$_{AO}$ 200–202, 204–206
– ACP$_T$ 196, 200, 201, 205, 206
– membrane-based factors 196–205
antigen-presenting cells (APCs) 14, 15
AP1 **163**
APD2 database **5**, 184, 187, 192
apidaecins 58
apolipoproteins 96, 117
arginine 42, 45, 47, 48, 51, 57, 58, 61
– α-ACPs 189, 190
artificial neural networks (ANNs) 135
Ascomycota 49
asparagine 54
aspartic acid 47, 54, 85, 99, 100, 189
astrocytoma 204
atopic dermatitis (ATD) 18, 92, 102
ATP 56, 153
ATPase 59, *59*
aurein-1.2 152, **155**, **163**, 163, 204
– amphiphilicity 117, 119–122
aureins 43, 181, **185**, 194, 195

b

Bacillus spp 148
Bacillus brevis 1
Bacillus megaterium 88
Bacillus subtilis 150
backbone 53, *120*, 193, *196*
bactenecins 48, 58, 61
bacteria 4, *5*, 8, 10–12, 15, 17, 184
– AAMPs 83–87, 102
– – blood components 96, 97
– – brain 87–91

– – epidermis 92–94
– – epididymis 95
– – food proteins 98, 99
– – respiratory tract 85, 86
– – structure – function relationship 99
– amphiphilicity 116, 119
– anticancer 199, 204
– CAMPs 39, 42, 44–50, 51–55, 57, 60
– MDR 39, 40
– membrane interaction models 145–150, 153–168
– PACT 41
– *see also* Gram-negative bacteria; Gram-positive bacteria
bacterial chaperone (DnaK) 59, *59*, 60
bactericons 83
bacteriophages 40
bacteriorhodopsin 122
barrel-stave pore model 145, 149, **155**, 155–159, 167
B-cell epitopes 183
β-AAMPs 91, 95
β-ACPs 184, **186**, 190, 192, 195
– β-ACP$_{AO}$ **186**, 202
– β-ACP$_T$ **186**, 190, 195
β-barrel TM porins 116
β-CAMPs 43, *44*, 51–55
β-defensins 1, 3, 7–16, 18, 51, 52, *117*
– AAMPs 83, 95, 96
β-glucan 146, *146*, 147
β-lactoglobulin (β-LG) 97, **98**
β-sheet structures 7, *8*, **8**, 9, 49, 148, 154
– amphiphilicity 115, 116, *117*
– amyloidogenesis 163–166
big defensins (BDs) 9, 10
biosynthetic pathways 40
blood–brain barrier 87
blood components 83, 96, 97, 101
blood–epididymal barrier 95
BMAP-27 and BMAP-28 **185**, 195, 201
bombinins 2, 42
bombolitin II 163
bovine milk 97, 98
bracelet cyclotides *196*
bracelet peptides 53, *53*, 55
brain 83, 87–91, 101
Brasseur approach 132, 133
breast cancer 184, 192, 193, 195, 200, 201, 204
bromotryptophan 42
buforin IIb 154, **155**, 156, 159, 168
– anticancer **185**, 193, 201, 202

c

C9 4
caerins 5
caerwein neuropeptides 6, 7
calmodulin 127
cancer 18, 41, 151, 153, 154, 168, 204
- AAMPs 99, 101
- CAMPs 42, 53
- *see also* anticancer peptides (ACPs)
Candida albicans 55, 90, 92, *150*
Candida tropicalis 150
Candida utitis 150
canine β-defensins (CBDs) 17
canine coat color 17, 18
CAP-18 148
cardiolipin (CL) 56, 149, 150, *150*, 152, 201
carmustinine 182
carpet-type mechanism 133, 158, 159, *159*, *160*
- AAMPs 83
- ACPs 181, 187
- CAMPs 46, 52
- membrane interaction models 145, 149, 152, **155**, 158–160, 167
casein 97, 98, **98**
cateslytin 51, 117
cathepsin D 91
cationic ACPs 205
cationic AMPs (CAMPs) 2, 7, 39, 42–62, 102, 154
- blood components 96, 97
- membrane interaction models 145, 146, 148–154, 166, 167
- respiratory tract 86
- structure–function relationship 99–101
CB1a **185**, 202
CD4[4] memory T cells 12, 13
cecropin A 157
cecropin B 199
cecropins 1, 2, 17, **155**, 159
- anticancer 181, **185**, 199, 200
cervical epithelial carcinoma 184, 204
cervix carcinoma 192
channel-blocking drugs 168
channel-forming peptides 117
ChBac3.4 **186**, 190
chemokines and chemokine receptors 12–14, *14*, 118
chemotherapy (CT) 182
chitins 146, *146*, 147
cholesterol 54, 194, 199, 200, 206, 153
choline 57
chondroitin sulfate 202
chromogranin A 51

chymosin 97, 98
cinnamycin 199
cis-proline 53, *88*
citrate synthase 124
citropin-1.1 119, *119*, 152, **163**, 163, 204
citropins 43, **185**
clavanin A 154, 168
clavanins 55, 194
ClC chloride channel 122
colicin E1 **163**, 163
colon cancer 184, 195, 201, 203
colorectal cancer 192, 197
comparative genomics 40
computational techniques 115, 116, 118
- qualitative methods 118–121
- quantitative methods 121–133
Cr-ACP1 197
Crohn's disease 18
Cryptococcus neoformans 97
C-terminus 6, 9, 18, *162*, 162, 163
- AAMPs 89, *90*, 91, 92, 94
- ACPs 185, 186, 195
- amphiphilicity *117*, 118, 129, 133
- CAMPs 42, 51, *58*, 59
CyBase database 184
cyclic lipopeptides 39
cyclic phakellistatins 193
cyclotides 4, 100, 118, 184, 195, *196*, 199
- β-ACPs **186**
- CAMPs 43, 52, 53, *53*, 54, *54*, 55
cycloviolacins 53, *53*, 148, **186**, *196*, 200
cysteine 7, *58*, 192
cysteine knot structures 4, 43, 53, 195, *196*
cysteine-stabilization 2, 7, 8, 101, 118
- CAMPs 43, *44*, 51
cystic fibrosis (CF) 86, 87, 102, 132
cytokines 4, 11, 13, 15, 16, 92
cytoplasmic membrane (CM) 145, 146, *147*, 148, 149
- α-CAMPs 149–154
- CAMPs 166, 167
cytotoxic peptides 117

d

D1 (K13) 119, *120*
daptomycin 84, *84*, 85
datasets and databases 5, 18, 43, 46
- ACPs 184, 189, 189–196, *197*
DDK 153
decontaminants 83
dermaseptins 159, **164**

dermcidin (DCD) 3, 87, 91–94, 102
– DCD-1 92–94, 99–101
– DCD-1L 92–94, 99–101
DEFB118 100–102
defense peptides 183, 184, 187, 192, 194
defensins 1–3, 5, 7–12, 18, 86, 118, 205
– CAMPs 42, 49, 51, 52, 61
– canine coat color 17, 18
– effectors of immunity 12–16
– membrane interaction models 148–150
– wound healing 16, 18
δ-hemolysin 157, 158
δ-lysin **155**
dementia 102, 168
dendritic cells 12, 13, 15
dental and periodontal disease 83, 98, 102
diabetes 16
disordered toroidal pore 145, **155**, 158, 167
disulfide bridges 7, 8, 9, 51, 52, **98**, 192
dogs 10, 17, 18
drosocin 58, 59
duramycin 199

e
Eisenberg consensus scale 125, 127
enantiomers 199
endocarditis 97
endocytosis 56
endostatins 205
endotoxins 4
enkelytin 96, 98, 100–102
– brain 87, 88, 88, 89, 90
Enterococci 85
eosinophils 3, **164**, 204
epidermal growth factor (EGF) 16
epidermoid cancers 197
epidermis 83, 91–94
epididymis 83, 94–96, 101
epinecidin-1 **185**
ergosterol 51, 60, 153, 154
erythrocytes 47, 146, 184, 200–202
erythroleukemia 204
Escherichia coli 95, 97, 150, 150, 163
eukaryotes 1, 4, 7, 8, 41, 129, 148
– AAMPs 85, 101
– ACPs 181, 185, 205, 206
– α-ACPs 191, 199, 200
– CAMPs 46, 49, 50, 60
– membrane interaction models 149, 153, 159, 168
exoploysaccharides 148

extended ACPs (E-ACPs) 43, *44*, 55–60, 184, **186**, 189
– E-ACP$_{AO}$ **186**, 190
– E-ACP$_T$ **186**, 190, 194

f
FALL-39 47
fertility control 18, 83, 96, 102
fibrils 163
fibrinogen 96, 97
fibrinopeptide A (FPA) 96, 97
fibrinopeptide B (FPB) 96, 97
fibroblasts 11
FLAK50 Z5, FLAK 98 and FLAK 50 T4 197
foliate biosynthesis 40
food preservatives 83
food proteins 97–99
Fourier transforms 123, 124, 126, 133
fowlicidin-2 100
FsAFP2 149
fungi 3, 5, 7, 8, 11, 12, 41
– AAMPs 85, 97, 99–102
– CAMPs 39, 42, 44, **46**, 46, 48–53, 55–57, 60, 62
– membrane interaction models 145–149, *150*, 152–154, 156
fusion peptides 161–163

g
gaegurins **185**, 195
γ-core motif 8
gangliosides 201, 202
gastrointestinal tract 83, 97–99, 199, 202, 204
gene-encoded AAMPs 83
glioblastoma 204
glioma 197, 204
glutamic acid 47, *88*, 88, 100, 148, 189
glycine 55, *93*, *119*, 119, 129, 193
glycocalyx of fungi 46
glycolipids 199, 201, 202
glycopeptides 84, 85
glycophorin A 122, 129
glycoproteins 15, *146*, *147*, 199, 201, 202
glycosaminoglycan (GAG) 202, 203, 206
glycosphingolipids 201, 202
gomesin **186**, 192, 195, 199, 205
G-protein-coupled receptors (GPCRs) 13, 15, 122
"graded release" model 157
gramicidins 1
Gram-negative bacteria 9, 12, 39, 41, 119
– AAMPs 83, 84, 86, 93, 98
– CAMPs 44, *45*, **46**, 46–51, 60

– membrane interaction models   145, *146, 147*, 148–150, 152, 154, 166
Gram-positive bacteria   1, 2, 9, 41
– AAMPs   83, 84, 86, 88, 93, 98
– CAMPs   44, *45*, **46**, 46, 48–51, 60
– membrane interaction models   *146, 147*, 148–150, 152, 154
granulocytes   3

h
HA2   163
Haemophilus influenza   52
halocins   83
HE2C   95, 101, 102
head-groups   151–153, 155, 156, *166*, 190, 194
hedyotide B2   54
Hedyotis biflora   54
helical-net diagram (Dunhill)   119, 120, *120*
helical wheels   118–120, 127, 133, 197
HeliQuest database   131
hemoglobin   117
hemolymphs   2
hemolytic activity   128, 133, 153
– CAMPs   47–51, 53–55
heparan sulfate (HS)   202, 203, 205
heparin   202
hepatoma   198
hepcidin TH2–3   184, **186**
hexameric repeat peptides   51
hidden Markhov models (HMMs)   134, 135
histatins   43, 55–57, 60, 100
histidines   55, 57, 62, 191, 198
HIV   10, 53
HNP-1, HNP-2 and HNP-3   52, **186**, 199
homology modeling   115
hospital acquired infections   40
Huge Toroidal Pore model   157
human β-defensins (HBDs)   12–16, 51, 52
human prion protein   **163**
hyaluronan   203
hydropathy analysis   115, 121, 122, *122*, 123, 130
hydrophilic   93, 116–118, *119*
– ACPs   190, 195, *196–198*
– CAMPs   42, 46, *53*
– membrane interaction models   156, 156, *162*, 165, *166*
– qualitative methods   119, 120, *121*, 121
– quantitative methods   129, 131
hydrophobic arc   181, **188**, 192

hydrophobic moment (<μ_H>)   124–134
– ACPs   **188**, 195, 196
– α-ACPs   186
– CAMPs   49, *50*, 50
hydrophobicity (<H>)   134, 165, *166*, 191–195
– AAMPs   91, *93*, 99, 101
– ACPs   184, 185, 187, 190–195, *196*, *197*, *198*, 207
– α-ACPs   **188**, 192
– amphiphilicity   115, 116, 118
– – qualitative methods   119–121
– – quantitative methods   121–131
– CAMPs   42, 43, 46–50, 52–55, 57, 61, 99
– qualitative methods   119–121
– quantitative methods   121–131
– hydrophobicity scales   126, 131
hyperglycemia   16

i
imidazole   191
immune systems   12–16, 18, 154, 182, 183
– AAMPs   83, 85–87, 89, 91, 92, 96, 97, 102
– CAMPs   39, 47
immunocytes   89, 96
immunosuppression   15, 182
immunotherapy   183
indolicidin   4, *44*, 116, 135, 152, **186**
– CAMPs   57, 60, 61
inflammation   1, 3, 11–13, 15, 18
influenza   52, 163
innate immune system   1, 13, 15, 18, 154, 181
– AAMPs   83, 85–87, 92, 96, 97
– – brain   87, 89, 91
– CAMPs   39, 41
insulin growth factor-1   16
interleukins (ILs)   15, 16
isoleucine   45, 61, 194, 195

k
kalata B1   53, 54, 148, 199, 200
kalata B2   *196*, 200
kalata-kalata   52
kappacins   98–101
Keller approach   130, 131
keratinocytes   11, 15, 16, 91, 92, 204
Klebsiella pneumonice   150
KvAP voltage-gated potassium channel   122
KW5   202, 203

l
lactic acid   198
lacticin Q   **155**, 157

Lactobacillus spp 97
Lactococcus lactis 157
lactoferricin B 57, **186**, 199, 202, 203, 205
lactoferrin 2, 4, 11, 48, **164**, 184
lantibiotics 83, 199
lasioglossins 184, **185**
"leaky slit" model 165, 165, *166*
leishmaniosis 41
lethal dose (LD) 184
leucine 45, 61, 194, 195
leukemia 193, 199, 202, 204
leukocytes 2, 3, 12, 57, 96
linezolid 85
lipid-binding proteins 117, 121, 131
lipopeptides 84, 85
lipophilicity 194
lipopolysaccharide (LPS) 148
lipoprotein *146*, *147*
lipoteichoic acid 148
liver cancer 193
LL-37 4, 148, 154, **155**, 164, **164**
– anticancer **185**, 200, 202
"lock and key" model 206, 207
lung cancer 184, 192, 195, 198, 201, 202, 204
lymphomas 198, 204
lysenins 46
lysine 42, 45, 47, 48, 61, *120*, 189–190
lysophosphatidylcholine (LPC) 151, 152
lysosomal phospholipase inhibition 132
lysozyme 2, 3
lysylated PG (LysylPG) 149, *150*
lytic channel-forming peptides 128
lytic peptides 115, 128, *128*, 129

m

macrophage inflammatory protein-3α (MAP-3α) 14, 15
macrophages 12–15
magainin-2 158, **164**, 195, *197*, *198*, 202
magainins 1, 3, 4, 17, 43, *44*
– anticancer 181, **185**, 199, 200
– membrane interaction models 153, **155**, 157, 158, 161, 164, 167
malaria 41
Malassezia folliculitis 41
mast cells 13, 15
mastopran 184
maximins 4, *129*, 130, *131*
McCaldon and Argos dataset 45, 47, *189*, 189–191
m-calpain 163
MCoCC-1 and MCoCC-2 **186**

melanocortin 1 receptor (Mc1r) 17
melanocytes 11, 17, 204
melanoma 184, 192, 202–204
melittin 128, **155**, 158, 159, **164**
– anticancer **185**, 193, *197*, 199, 200, 202
"membrane discrimination" model 159
membrane protein SbmA 59, *59*
membranes 4, 11, 15, 18, 145–168, 196–205
– AAMPs 83–86, 89–96, 98–102
– ACPs 181, 182, 187, 189, 190–205
– amphiphilicity 115–119, 121, 122, 125, 126
– CAMPs 39, 42, 43, 46, 51, 52, 54–57, 59–62
– disruption model 14, *14*
– interaction models 145–149, **155**, 166–168
– – α-CAMPs 149–161
– – amyloidogenesis 163–166
– – established 155–159
– – novel 159–161
– – tilted peptides 161–163
meninges 87
M-enk *90*
M-ent-RF *90*
Met-Enk 89
methicillin-resistant *Staphylococcus aureus* (MRSA) 41, 85, 92, 135
microcins 83, **164**
Micrococcus luteus 88
minimum inhibitory concentrations (MICs) 3, 88, 92
– CAMPs 45, 46, **46**, 48, 49, 59, 60
Möbius cyclotides *196*
Möbius peptides 53, *53*, 55
molecular dynamics (MD) simulations 149, 158
molecular electroporation model 160, 161
molecular hydrophobic potential (*MHP*) 132
molecular shapes 151, *151*, 159
monocytes 12, 13, 15, 154
mucin-1 201
multidrug resistance (MDR) 39–41, 135, 182, 183
– AAMPs 85, 87, 92, 94, 102
multiple sequence alignments 115
MX-226 135
Mycobacterium tuberculosis 92
myeloma 204
myoglobin 117
myxobacteria 8

n

net positive charge 187, **188**, 189–191
– ACPs 184, 185, 187–191, 206, 207
neural networks 115
neural tumors 198
neuroblastoma 204
neurocandidiasis 90
neurodegenerative disorders 163
neuronal nitric oxide synthase inhibitors 4
neuropeptides 4, 87, 89, 102
neutral endopeptidase (NEP) 90
neutrophils 3, 12, 13, *117*, 118, 154
Newcastle disease virus 161
nisin bind lipid II 199
NK-2 **185**, 205
NK-lysin **155**, 161
non-ribosomal prokaryotic AAMPs 83, 84
non-stereo-selective antimicrobial mechanism 59, 60
N-terminus 6, 9, 13, 17, 193, 195
– AAMPs in epidermis *93*, 93, 94, *94*
– amphiphilicity *117*, 118, 122, 133
– CAMPs 48, *58*, 58
– membrane interaction models *162*, 162
nucleic acids 154
nucleosides 56
nucleotides 56

o

Oldenlandia affinis 52, 53
oligomers 164–166, *166*
oligopeptides 45, *45*, 61, 189
oncocin 60
oncolytic AMPs 181–183, 205–207
– membrane-based factors 196–205
– peptide-based factors 183–196
onychomycosis 41
opioid peptides 87, 89, *90*, 96
opioid proteins 102
ovalbumin (OA) 97, **98**
ovarian cancer 195, 201, 204
oxazolidinones 39

p

P9A and P9B 2
paclitaxel 182
pancreatic cancer 201
pancreatin 97
paradaxin 129
parasin 48
parasites 12, 40, 41
– CAMPs 9, *50*, 50, 51, 52, 55
penaeidin 58

penicillin 2
pepsin 97
peptaibols 156
peptide B 96, 98, 100–102
– brain 87, 88, *88*, 89, *90*
peptidoglycan *146*, *147*, 148
perforin 4
pests and insects 39, 53, 54, 85, 118
PGLa 3
phage therapy 40, 41
phagocytes 13, 160
phenothiaziniums 41
phenylalanine 45, 61, 194, 195
pheochromocytoma 204
pheromones 4
phosphatidylcholine (PC) 149, 151, *151*, 160
phosphatidylethanolamine (PE) 51, 54, 148–152, 199
phosphatidylglycerol (PG) 149, *150*, 150, 152
phosphatidylserine (PS) 201, 203–206
phospholipids 149, *150*, 199, 201, 203–205
photodynamic antimicrobial chemotherapy (PACT) 40, 41
phylogenetic analyses 1, 6, 8, *9*, 18
PhytAMP 184
piscidin-1 158
planktonic bacteria 41
plantarcin A 164, **164**, 204
plasma 96, *146*, *147*
platelets 3, 96, 97
platypus venom 11, 42
plectasin 149
Plate, lung, nasal epithelium clone (PLUNC) 4
polybia-MP1 **185**
polysaccharides 46, *146*, *147*
pore formation 51, 52, 55
porins *146*, *147*
Porphyromonas gingivalis 98
potassium 84, 122
PR-39 58, 61, **186**, 190, 194, 197
PR-AMPs 58, 59, *59*, 60, 61
prediction 163, 186, 193
– amphiphilicity 119, 124, 125, 130, 133, 134
premature stop codons 10
pre-prodermaseptins 5–7, 18
proenkephalin A (PEA) 87, 88, *88*, 89, *90*
prokaryotes 1, 8, 50, 83–85, 146, 148
proline *88*, 88, 100, 193
– CAMPs (PR-CAMPs) 53, 58, *58*, 62
prophenin PF2 194

prostate cancer   184, 191, 195, 200, 204
protegrin-1   *44*, 48, 100, **164**, 190
protein anchor regions   49
protein synthesis   40
proteoglycans (PGs)   199, 201–203
Pseudomonas aeruginosa   2, 87, 92, 119, *150*
Pseudomonas solanacearum   2
Pseudoplectania nigrella   149
psoriasis   15, 18
psyle A – psyle F   **186**
puroindoline A   57
purothionin   2
pyrrhocoricins   58

q

quantitative structure – activity relationship (QSAR) models   134
quinupristin-dalfopristin   85

r

radiation therapy (RT)   182
rat β-defensins (RBDs)   16
reactive oxygen species (ROS)   56, 62, 160
receptor tyrosine kinases   122
residue composition   187, **188**, *189*, 189–191, 206, 207
respiratory tract   83, 85–87
retrocyclins   10, 101
RGD-tachyplesin-1   205
rhabdomyosarcoma   192, 204
rhabdosarcoma   202
rifampicin   85
rifamycin   85

s

Saccharomyces cerevisiae   *146*, *147*
sakacin P   **164**
Salmonella typhimurium   52, 92
salt bridging   86, 99, 100
"sand in a gearbox" model   161
Segrest method   *127*, 127–129
selectivity of ACPs   184, 185, **185**, 187, 189–193, 198–202, 204–207
sepsis   148
sequence length   184, 187, 206, 207
serine   88, 89, 100
sessile bacteria   41, 87
sexually transmitted diseases   10, 40, 96
Shai-Huang-Matsazuki (SHM) model   95, 158, 159, *160*, 167
sialic acid   201, 202, 206
SK84   193
SMAP-29   148
snorkeling mechanism   190

specific residues   *44*, 55–60
sphingolipids   149
sphingomyelin (SM)   149
spider β-ACP$_T$ 192
Staphylococcus spp   52, 85
Staphylococcus aureus   88, 95, 97, 149, *150*, 158
– see also MRSA
Staphylococcus epidermidis   85, 92, *150*
STAT proteins   16
stereo-selective antimicrobial mechanism   59
sterols   60, 145, 153, 166, 199, 200, 206
S-thanatin   153
stomach carcinoma (AGS)   199, 202, 204
Streptococcus   148
Streptococcus mutans   98
Streptococcus pneumonice   150
Streptomyces roseosporous   84
streptomycin   2
structure – function relationships   99–101, 115, 181, 184, 187
surfactant associated anionic peptides (SAAPs)   83, *86*, 86, 87, 99–102
styelin D   42
synovial sarcoma   191

t

T-cell epitopes   183
T-cells   87, 154
TA-CAMPs   57, 61
tachyplesin-1   **186**, 203
teichoic acid   *146*, *147*, 148
teichuronic acid   *146*, *147*
teicoplanin   84, 85
temozolomide   182
temporin A   159
temporin B   **155**, 160, **164**, **164**
temporin-1DRa   **185**
temporin L   **155**, 159, 160, **164**, 164, 166, **185**, 200
θ-defensins   3, 7, 10, 101
thiamine biosynthesis   40
thiol   192
thionins   2
thrombin   96
thymosin-β$_4$ 87, 97, 102
tilted α-helices   122, 123, 128, 129, 130, *131*
tilted peptides   49, 145, 149, **155**, 161–163, 167
– α-ACPs   187, 190, 191, 192, 193
– amphiphilicity   115, 119, 130, *131*, 133, 134
Toll-like receptors (TLRs)   14, 15

Tomich model 128, 129
topical biocides 83, 102
toroidal pore model 155, 156, 156–158, 181
– AAMPs 94, 94
– membrane interaction 145, 149, 152, 154–159, 166, 167
toxicity of ACPs 184, **185**, 185–187, 189–193, 198–207
transactivation model 14, 14
transforming growth factor-α 16
translocation 154, 157–159, 160, 167, 168
– AAMPs 83, 86, 95, 96
– anticancer 196, 202, 204
transmembrane (TM) α-helices 116, 121, 122, 125–128, 130, 132, 133
transmembrane (TM) pores 46
transmembrane (TM) potential (ΔΨ) 145, 150, 151, 166, 198, 205
– CAMPs 56, 60
trans proline 88
Trichoderma 156
Triticum aestivum 2
tritrpticin 48, 57, 116
tropomyosin 124
trypanosomiasis 41
trypsin 97
tryptophan 57, 61, 193, 194
tryptophan and arginine CAMPs (TA-CAMPs) 57, 61
tumor necrosis factor-α 15

v
V13K **155**, 192
vaccines 183
valine 194, 195
vancomycin 84, 85
varv A and varv F **186**
varv A, varv E and vitri A **186**
vhl-2 196
vibi D, vibi E, vibi G and vibi H **186**
Vigna radiate 150
Viola odorata 53, 54
violacin A 54
viruses 10, 12, 15, 41, 117, 161–163
– AAMPs 85, 87, 92
– CAMPs 39, 46, 51, 52, 54
VP1 **163**, 163, **164**

w
Wenxiang diagrams 120, 121, 121
World Health Organization (WHO) 39, 40
wound healing 1, 16, 18, 97

x
Xanthomonas campestris 2

y
yeast 55

z
zwitterionic lipids 50, 149
zwitterionic membranes 190
Zygomycota 49